CALCULUS 2
SIMPLIFIED

Other Books by Oscar Fernandez

Everyday Calculus: Discovering the Hidden Math All Around Us

*The Calculus of Happiness: How a Mathematical Approach
to Life Adds Up to Health, Wealth, and Love*

Calculus Simplified

CALCULUS 2
SIMPLIFIED

Integration and Infinite Series

OSCAR E. FERNANDEZ

Princeton University Press
Princeton and Oxford

Published by Princeton University Press
41 William Street, Princeton, New Jersey 08540
99 Banbury Road, Oxford OX2 6JX

press.princeton.edu

Library of Congress Cataloging-in-Publication Data

Names: Fernandez, Oscar E. (Oscar Edward), author.
Title: Calculus 2 simplified : integration and infinite series / Oscar E.
    Fernandez.
Description: Princeton : Princeton University Press, [2025] | Preceded by:
    Calculus simplified / Oscar E. Fernandez. [2019]. | Includes
    bibliographical references and index.
Identifiers: LCCN 2024035585 (print) | LCCN 2024035586 (ebook) | ISBN
    9780691263755 (pbk.) | ISBN 9780691263762 (ebook)
Subjects: LCSH: Calculus–Problems, exercises, etc. | Calculus–Textbooks.
    | BISAC: MATHEMATICS / Calculus | STUDY AIDS / Study Guides | LCGFT:
    Textbooks.
Classification: LCC QA303.2 .F458 2019 (print) | LCC QA303.2 (ebook) |
    DDC 515–dc23/eng/20240929
LC record available at https://lccn.loc.gov/2024035585
LC ebook record available at https://lccn.loc.gov/2024035586

British Library Cataloging-in-Publication Data is available

Editorial: Diana Gillooly, Whitney Rauenhorst
Production Editorial: Elizabeth Byrd
Production: Jacqueline Poirier
Publicity: William Pagdatoon
Cover and interior design: Wanda España

Printed in the United States of America

10 9 8 7 6 5 4 3 2 1

To Alicia and Emilia

Thank you for filling our hearts with joy, every day.

To Zoraida

Estando juntos mi mundo se llena de luz. Bendita la luz de tu mirada.

Contents

Preface

Hi! Welcome to *Calculus 2 Simplified*. My name is Oscar Fernandez, professor of mathematics and former faculty director of the Pforzheimer Learning and Teaching Center at Wellesley College, and I will be your instructor.

Who Is This Book Intended For?

Here are three questions that will help you determine if this book is for *you*.

1. **Do you have a background in Calculus 1 (limits, differentiation, and some integration)?** If so, this book is for you.

2. **Are you currently enrolled in a Calculus 2 course (or soon will be)?** If so, this book is for you.

3. **Did you learn calculus long ago and are now seeking a quick refresher on the subject?** If so, this book is for you.

If you answered "no" to the first question, this book *might* not be for you. I encourage you to skim it first to see if it may still be an appropriate resource for you. If you end up deciding that you need more Calculus 1 background, I encourage you to check out *Calculus Simplified*. If you answered "yes" to any of the questions above, great! Read on.

Why You Should Use This Book

Reason 1: Its Goldilocks Approach to Calculus 2

Studies in cognitive science show that we learn best when content is taught at just the right level of challenge and complexity—not too much, and not too little. This is the "Goldilocks effect" in learning.

Consider now the challenge of learning calculus. The typical calculus student turns to three particular resources for help: a calculus textbook, a calculus professor, and a calculus supplement. Each of these resources has its strengths and weaknesses. Figure P.1(a) highlights three dimensions along which to understand those strengths and weaknesses: level of detail, personalization of content, and depth of insights.

Compare the aforementioned three calculus resources along those three dimensions and here are some things you'll notice:

- **Level of detail.** Most calculus supplements (e.g., *Calculus II for Dummies*) lack formal statements of theorems, making it unclear when one can apply the formulas and techniques discussed (the hypotheses of a theorem clearly articulate this). Most calculus textbooks, by contrast, are replete with formal statements

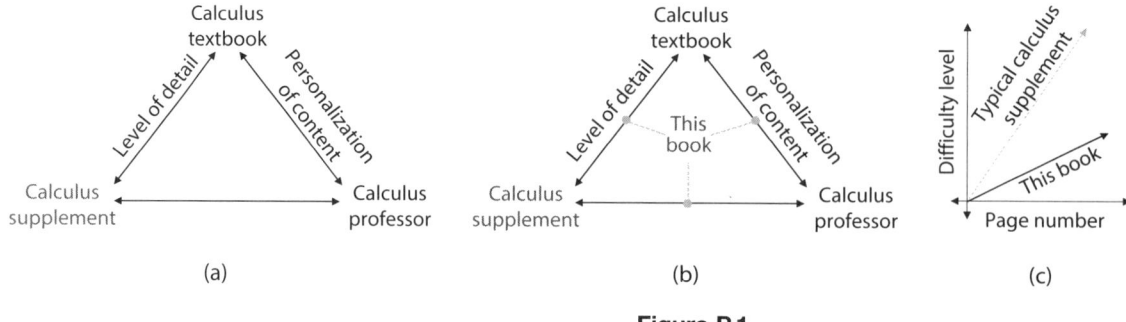

Figure P.1

of theorems and their proofs. This often obscures the intuition behind the concepts and results in a calculus learning experience that feels too formal. Conclusion: Too little detail can give you a false sense of confidence in your calculus understanding; too much detail can turn you off from calculus altogether.

- **Depth of insights.** Most calculus supplements offer only superficial mathematical insights, focusing instead more on teaching computational skills, procedures, and techniques. (Example: "Do *this* when you see *that*.") In fairness, most calculus supplements are *supplements*—they're intended to be used alongside a calculus textbook and/or calculus professor to furnish those deeper mathematical insights. But often *those* resources go too far, as evidenced by the familiar plea from calculus students, "Less theory, more examples, please!" Conclusion: Insufficient depth of insight can make calculus feel like rote learning; excessive emphasis on depth can make it seem overly theoretical and impractical.

- **Personalization of content.** Most calculus textbooks are thousand-page tomes containing way more content than a professor can cover in a Calculus 2 course. The average calculus textbook, therefore, is not at all personalized to your interests. We calculus professors do our best to distill all that content into roughly 30 hour-long lessons, ideally incorporating the particular interests of the students in the course. But while this is an improvement over the textbook, in a class of *many* students it's hard to personalize the content to *each* student. Conclusion: Too little personalization of content is a wasted opportunity to engage you in learning calculus, but the amount of personalization the average calculus instructor provides, while commendable, is still not enough.

Conclusion: *None of the resources just discussed is "just right" for learning calculus.* That's where this book comes in.

This book takes a "Goldilocks approach" to learning Calculus 2.

As figure P.1(b) is meant to illustrate:

- **This book balances intuition with theory to provide you with just the right level of detail.** Chapter 1 teaches you the core ideas of Calculus 2. It focuses on developing the intuition behind Calculus 2's main concepts, the mindset of Calculus 2, and Calculus 2's overarching framework. Subsequent chapters discuss the math of Calculus 2, with just the right balance of formal statements of definitions and theorems. With this approach you learn the terminology of Calculus 2 and understand the *full* story, complete with when it works and why.

- **This book empowers you to customize your Calculus 2 adventure.** Most Calculus 2 books jump right into techniques of integration content. But that requires a solid grasp of integration and differentiation. That's a big ask up front. *This book is different.* It gives you the choice to follow that traditional ordering of topics, if you'd like, *and* gives you an alternate route that delays integration content, asking only that you remember a few things about limits and basic differentiation up front. In other words,

  This book gives **you** *the choice to design your own Calculus 2 journey.*

  To experience that traditional Calculus 2 topics ordering—integration techniques, then applications of integration, then sequences and series—work through the chapters in the order they're numbered:

  Chapter 1 → **Chapter 2** → **Chapter 3** → **Chapter 4** → **Chapter 5**

  To experience the lower-prerequisites ordering—sequences and series, then integration techniques, then applications of integration—work through the chapters in this order:

  Chapter 1 → **Chapter 4** → **Chapter 2** → **Chapter 3** → **Chapter 5**

  You'll learn the same Calculus 2 content regardless of the chapter sequence you choose. My recommendation, however, is to choose this second path. It requires fewer up-front prerequisites compared to the traditional topics sequencing. Nevertheless, the choice is yours. And that's the point: *With this book,* **you** *get to choose how you learn Calculus 2.*

  This book also lets you choose the balance between theory and intuition. Sections *do* include discussions of theory, but they focus on intuition rather than proof. More technical discussions and proofier content are relegated to the chapter appendixes and the end-of-chapter exercises. Include that content in your calculus journey if you'd like, or skip it if you want. The choice is yours.

  Finally, the same freedom applies to the real-world applications discussed in this book. Most of these are discussed in the applications-oriented exercises. Include them in your calculus journey if you'd like; skip them if you'd rather not. The choice is yours.

  The net effect of all this is a gentler learning experience—as illustrated by figure P.1(c)—that's customizable and emphasizes intuition (hence the title *Calculus 2 Simplified*).

- **This book provides just the right amount of depth to the mathematical insights unearthed.** You will learn both the "how" *and* the "why" of Calculus 2. You will understand why its core concepts are important. And you will be exposed to the various other places (e.g., real-world contexts) Calculus 2 concepts show up. Together with the accompanying videos, animations, and interactive online applets (described below), this gives you ample opportunity to customize how broadly and deeply you'd like to learn the subject.

I cannot overemphasize how important these features of the book are. Calculus 2 has a reputation for being a hodgepodge of content. This is primarily due to the prevailing assumption among authors and instructors that Calculus 2 is merely a continuation of Calculus 1. But it's not. Calculus 2 introduces new major, beautiful, and important concepts and results that are unrecognizable to a Calculus 1 student. And, the methods used to discover and investigate these concepts are subtly but distinctly different from those used in Calculus 1. That's the point of chapter 1—to frame the course around those new results and methods, to set the stage, to *motivate you*.

Reason 2: The Bonus Features

The following additional features are designed to supercharge your learning of Calculus 2.

- **A focus on conciseness.** Excluding the exercises and appendixes,

  This book teaches you Calculus 2 in 137 pages.

  This counts the content of all five chapters. There are an additional 20 pages of exercises, plus the appendixes. A typical calculus textbook, by contrast, spends roughly 200–300 pages to teach Calculus 2. (To be fair, textbooks provide a more comprehensive treatment of the subject.)

- **Over 170 solved examples.** I've included 178 distinct solved examples in this book. I also included more than just the calculations for many of them—I wrote out my thought process, too. This will help you learn to think like a mathematician thinks about calculus.

- **Answers to *all* nonproof exercises.** There are 445 distinct exercises in this book, counting all subparts. I've included answers to *all* the exercises that aren't proofs or derivations, and solutions or hints to almost all that are; answers are at the back of the book.

- **Use of contrast and boxes to separate content.** Definitions, theorems, and alerts to applied exercises and explorations appear in chapters either in boxes or with thick vertical lines to help you spot them easily. End-of-chapter exercises with a bold title (e.g., "**Population density**") are applied exercises and are numbered **A1**, **A2**, etc.

- **Interactive online content linked to book content.** In mathematics, an interactive graph—or a video lesson—is sometimes worth a million words. That's why I've created interactive applets, animations of core concepts, and video explanations to accompany this book. You'll find those referenced throughout the book, sometimes via links or QR codes. All this content can also be accessed via the book's website:

  sites.google.com/view/fernandezmath/books/calc2s

- **Inclusion of references.** References appear in text as brackets and use a number (e.g., [1]) to identify the entry in the bibliography being referenced.

- **A direct line to the author.** I wrote this book to help you learn Calculus 2. And I stand by this goal. So, *feel free to email me with any questions, comments, or suggestions.* Seriously. Here's my email:

  ofernandezmath@gmail.com

  I also encourage you to submit feedback on the book via the link below.

  https://forms.gle/WbD4HNEjAEUhkf6n6

  Your submission will be anonymous. Your feedback will help me improve the book and will be incorporated into future editions.

Parting Thoughts

Calculus 2 Simplified is a resource for anyone interested in learning (or relearning) Calculus 2. It aims to strike that "just right" balance on the level of detail, depth of insights, and personalization of content. It also empowers you to take control of your learning by providing you with multiple pathways to mastering Calculus 2.

In terms of rigor, though *Calculus 2 Simplified* is specifically designed to streamline your learning experience, don't confuse "streamline" with "water down." This book is not a collection of Calculus 2 formulas or a brief review of the subject's main concepts (which presumes you already know Calculus 2). It's not an idiot's or dummies guide to Calculus 2 (you are neither). *Calculus 2 Simplified* is a college-level Calculus 2 course based on the notes I use to teach the same content in my college courses. It's a streamlined experience that removes excess content to accelerate your mastery of Calculus 2.

In terms of content, this streamlining means that *Calculus 2 Simplified* is not intended to be a comprehensive treatment of Calculus 2. In the present conception of what a calculus textbook consists of, this book is also not intended to be a textbook (though it can certainly be used as such in some settings). At the same time, *Calculus 2 Simplified* is much more than the run-of-the-mill calculus supplement. Explained within the theme of this preface, I think of *Calculus 2 Simplified* as occupying the "Goldilocks zone" between a calculus textbook and a calculus supplement. You'll find the vast majority of Calculus 2 topics covered herein, ideally in more accessible, customizable, and engaging ways than in a calculus textbook.

I'm excited to be working with you, and thank you for trusting me to be your guide on your Calculus 2 adventure. Once you're done learning the calculus in this book, I encourage you to read the epilogue; it contains some useful advice and encouragement for navigating mathematics beyond calculus. As your next step, I recommend reading the **Before You Begin** and **To the Student** sections before you dive into chapter 1. These two sections describe important information, such as what the icons in the margins mean and how to know which examples have videos associated with them, that you'll want to know before starting your Calculus 2 journey. See you soon!

Oscar E. Fernandez
Brookline, MA

Before You Begin . . .

Here's some useful information that will help you navigate this book.

Numbering Scheme

- Equations in the chapters are numbered in (chapter.equation) format. Example: Equation (3.17) refers to the 17th numbered equation in chapter 3. Equations in the appendixes replace the chapter number with the appendix identifier. Example: Equation (A2.7) refers to the 7th numbered equation in appendix A2, the appendix to chapter 2. Figures and tables follow the same numbering scheme except without parentheses. Example: Figure B.5 refers to the 5th numbered figure in appendix B.

- Equation numbers appear flush right in the book, like this: (3.17)

Typography

- Definitions and theorems appear next to gray vertical rectangles, like this:

  THEOREM 3.2: Theorem text . . .

- Superscripts, like this,[1] point to a footnote.

- Comments appear in the margins, referenced with asterisks like this.*   * Here's a comment.

- Suggested end-of-chapter exercises are referenced in two ways. Most are referenced at the end of an example, like this:

  EXAMPLE 0.1   Example prompt here.

  Solution Here would be the solution. The rightmost square signals the end of the solution. ■

  Related Exercises   1, 5(a), A5, D4, etc.

  Applied exercises have an "A" prefix, and exercises involving derivations, proofs, or explorations have a "D" prefix. Some exercise suggestions are not connected to particular examples or are delayed until I make some comments on an example's solution.

---

[1] This is a footnote.

- Interactive Figure icons in the margin—like the one on the left—indicate that I've created an interactive online supplement to the graph(s) or content being discussed. I've also recorded videos for many of the examples in the book. (All these resources are accessible via the book's website, listed in the preface.) These videos are referenced right after the example number with the word "Video," like this:

  EXAMPLE 0.2 (Video)   An example with an accompanying video.   ■

- In most sections I suggest tips and summarize takeaways. Sometimes I give a few more in a special subsection that appears under the subheading **Additional Tips, Tricks, and Takeaways**.

- At times you'll see one of the four icons below placed in the outer margin of a page. These are designed to communicate information about the content. The first icon below—the upward-sloping mountain—signals that a long calculation or derivation is ahead, and the second icon—the downward-sloping mountain—signals the end of that calculation or derivation. The light bulb icon alerts you to an insight. The map icon signals a check-in. At these points I discuss where we are in the Calculus 2 journey and where we're heading.

- Finally, QR codes are included on various pages. Scanning these with a phone camera will take you to an animation or interactive graph of the concept or formula referenced. Try this with the QR code below. You should see a GIF animation of a graph being rotated about the x-axis to produce a "surface of revolution."

To the Student

Welcome to *Calculus 2 Simplified!* Before you embark on your adventure through this book, I thought I would give you a few practical tips intended to help you conquer calculus.

How This Book Organizes the Content

Research shows that students learn best when they know *beforehand* what they are about to learn, and when that learning is scaffolded (a term that basically means "chunked into digestible bits that are sequenced appropriately to maximize learning"). You've picked up a Calculus 2 book, so you know you're going to learn Calculus 2 from this book. But let me now give you a quick tour of how the book introduces you to that content, and how it's scaffolded.

- **Chapter 1: The Bird's-Eye View.** The first chapter describes the driving questions behind Calculus 2, how we'll tackle them in this book, and what the main results will be. After completing that chapter you'll have a bird's-eye view of the subject *and* a sense of where we'll be heading in subsequent chapters.

- **Chapters 2–5: The Details.** These chapters drop down from that bird's-eye view to work out the details of the main results previewed in chapter 1. Each chapter begins with a preview that overviews the path the chapter takes to developing those main results.

- **In-Chapter Excursions.** Throughout each chapter you'll periodically see me suggesting you try out one of the applied exercises (to see how Calculus 2 applies to real-world problems) or one of the exploration exercises (to deepen your understanding of the subject).

This structure is designed to help you know, at all times, where we are in our Calculus 2 journey, where we've come from, where we're heading, and why we're heading there.

How to Read a Math Book (Including This One)

Though I've done my best to infuse my writing with the elements of a novel—narrative, plots, etc.—this book is not a novel. One thing this implies is that you need to read this book differently from a novel. For example, simply *reading* this book won't help you understand Calculus 2. Rather, I recommend you *work through* this book—work out the examples, work out the solutions to the exercises, work through the supplemental content. By *doing* mathematics you'll be helping yourself *learn* mathematics. Moreover, jot down questions and comments as you read and work through this book. This will ensure you are learning *actively* rather than passively.

Lastly, let me mention the special role that theorems play in mathematics and how to ensure you're getting the most out of them. Loosely speaking, a theorem is a statement that has been proven true by following the rules of logic. A typical theorem has the following structure: preamble, hypotheses, conclusion. Example:

THEOREM (PYTHAGOREAN THEOREM): Consider a right triangle in the plane. Let c denote the length of the hypotenuse of the triangle, and a and b the lengths of the other two sides. Then $c^2 = a^2 + b^2$.

In this theorem the first sentence is the preamble; its role is to provide context for what the theorem says. The second sentence in the theorem contains some assumptions (as happened here, sometimes the preamble also contains assumptions). The last sentence contains the conclusion.

Echoing my earlier advice to *work through* this book, do the same with theorems—whenever you come across a theorem, take a moment to understand what it's saying. Try drawing pictures, explaining the theorem in words, and imagining removing some of the hypotheses to see how the conclusion might be affected. Doing all this will help you appreciate what the real use of the theorem is, help you remember it, and help you learn when it can (and cannot) be applied.

How to Become a Better Student

I have one last recommendation for you: *Use the latest research from the science of learning while you study.* Study strategies like retrieval practice and interleaving— both backed by cognitive science research—can supercharge your studying. You can read more about these and other research-backed study strategies in the supplemental document I wrote for *Calculus Simplified* titled *Evidence-Based Study Strategies*, available on the present book's website.

Alright, let's get started with your calculus adventure!

To the Instructor

You might be thinking: "Not another calculus book!" But this one is different. As I wrote in the preface, *Calculus 2 Simplified* is an attempt to strike a "just right" balance in the level of detail, depth of insights, and personalization of content for what we typically teach in a Calculus 2 course.

Calculus 2 Simplified is also a sign of the times. In the age of sound bites and social media, shorter and more succinct treatments of calculus resonate more with students. That's what drove my earliest attempts to simplify the treatment of calculus at this level (e.g., not Spivak). The resulting byte-sized explanations utilized in this book cut down the time investment necessary for students to quickly familiarize themselves with a calculus concept and to pinpoint their areas of confusion. That can lead to enhanced learning and shorten the delay between student confusion and instructor support. I've experience all this in practice while using this very text as the textbook for the Calculus 2 course I teach at Wellesley College. My Wellesley colleagues, who regularly use *Calculus Simplified* as the textbook for their Calculus 1 courses, relay similarly positive experiences with these book's "simplified" approach. (I encourage you to flip through this book and also *Calculus Simplified* to see if they might serve similar purposes for your courses and your students.)

Students also appreciate that my calculus books aren't thousand-page tomes costing hundreds of dollars. And though there are plenty of open-source calculus resources out there—including many excellent ones—my calculus books are among the few (only?) to offer both the student and the instructor options for customizing the calculus journey. In this book, in particular, you and your students have the option to learn Calculus 2 using a "sequences first" approach. To do that, simply start with chapter 4 after working through chapter 1. As I mention in the preface, this nontraditional ordering of topics asks less up front of students (no need to remember what definite integrals are, or get up to speed with u-substitution by Day 3 of the course). This would translate to a gentler on-ramp for students in your course, allowing those with gaps in their background knowledge more time to reach out to you and work to fill those gaps.

Thanks for taking a look at *Calculus 2 Simplified*. I hope you find its contents useful for you and your students.

1   The Fast-Track Introduction to Calculus 2

Chapter Preview. Calculus is a new way of thinking about mathematics. And Calculus 2 builds on that new perspective in new ways. This chapter introduces you to the calculus mindset, the core concepts of Calculus 2, and the sorts of problems these innovations help solve. The focus throughout is on the *ideas* behind Calculus 2 (the big picture of Calculus 2); the subsequent chapters discuss the *math* of Calculus 2. After reading this chapter, you'll have an intuitive understanding of Calculus 2 that'll ground your subsequent studies of the subject. Alright, let's start the adventure!

1.1   First Things First: What Is Calculus?

In *Calculus Simplified* [2] I gave this two-part answer to that question:

> *Calculus is a mindset—a dynamics mindset. Contentwise,*
> *calculus is the mathematics of infinitesimal change.*

This frame on calculus applies as much to Calculus 2 as to Calculus 1 (and any mathematics that also uses calculus). So, let's unpack that answer, now in the specific context of Calculus 2.

Calculus as a Way of Thinking

The mathematics that precedes calculus—often called "precalculus," which includes algebra and geometry—largely focuses on *static* problems: problems lacking change. By contrast, change is central to calculus—calculus is about *dynamics*. Example:

- *Precalculus question:* Find the pattern in the sequence of numbers

$$1, 1, 2, 3, 5, 8, 13, 21, \ldots$$

- *Calculus question:* Does the ratio of consecutive numbers in the sequence above *approach* a specific number?

The sequence of numbers above is the famous **Fibonacci sequence**. In this sequence, the nth number (let's denote that F_n) is the sum of the two preceding numbers (F_{n-1} and F_{n-2}), starting with $F_1 = 1$ and $F_2 = 1$. That's the precalculus answer to the precalculus question—a (static) formula. But the *calculus* answer to the *calculus* question reveals something magical and enlightening:

$$\lim_{n \to \infty} \frac{F_{n+1}}{F_n} = \varphi = \frac{1 + \sqrt{5}}{2} \approx 1.618\ldots,$$

the **golden ratio** (figure 1.1(a)). Translation: The ratio of consecutive Fibonacci numbers tends to φ ("phi") as we get further into the sequence. "Tends" and "further into" here convey the *dynamics* of this calculus answer. And what about the

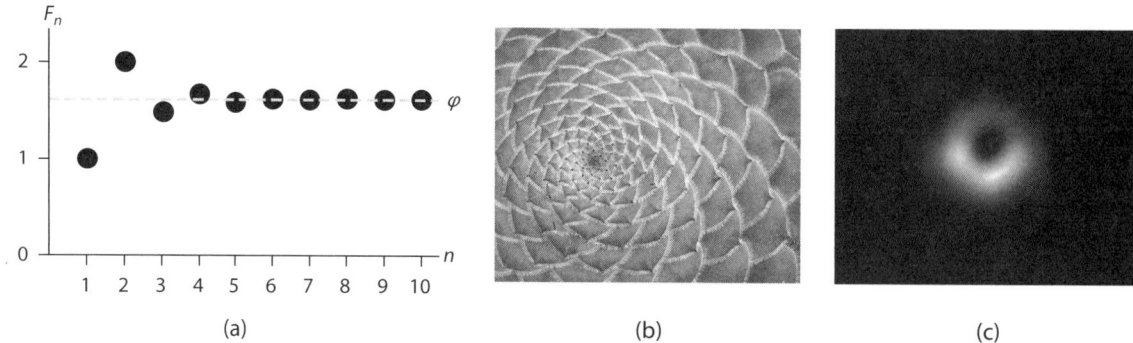

(a)   (b)   (c)

Figure 1.1: (a) The ratios of consecutive Fibonacci numbers F_{n+1}/F_n (black) approach the golden ratio φ as n increases. (b) Close-up of an *Aeonium* succulent by Max Ronnersjö showing spiral phyllotaxis, where successive leaves grow at approximately the "golden angle" $2\pi/\varphi^2$. Retrieved from Wikipedia Commons. (c) The black hole at the center of Messier 87, a galaxy in the constellation of Virgo. By the European Southern Observatory; retrieved from Wikipedia Commons.

magical and enlightening aspects I alluded to? Well, it turns out that *the golden ratio is hidden in many of Nature's patterns*. It's encoded in the spiral arrangements of leaves on plant stems (figure 1.1(b)), in the proportions of components in human hearts and brains, and in theoretical models of black hole physics (figure 1.1(c)) [3]. Takeaway: Calculus is hidden in Nature, *in you*.

This statics versus dynamics distinction between precalculus and calculus runs even deeper—change is the *mindset* of calculus. The subject trains you to approach a problem from a dynamics (versus statics) perspective. We saw this in Calculus 1 when we studied differentiation and interpreted derivatives as instantaneous *rates of change*. And we saw it again when we studied integration and *accumulation* functions. (Appendixes C–D review Calculus 1, in case you'd like a refresher.) This dynamics mindset carries over into Calculus 2. Let me illustrate this—and the continuing role of "infinitesimal change" in calculus—via Zeno's paradox.

The Continuing Role of Infinitesimals in Calculus 2

Zeno of Elea (ca. 490–430 BC) was a Greek philosopher who devised a set of paradoxes arguing that motion is not possible. (Clearly, Zeno did not have a calculus mindset.) One such paradox—the Dichotomy Paradox—can be stated as follows:

To travel a certain distance you must first traverse half of it.

Figure 1.2 illustrates this. Therein Zeno is trying to walk a distance of 2 feet. But because of Zeno's mindset, with his first step he only walks half the distance: 1 foot (figure (b)). He then walks half of the remaining distance in his second step (0.5 foot) and reaches the 1.5-foot mark (figure (c)). After Zeno has taken n steps, the distance d_n he's traveled is given by

$$d_n = 1 + \frac{1}{2} + \frac{1}{4} + \cdots + \frac{1}{2^{n-1}}. \tag{1.1}$$

As Zeno continues his walk, the total distance walked d_n always gets closer to 2 yet never reaches 2 (because Zeno's steps always traverse *half* the remaining distance).

If we checked back in with Zeno after he's taken an *infinite* number of steps, however, his total distance traveled d would be ... drum roll please ... *infinitesimally close* to 2—as close to 2 as you can imagine but not equal to 2.

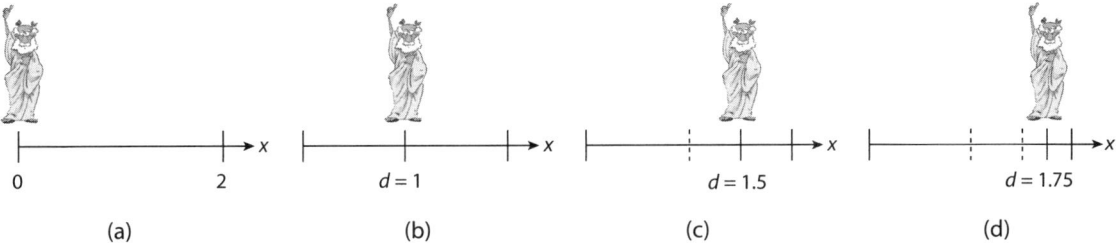

(a)   (b)   (c)   (d)

Figure 1.2: Zeno trying to walk a distance of 2 feet by traversing half the remaining distance with each step.

This example illustrates the dynamics mindset of calculus. We discussed Zeno *walking*; we thought about the *change* in the distance he traveled; we visualized the situation with a figure that conveyed *movement*. (Calculus is full of action verbs!) But the example also challenges us. What seems intuitively clear to us is that after an infinite number of steps Zeno would've covered a distance equal to

$$1 + \frac{1}{2} + \frac{1}{4} + \cdots + \frac{1}{2^{n-1}} + \cdots = 2. \tag{1.2}$$

That is, *the infinite sum of the distances Zeno covers with each step should equal 2.* But how can we add up an infinite number of numbers? And how/why does the particular infinite sum above "yield" 2? (Those are the challenges.) To tackle these new calculus questions requires new calculus concepts and methods that leverage their inherently *dynamic* nature. Luckily, one of the pillars of Calculus 1 provides a stable foundation on which to build these new concepts and methods: limits.

1.2   Limits: (Still) The Foundation of Calculus

Let's return to equation (1.1). It turns out that we can express the sum therein much more compactly as

$$d_n = 2 - \frac{1}{2^{n-1}} = 2\left(1 - \frac{1}{2^n}\right). \tag{1.3}$$

(We'll learn how to do this in chapter 4.) With the help of the Limit Laws from Calculus 1,* it follows that

$$\lim_{n \to \infty} d_n = \lim_{n \to \infty} \left(2 - \frac{2}{2^n}\right) = 2 - \lim_{n \to \infty} \frac{2}{2^n} = 2. \tag{1.4}$$

* These are reviewed in appendix C.

This is Calculus 2's answer to the mystery of (1.2). It expresses the intuitive idea that the 2-foot mark is the *limiting* value of the total distance Zeno's traversing. Equation (1.4), therefore, is a statement about the *dynamics* of Zeno's walk, in contrast to (1.1), which is a statement about the *static* snapshots of how far Zeno has traveled after n steps. Moreover, (1.4) reminds us that d_n is always *approaching* 2 yet never *arrives* at 2. Indeed, as you may recall from Calculus 1:

Limits approach indefinitely (and thus never arrive).

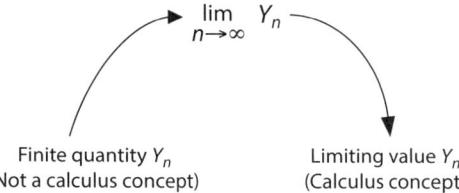

Finite quantity Y_n
(Not a calculus concept)

Limiting value Y_n
(Calculus concept)

Figure 1.3: The Calculus 2 workflow.

We'll learn much more about infinite sums in chapter 4. For now, the Zeno example is sufficient to illustrate one key idea: *Limits are also the foundation of Calculus 2.* The Calculus 1 mansion was built on limits, and so will be the Calculus 2 mansion. And the workflow we'll use for building new Calculus 2 concepts from limits will be similar to the one introduced in *Calculus Simplified* [2]: *Start with a finite quantity Y_n that depends on an integer n, and then let n tend to infinity (i.e., take the limit as $n \to \infty$) to arrive at a calculus result* (see figure 1.3). Working through this process—like we just did with the Zeno example—for various quantities Y_n of real-world and mathematical interest is part of what *doing* Calculus 2 is all about.

1.3 The Three Difficult Questions That Drove the Development of Calculus 2

Calculus 2 developed out of a need to answer the following three Big Questions.

1. ***The Geometry Question:*** *Can we calculate the length of any curve, area of any surface, and volume of any solid?* The ancient Greeks (and other ancient civilizations) could calculate lengths, areas, and volumes for polyhedra and some curvy shapes (e.g., spheres), but that was about it. For example, for thousands of years mathematicians struggled to calculate the surface areas and volumes of objects as simple as the flower vase shown in figure 1.4(a). They could paint them and fill them with water but not know beforehand how much paint or water they'd need.

2. ***The Infinite Sum Question:*** *Does an infinite sum have a sum, and if so, what's the sum?* For centuries mathematicians have used clever geometric arguments to tackle infinite sums. For example, adding up all the areas inside the square in figure 1.4(b) shows that

$$\frac{1}{2} + \frac{1}{4} + \frac{1}{8} + \frac{1}{16} + \cdots = 1. \tag{1.5}$$

Notice that if we multiply this equation by 2 we get (1.2), yielding a second (this time geometric) verification of that sum. But does a general infinite sum, like the one shown in the cloud in figure 1.4(b), have a sum? And if so, how can we calculate that sum? Tough questions.

3. ***The Approximation Question:*** *Without knowing the exact value of a function, can we accurately approximate it?* Before calculators and computers, accurately approximating quantities like the ones shown in figure 1.4(c) was a difficult

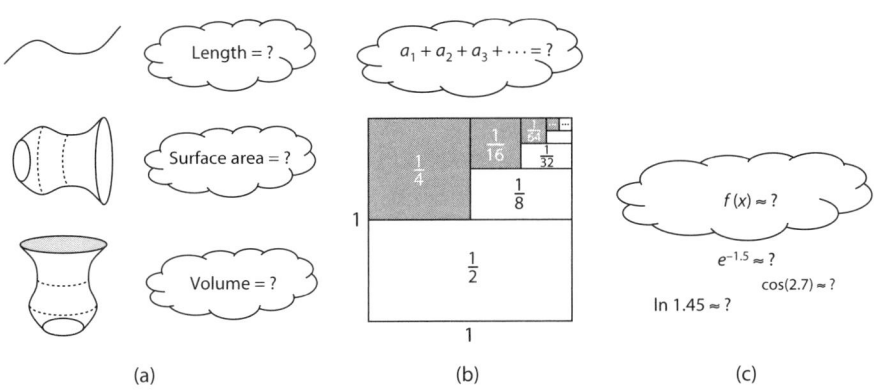

Figure 1.4: The three sets of Big Questions that drove Calculus 2's development.

problem. And though today's technology makes that easy, the algorithms that technology uses to produce those accurate approximations trace back to the early work on this third Big Question. For example, how can we accurately approximate quantities like $\sqrt{1.05}$, or the other quantities under the cloud in figure 1.4(c), without using a calculator?

I hope these descriptions impart the mystery and difficulty involved in trying to answer these Big Questions. Indeed, *the resolution of these questions took millennia*. But it won't take that long for you. By the time you're done reading the next three pages you'll have developed an intuitive understanding of how to resolve each of those questions. That understanding is grounded in my earlier answer to the question, What is calculus? *A dynamics mindset*. Let's see how.

First, note that nothing about figure 1.4 says "dynamics." Every image is a static depiction of something (e.g., a volume). Yet in the real world we *fill* beakers with liquids to measure volume and *add up* numbers to obtain a sum. (There are those action verbs again.) We've turned on our calculus (dynamics) mindset. The next step is to look for the Y_n we'll need so we can apply the Calculus 2 workflow (figure 1.3). To illustrate this process, let's *calculus* three of the questions in figure 1.4— yep, I'm encouraging you to think of calculus as a verb—and search for the finite quantities Y_n that, in the limit as $n \to \infty$, yield the desired quantities (e.g., volume). Figure 1.5 illustrates the results. Let me give you a tour of that figure now.

- *Row 1:* The length of the curve in the third column—called the **arc length** of that curve—is realized as the infinite limit (second column) of the total length s_n of n line segments (the gray ones in the figure) created by choosing suitable points on the curve (first column). Thus, here $Y_n = s_n$.

- *Row 2:* The infinite sum of a set of numbers (third column) is realized as the infinite limit (second column) of the sum of the first N of them, S_N (first column).[1] Thus, here $Y_N = S_N$.

- *Row 3:* The value $f(x)$ of a function for x-values near $x = 0$ (third column) is realized as the infinite limit (second column) of polynomials $T_n(x)$ of increasing degree n (first column). So, here $Y_n = T_n(x)$.

[1] This is exactly what we did in the Zeno example in equation (1.4).

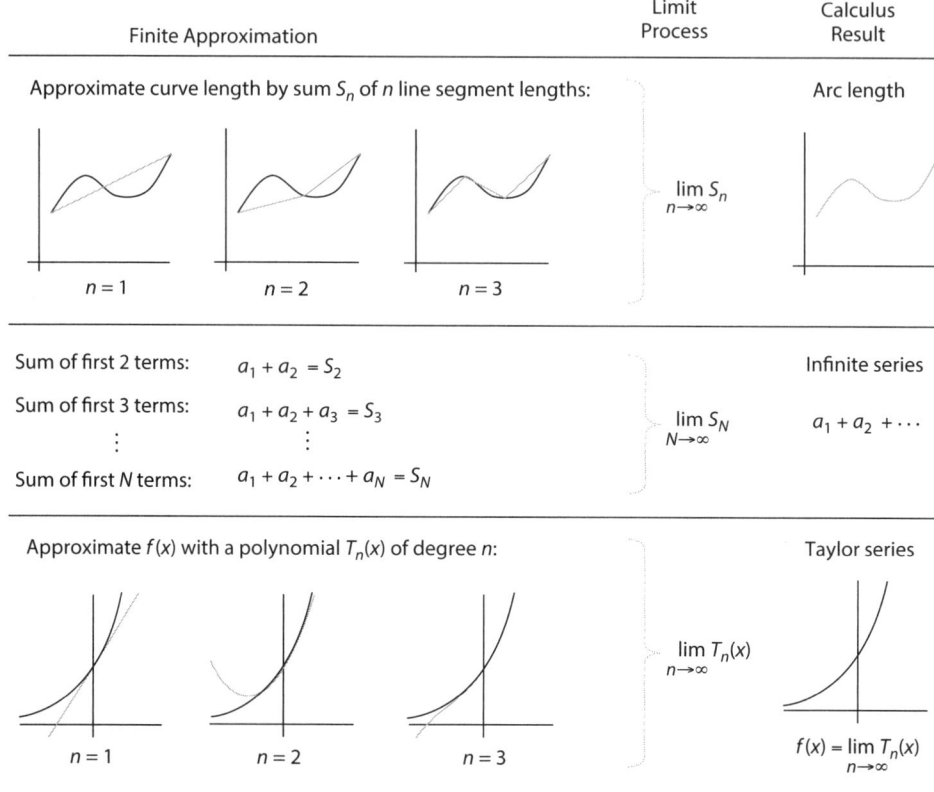

Finite Approximation	Limit Process	Calculus Result

Approximate curve length by sum S_n of n line segment lengths: — Arc length

$n = 1$ $n = 2$ $n = 3$

$\lim\limits_{n\to\infty} S_n$

Sum of first 2 terms: $a_1 + a_2 = S_2$ Infinite series

Sum of first 3 terms: $a_1 + a_2 + a_3 = S_3$

$\vdots$ $\vdots$ $\lim\limits_{N\to\infty} S_N$ $a_1 + a_2 + \cdots$

Sum of first N terms: $a_1 + a_2 + \cdots + a_N = S_N$

Approximate $f(x)$ with a polynomial $T_n(x)$ of degree n: — Taylor series

$n = 1$ $n = 2$ $n = 3$

$\lim\limits_{n\to\infty} T_n(x)$

$f(x) = \lim\limits_{n\to\infty} T_n(x)$

Figure 1.5: The Calculus 2 workflow (figure 1.3) applied to the three Big Questions.

We'll learn how to calculate arc length in chapter 3. The result will be a definite integral that in many cases will require advanced integration methods. We'll learn those at the end of chapter 2. At the beginning of that chapter we'll learn how to approximate definite integrals, which comes in handy when we can't evaluate them exactly. In chapter 4 we'll then revisit the last two rows of figure 1.5. The infinite sum of numbers in the second row of the figure is called an **infinite series,** and through our attempts to approximate the sums of certain infinite series we'll end up circling back to the third row of the figure. We'll call the polynomials $T_n(x)$ illustrated therein (the gray curves) **Taylor polynomials**. And by observing that as n increases they approximate $f(x)$ better, we'll build up to the climax of the chapter: the *remarkable* result that sometimes the limit as $n \to \infty$ of the Taylor polynomial $T_n(x)$ is $f(x)$. Translation: *Some functions can be represented as "infinite degree" polynomials!* (We'll call these **Taylor series**.)

I don't expect you to understand everything I've just described. My intent was instead to provide you with a roadmap of the main stops and a preview of the highlights of our upcoming calculus adventure. I also hope that the preceding discussion helps you appreciate this book's approach to Calculus 2. Figure 1.5 originated from switching to a dynamics mindset to tackle our three Big Questions. We then applied the Calculus 2 workflow to realize each of the calculus objects in the third column of the figure as infinite limits of appropriate Y_ns. Takeaway: *In*

scanning figure 1.5 from column to column (left to right) you're following the Calculus 2 workflow.

This completes my big-picture overview of Calculus 2. We've by no means resolved the Big Questions illustrated in figure 1.4, but we've created a roadmap for tackling those Big Questions. What's left now is to apply the Calculus 2 workflow to our Big Questions to develop the mathematics of Calculus 2. That's what we'll do in the rest of the book. See you in the next chapter!

2 Integration Techniques and Approximations

Chapter Preview. In Calculus 1 you learned a few techniques for evaluating definite integrals. But we'll need more advanced methods for Calculus 2 purposes. This chapter will teach you those, grouped into two categories: approximation methods and analytic methods. We'll begin by discussing Riemann sums and the trapezoidal approximation. These help us approximate a definite integral using areas of rectangles and trapezoids, respectively. By studying how these approximations increase in accuracy, we'll then develop theorems that can be used to make those approximations as accurate as we want. The remainder of the chapter then shifts gears to evaluating integrals exactly (i.e., analytically). There I'll teach you advanced integration techniques (beyond u-substitution), such as integration by parts. Along the way I'll give you lots of tips, tricks, and takeaways to help you master all this content. I'll assume you're comfortable with the content in appendixes A–C (these cover precalculus, limits, and differentiation), so skim those first if you haven't already. Alright, let's start the adventure!

2.1 Integrating, Leibniz's Way

Let's get started by reviewing the notation for the definite integral.[1] Gottfried Leibniz, the co-inventor of calculus, took an "infinitesimals" approach to calculus. Indeed, the modern notation for the definite integral,

$$\int_a^b f(x)\,dx, \tag{2.1}$$

is largely due to him. Leibniz thought of a definite integral as a sum of areas of *infinitesimal* widths. Figure 2.1 illustrates what I mean.

At $t = x$ I've drawn one of Leibniz's "infinitesimal rectangles." Or, rather, I've tried to. The true infinitesimal rectangle's thickness is dx, but depicting that would make the rectangle look like a line.[2] And that was part of the genius of Leibniz's insight. You see, the y-value of $y = f(t)$ at $t = x$ is $f(x)$. So, the area of Leibniz's

---

[1] A bit of history, if you're interested. The 1600s was an exciting time in mathematics. It was a period of rapid discovery and innovation. Newton, Leibniz, and others were formulating and discovering calculus; their colleagues were applying the results to make new advances in other fields of mathematics; and all that work was stimulating the development of new branches of mathematics. The 1700s saw those calculus-based results applied by famous mathematicians like Euler and Lagrange to solve problems in the sciences. The math worked—it explained real-world phenomena—but many feared calculus was built on a house of cards. "Infinitesimals," in particular, were too vague a concept for many to embrace. In the 1800s mathematicians addressed those criticisms head on and began a campaign to "rigorize" calculus. A big chunk of this work was done by Augustin-Louis Cauchy, whose 1821 book *Cours d'Analyse* gave us the rigorous definition of the limit we use today, among other contributions. In 1868 Riemann "rigorized" integration (using Riemann sums). Since then, these rigorous treatments of calculus have slowly trickled down the textbook chain and eventually landed in the typical calculus textbook. That's not a bad thing—we want to ensure what we're doing is correct (i.e., can be proven true)—but we also want to preserve the intuition and insights of the founders of calculus. Hence my approach: *teach you both.*

[2] To read Leibniz in his own words writing about dx, see [1], pp. 134–44.

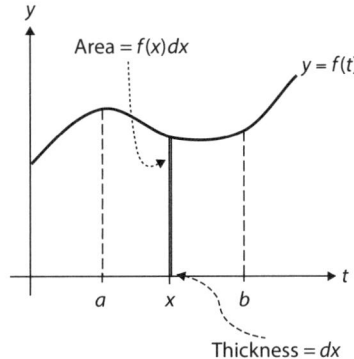

Figure 2.1

"infinitesimal rectangle" is $f(x)\, dx$. Leibniz wrote the sum of the areas of all such "infinitesimal rectangles" between $t = a$ and $t = b$ as

$$\text{S } f(x)\, dx \quad \text{from } t = a \text{ to } t = b,$$

where the "S" was shorthand for "sum." Over time that "S" got more elongated and eventually morphed into the symbol $\int$ we're now used to:

$$\int f(x)\, dx \quad \text{from } t = a \text{ to } t = b$$

Today we append the bounds $t = a$ and $t = b$ to the integral sign to arrive at the standard notation for the definite integral:

$$\int_a^b f(x)\, dx$$

Takeaway: *The definite integral is an (infinite) sum of "infinitesimal rectangles."*

   As enlightening and intuitive as Leibniz's approach to the definite integral is, using it to compute definite integrals in practice would require summing an infinite number of areas of "infinitesimal rectangles." This fact hints at a historical fact: The Infinite Sum Question from chapter 1 has always been intertwined with integration theory—mathematicians have always turned to infinite series to help them calculate "the area under the curve." And just like we previewed in chapter 1 that the most successful approach to tackling the Infinite Sum Question is to first study *finite* sums and then eventually take a limit, the same approach was used by Bernhard Riemann in the late 1800s to "rigorize" integration. (Read the footnote on the first page of this chapter for more.) Over the next few sections we'll study Riemann's approach, pausing intermittently to discuss one side benefit—that it helps us approximate definite integrals in cases when we can't calculate them exactly. In section 2.4 we'll then discuss how Riemann made sense of Leibniz's approach that "the definite integral is an (infinite) sum of infinitesimal rectangles."

2.2 Approximating Integrals, Riemann's Way

Take a look at figure 2.2. In Calculus 1 you learned that the area A of the shaded region is

$$A = \int_0^3 (x^2 + 1) \, dx.^*$$

You also learned how to evaluate integrals like these using the Evaluation Theorem, reprinted here to save you the page flipping to appendix D.

THEOREM 2.1: THE EVALUATION THEOREM. Suppose that f is continuous on $[a, b]$ and that F is an antiderivative of f (i.e., $F'(x) = f(x)$). Then,

$$\int_a^b f(x) \, dx = F(b) - F(a).$$

Recall that to use this theorem we need to (1) verify that f is continuous on $[a, b]$; (2) find an antiderivative F of f; and (3) evaluate "$F(b) - F(a)$." For the area A above, $f(x) = x^2 + 1$ is continuous, and $F(x) = x^3/3 + x$, so $A = F(3) - F(0) = 12$.

The Evaluation Theorem is fast, but it requires an antiderivative of f. Sometimes even after trying advanced integration techniques—we'll cover those later in this chapter—one is unable to find that precious antiderivative required by the theorem. But not all hope is lost:

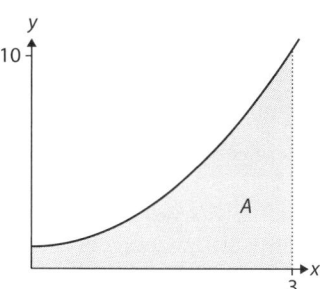

Figure 2.2: $f(x) = x^2 + 1$.

We can still approximate the definite integral. In this section we'll work on approximating definite integrals using areas of rectangles. We'll then extend that to using trapezoids in section 2.3. The best part of all that: no antiderivatives required!

Left- and Right-Hand Riemann Sums

To begin, let's return to the area A in figure 2.2 and calculus the figure. The area pictured in that figure is a static image; it's just a gray blob on printed paper (or on a screen). But both the printer and the screen needed to *shade in* the gray-colored area to produce the result you see. And *how* did the printer/screen know where on the paper/screen to shade? It used a grid. So let's add that to the figure. Figure 2.3(a) shows the result.

Let's now use our grid to approximate the area A. Notice first that the grid in figure (a) subdivides the interval $[0, 3]$ into three subintervals of equal width: $[0, 1]$, $[1, 2]$, and $[2, 3]$. Such equal-width partitions of an interval are called **equipartitions**, and the common width of their subdivisions is denoted Δx. (In figure (a), $\Delta x = 1$.) We're now going to erect rectangles from these subdivisions and add up their areas to obtain our approximations to A. We've settled on the base of those rectangles—the subdivisions $[0, 1]$, $[1, 2]$, and $[2, 3]$—but not on their heights. Figure (b) illustrates one choice: make the rectangles' heights the y-values at the *right endpoints* of each subdivision. The height of the first rectangle is the y-value of $f(x)$ at $x = 1$; that is, $f(1) = 1^2 + 1 = 2$. The same goes for the heights of the second and third rectangles: $f(2) = 2^2 + 1 = 5$ is the height of the second, and $f(3) = 3^2 + 1 = 10$ the height of the third. Since a rectangle's area is "base times

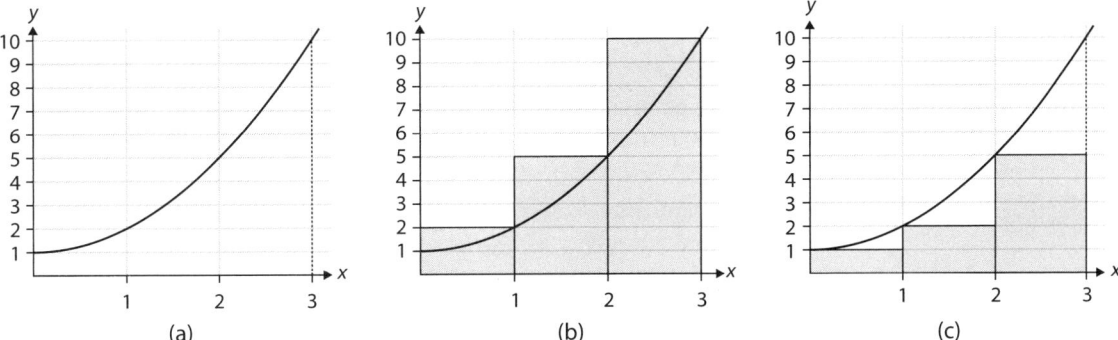

Figure 2.3: Part of the graph of $f(x) = x^2 + 1$ (a) along with coarse approximations to the area A between it and the x-axis in (b)–(c). The total area of the shaded region in (b) overapproximates A; the one in (c) underapproximates it.

height," the area of the first rectangle is therefore $\Delta x \cdot f(1)$, that of the second is $\Delta x \cdot f(2)$, and that of the third is $\Delta x \cdot f(3)$. Their total area is therefore

$$f(1)\Delta x + f(2)\Delta x + f(3)\Delta x = \sum_{i=1}^{3} f(i)\Delta x, \qquad (2.2)$$

where I've introduced **sigma notation**. Briefly,

$$\sum_{i=1}^{n} (\text{stuff involving } i)$$

means "sum whatever the quantity involving i is, starting with its value when $i = 1$ and ending with its value when $i = n$." For example,

$$\sum_{i=1}^{4} i^2 = 1^2 + 2^2 + 3^2 + 4^2.$$

Returning to (2.2) and using $f(x) = x^2 + 1$ and $\Delta x = 1$ yields

$$\sum_{i=1}^{3} f(i)\Delta x = [f(1) + f(2) + f(3)]\Delta x = [2 + 5 + 10](1) = 17. \qquad (2.3)$$

This is the total area of the shaded rectangles in figure 2.3(b), and it yields our first approximation to the area A in figure 2.2: $A \approx 17$.

The quantity on the right-hand side of (2.2) is an example of something called a **right-hand equipartition Riemann sum.**[*] We could also approximate A using a **left-hand equipartition Riemann sum**. The procedure is the same, except that we use the *left endpoints* of each subdivision in the equipartition to generate the rectangles' heights. Figure 2.3(c) illustrates the three rectangles that approach generates. The resulting (left-hand) Riemann sum is

$$f(0)\Delta x + f(1)\Delta x + f(2)\Delta x = \sum_{i=0}^{2} f(i)\Delta x.$$

Since $f(x) = x^2 + 1$ and $\Delta x = 1$, this yields

$$\sum_{i=0}^{2} f(i)\Delta x = [f(0) + f(1) + f(2)]\Delta x = [1 + 2 + 5](1) = 8. \qquad (2.4)$$

[*] Bernhard Riemann gave us the first rigorous definition of the definite integral via exactly the method we're discussing now.

This is the total area of the shaded rectangles in figure 2.3(c), and it yields our second approximation to the area A in figure 2.2: $A \approx 8$.

We've succeeded in approximating A in two ways. Unfortunately, both of these approximations are pretty far from the actual value of A (recall we calculated $A = 12$). We used only three rectangles, though. Would our approximation get better if we used more? Figure 2.4 provides a clue. Therein we see the left- and right-hand Riemann sum approximations with $n = 10$ and $n = 30$ rectangles. (I've also introduced the notation "L_n" for the left-hand equipartition Riemann sum with n rectangles, and "R_n" for the right-hand one.) Those pictures suggest that as n gets larger, L_n and R_n better approximate A.* How can we verify this? Well, we could calculate L_n and R_n explicitly. But doing the computations for $n = 100$, for example, would be tedious. So let's systematize what we did in figure 2.3 and derive general formulas for L_n and R_n to see if that helps.

* Scan the QR code below to see this for L_n.

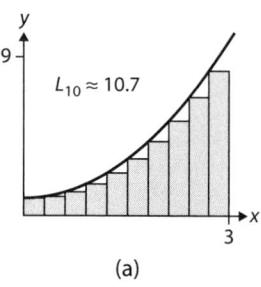
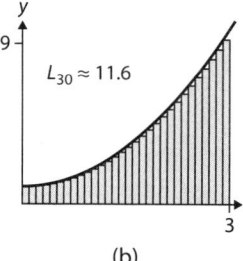
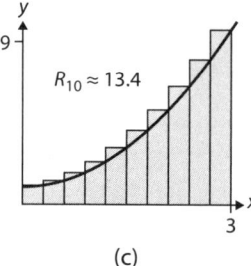
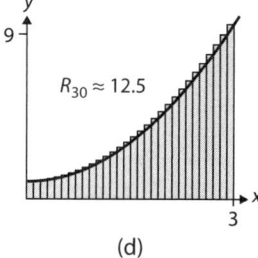

(a)   (b)   (c)   (d)

Figure 2.4: Part of the graph of $f(x) = x^2 + 1$. In (a) and (b) the area under the curve is approximated using left-hand approximations using 10 and 30 rectangles, respectively. In (c) and (d) right-hand approximations with 10 and 30 rectangles, respectively, are used. Access the interactive version of this figure at sites.google.com/view/fernandezmath/apps/cs2 to dynamically see how changing the number of rectangles changes the approximation.

Our first step is to equipartition the interval of interest. We're doing things more generally now, so we'll denote that interval by $[a, b]$. Now, chop up $[a, b]$ into n subintervals of equal width Δx; figure 2.5(a) illustrates that, where I've set $x_0 = a$ and $x_n = b$. Notice that because there are n subintervals, each of which has width Δx, then $n\Delta x$ is the total distance from a to b. That distance is also $b - a$, so

$$n\Delta x = b - a \implies \Delta x = \frac{b - a}{n}.$$

We've now got our equipartition, which I'll denote henceforth by $[x_0, x_1, x_2, \ldots, x_n]$, and a formula for Δx. Next up: write out the Riemann sums.

To do that, notice that each consecutive pair of x-values in the equipartition defines a subinterval (e.g., $[x_3, x_4]$). The right-hand equipartition Riemann sum R_n uses the right endpoints of each of those subintervals as the function inputs:

$$R_n = f(x_1)\Delta x + f(x_2)\Delta x + \cdots + f(x_n)\Delta x = \sum_{i=1}^{n} f(x_i)\Delta x.$$

This is the generalized version of (2.3). The left-hand equipartition Riemann sum L_n uses the left endpoints of each of those subintervals as the function inputs, so

$$L_n = f(x_0)\Delta x + f(x_1)\Delta x + \cdots + f(x_{n-1})\Delta x = \sum_{i=0}^{n-1} f(x_i)\Delta x.$$

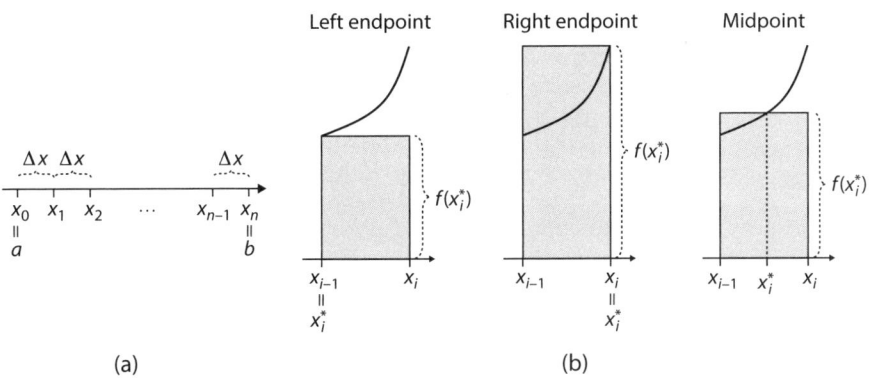

Figure 2.5: (a) An equipartition of $[a, b]$ into n subintervals of equal width Δx. (b) The rectangles that result from the different x_i^* choices.

This is the generalized version of (2.4). The following definition records our results.

DEFINITION 2.1: RIGHT- AND LEFT-HAND EQUIPARTITION RIEMANN SUMS. Let f be a function defined on $[a, b]$. Partition $[a, b]$ into n subintervals of equal width $\Delta x = (b - a)/n$. Then:

- The **right-hand equipartition Riemann sum**, R_n, is defined by

$$R_n = \sum_{i=1}^{n} f(x_i)\Delta x = [f(x_1) + f(x_2) + \cdots + f(x_n)]\Delta x. \qquad (2.5)$$

- The **left-hand equipartition Riemann sum**, L_n, is defined by

$$L_n = \sum_{i=0}^{n-1} f(x_i)\Delta x = [f(x_0) + f(x_1) + \cdots + f(x_{n-1})]\Delta x. \qquad (2.6)$$

EXAMPLE 2.1   Verify that the $n = 3$ left-hand Riemann sum for $\int_0^3 (x^2 + 1)\, dx$ is $L_3 = 8$.

Solution Here $a = 0$, $b = 3$, and $n = 3$, so $\Delta x = \frac{3-0}{3} = 1$. Starting at $a = 0$ and adding multiples of Δx until we get to $b = 3$ generates the equipartition

$$[0, 1, 2, 3].$$

The left-hand Riemann sum (2.6) uses the bolded x-values (the left endpoints of the subintervals) as the function inputs. And since $f(x) = x^2 + 1$ here,

$$L_3 = \left[\underbrace{(0^2 + 1)}_{f(0)} + \underbrace{(1^2 + 1)}_{f(1)} + \underbrace{(2^2 + 1)}_{f(2)} \right] \underbrace{(1)}_{\Delta x} = 8. \qquad \blacksquare$$

EXAMPLE 2.2   Approximate $\displaystyle\int_0^4 x^2\, dx = 21.\overline{3}$ with an $n = 10$ left-hand Riemann sum.

Solution Here $a = 0$, $b = 4$, and $n = 10$, so $\Delta x = \frac{4-0}{10} = \frac{2}{5} = 0.4$. Starting at $a = 0$ and adding multiples of Δx until we get to $b = 4$ generates the equipartition

$$\left[0, \frac{2}{5}, \frac{4}{5}, \ldots, \frac{18}{5}, 4 \right].$$

The left-hand Riemann sum (2.6) uses the bolded x-values (the left endpoints of the subintervals) as the function inputs. And since $f(x) = x^2$ here,

$$L_{10} = \left[0^2 + \left(\frac{2}{5} \right)^2 + \left(\frac{4}{5} \right)^2 + \cdots + \left(\frac{18}{5} \right)^2 \right] (0.4) = 18.24. \qquad \blacksquare$$

EXAMPLE 2.3   Set up, but do not evaluate, the $n = 24$ right-hand Riemann sum approximation to $\int_{-1}^{1} x^3 \, dx = 0$.

Solution Here $a = -1$, $b = 1$, and $n = 24$, so $\Delta x = \frac{1-(-1)}{24} = \frac{2}{24} = \frac{1}{12}$. Starting at $a = -1$ and adding multiples of Δx until we get to $b = 1$ generates the equipartition

$$\left[-1, -\frac{11}{12}, -\frac{10}{12}, \ldots, \frac{11}{12}, 1 \right].$$

The right-hand Riemann sum (2.5) uses the bolded x-values (the right endpoints of the subintervals) as the function inputs. And since $f(x) = x^3$ here,

$$R_{24} = \left[\left(-\frac{11}{12} \right)^3 + \left(-\frac{10}{12} \right)^3 + \cdots + \left(\frac{11}{12} \right)^3 + 1^3 \right] \left(\frac{1}{12} \right). \qquad \blacksquare$$

* Remember: Exercises with an A prefix are applications.

Related Exercises   1–3, A1–A2, A3(d).*

APPLICATIONS   Applied exercise A1 uses Riemann sums to estimate the *cardiac output* of a person's heart (the volume of blood it pumps per second). Applied exercise A2 explores the **normal distribution**, which is used in statistics and the physical and social sciences to describe the probability that an event occurs. Applied exercises A3 explore a population's survival function, life expectancy at birth, and life span inequality.

** We'll soon discuss figure 2.6(c).

Figures 2.6(a)–(b) illustrate the results of the last two examples we just did.** (In figure (b), the light gray-colored rectangles are below the x-axis because $f(x) = x^3 < 0$ for $x < 0$.)

We've succeeded in systematizing the calculation of L_n and R_n. But it still takes a while to carry out the calculation. Verifying the $R_{24} \approx 0.083$ in figure (b), for example, would entail cubing the 24 terms in the R_{24} equation in the previous example and then adding up the results. That's be a lot of work. Luckily, if $f(x)$ is a polynomial there are useful *summation formulas* that can help. Section A2.1 in the appendix discusses those, if you're interested. Those formulas speed up calculations but still don't answer our nagging question: Are we *sure* that as n increases, L_n and R_n better approximate the definite integral?

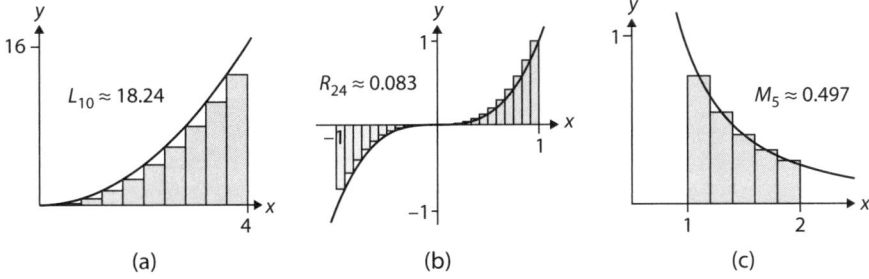

Figure 2.6: Illustrating the results from examples 2.2–2.4.

More General Riemann Sums

Maybe we've been too rigid in our approach. So let's try injecting some freedom of choice to see if that helps. Let's go back to the equipartition $[x_0, x_1, x_2, \ldots, x_n]$ of $[a, b]$ and now allow ourselves to choose *any* x-values in the subintervals $[x_0, x_1]$, $[x_1, x_2]$, etc., to form the Riemann sum (not just the left- or right-hand endpoints). If we denote by x_1^* the one chosen in the interval $[x_0, x_1]$, by x_2^* the one chosen in the interval $[x_1, x_2]$, etc., then the resulting Riemann sum is

$$S_n = f(x_1^*)\Delta x + f(x_2^*)\Delta x + \cdots + f(x_n^*)\Delta x = \sum_{i=1}^{n} f(x_i^*)\Delta x. \qquad (2.7)$$

This more general Riemann sum can spawn new Riemann sums beyond the left- and right-hand ones we've been working with. One notable example is the **midpoint Riemann sum**, illustrated in figure 2.6(c). In this Riemann sum one chooses each x_i^* to be the midpoint of the interval $[x_{i-1}, x_i]$:

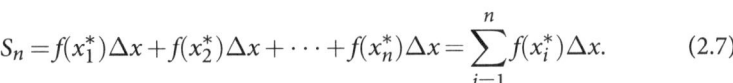

Figure 2.5 illustrates this, along with how we can now view the left- and right-hand Riemann sums as special cases of (2.7) corresponding to particular choices of x_i^*. The definition below records all this, and also provides us with an explicit formula for the midpoint Riemann sum.

DEFINITION 2.2: EQUIPARTITION RIEMANN SUMS. Let f be a function defined on $[a, b]$. Partition $[a, b]$ into n subintervals of equal width $\Delta x = (b - a)/n$. Let x_i^* be such that $x_{i-1} \leq x_i^* \leq x_i$. Then the **equipartition Riemann sum** S_n of f over $[a, b]$ is

$$S_n = \sum_{i=1}^{n} f(x_i^*)\Delta x = [f(x_1^*) + f(x_2^*) + \cdots + f(x_n^*)]\Delta x. \qquad (2.8)$$

When $x_i^* = \frac{x_{i-1}+x_i}{2}$ the resulting Riemann sum is called the **midpoint Riemann sum:**[3]

---

[3]The definition assumes an equipartition of $[a, b]$. One can generalize this to partitions with varying widths, Δx_i.

$$M_n = \sum_{i=1}^{n} f\left(\frac{x_{i-1} + x_i}{2}\right) \Delta x$$

$$= \left[f\left(\frac{x_0 + x_1}{2}\right) + f\left(\frac{x_1 + x_2}{2}\right) + \cdots + f\left(\frac{x_{n-1} + x_n}{2}\right) \right] \Delta x. \quad (2.9)$$

EXAMPLE 2.4   Approximate $\int_1^2 \frac{1}{x^2} \, dx = 0.5$ with an $n = 5$ midpoint Riemann sum.

Solution Here $a = 1$, $b = 2$, and $n = 5$, so $\Delta x = \frac{2-1}{5} = \frac{1}{5} = 0.2$. Starting at $a = 1$ and adding multiples of Δx until we get to $b = 2$ generates the equipartition

$$[1, 1.2, 1.4, 1.6, 1.8, 2].$$

The midpoint Riemann sum (2.9) uses the average of each consecutive pair of x-values as the function inputs (example: $\frac{1+1.2}{2} = \frac{2.2}{2} = 1.1$):

$$[1.1, 1.3, 1.5, 1.7, 1.9].$$

And since $f(x) = \frac{1}{x^2}$ here,

$$M_5 = \left[\frac{1}{(1.1)^2} + \frac{1}{(1.3)^2} + \frac{1}{(1.5)^2} + \frac{1}{(1.7)^2} + \frac{1}{(1.9)^2} \right] (0.2) \approx 0.497.$$

Figure 2.6(c) illustrates this result.   ■

Related Exercise 4.

Notice that the answer we just got for M_5 is *very* close to the actual value of the integral (0.5). Look back at the last few examples and you'll see that this is the most accurate definite integral approximation we've obtained thus far. We're on to something here. We still can't determine the accuracy of our approximations without knowing the value of the definite integral. But breaking out of the left/right Riemann sum mold has dramatically increased our accuracy. So let's continue following that trail of breadcrumbs and see where it leads.

2.3   The Trapezoidal Rule

We've varied the x_i^* values used to form the Riemann sum but kept the rectangles. *What if we used different shapes?* Rectangles' tops are flat. What if we allowed them to slope? The resulting shape is a *trapezoid*, and the associated definite integral approximation is called the *Trapezoidal Rule*.

To introduce the Trapezoidal Rule, let's return to $f(x) = x^2 + 1$ and try approximating the area between $x = 0$ and $x = 3$ (the shaded area in figure 2.7(a)) using trapezoids. Figure 2.7(b) uses the same equipartition of $[1, 3]$ we constructed before. But now—the new part—it constructs (three) *right trapezoids** from the associated subintervals. Now, the area of a right trapezoid with base length b and heights h_1 and h_2 is $\frac{1}{2}b(h_1 + h_2)$. The base length of the three trapezoids in the figure is the same; let's call it Δx again. What changes are the two heights of each trapezoid. For the first trapezoid, for example, the leftmost height is $f(0)$ and the rightmost is $f(1)$.

* A trapezoid with two right angles in it.

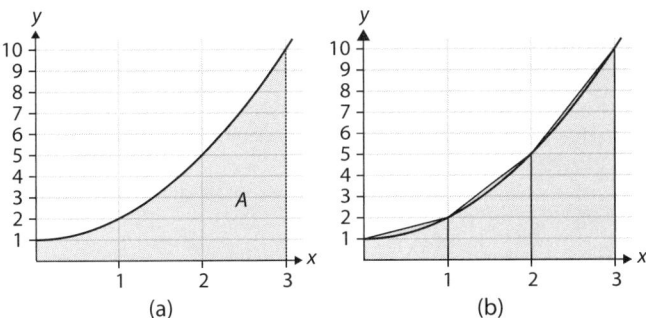

Figure 2.7: (a) $f(x) = x^2 + 1$ along with the area A under the curve and between $x = 0$ and $x = 3$. (b) The trapezoidal approximation (using three trapezoids) to the area under the graph in (a).

Therefore its area is

$$\frac{1}{2}[f(\mathbf{0}) + f(\mathbf{1})]\Delta x.$$

I've bolded the numbers here to highlight for you that 0 and 1 are the left and right endpoints of the base of that first trapezoid. Similarly, the areas of the next two trapezoids are

$$\frac{1}{2}[f(1) + f(2)]\Delta x, \quad \frac{1}{2}[f(2) + f(3)]\Delta x,$$

respectively. The "trapezoidal approximation," therefore, to the area A in figure 2.7(a) is

$$A \approx \frac{1}{2}[f(0) + f(1)]\Delta x + \frac{1}{2}[f(1) + f(2)]\Delta x + \frac{1}{2}[f(2) + f(3)]\Delta x.$$

Factoring out a $\frac{1}{2}\Delta x$ from each term yields

$$A \approx \frac{1}{2}[f(0) + f(1) + f(1) + f(2) + f(2) + f(3)]\Delta x$$

$$= \frac{1}{2}[f(0) + 2f(1) + 2f(2) + f(3)]\Delta x. \tag{2.10}$$

And since $f(x) = x^2 + 1$ and $\Delta x = 1$, this sum yields 12.5. (This is the total area of the shaded trapezoids in figure 2.7(b).) This is pretty close to the actual value $A = 12$ we calculated at the start of this chapter. It's also a *way* better approximation than the L_3 and R_3 approximations (the left- and right-hand Riemann sums with three rectangles) from the previous section. (Compare figure 2.7(b) and figures 2.3(b)–(c) to see that visually.)

Let's now generalize what we've done to handle more than three rectangles in our approximation. We'll start with figure 2.8. To create this figure I replaced the rectangle in figure 2.5(b) with a right trapezoid. The area of the shaded trapezoid is

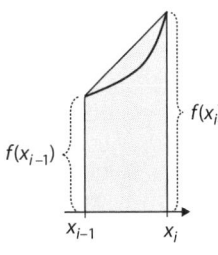

Figure 2.8

$$\frac{1}{2}[f(x_{i-1}) + f(x_i)]\Delta x,$$

where $\Delta x = x_i - x_{i-1}$. On an equipartition $[x_0, x_1, \ldots, x_{n-1}, x_n]$ of $[a, b]$, the sum T_n of the n associated trapezoidal areas would be

$$T_n = \frac{1}{2}[f(x_0) + f(x_1)]\Delta x + \frac{1}{2}[f(x_1) + f(x_2)]\Delta x + \frac{1}{2}[f(x_2) + f(x_3)]\Delta x + \cdots$$

$$+ \frac{1}{2}[f(x_{n-2}) + f(x_{n-1})]\Delta x + \frac{1}{2}[f(x_{n-1}) + f(x_n)]\Delta x$$

$$= \frac{1}{2}[f(x_0) + f(x_1) + f(x_1) + \cdots + f(x_{n-1}) + f(x_{n-1}) + f(x_n)]\Delta x$$

$$= \frac{1}{2}[f(x_0) + 2f(x_1) + 2f(x_2) + \cdots + 2f(x_{n-1}) + f(x_n)]\Delta x. \qquad (2.11)$$

This approximation is known as the *Trapezoidal Rule*.

> **DEFINITION 2.3: THE TRAPEZOIDAL RULE.** Let f be a function defined on $[a, b]$, and let $[x_0, x_1, \ldots, x_{n-1}, x_n]$, where $a = x_0 < x_1 < x_2 < \cdots < x_n = b$, be an equipartition of $[a, b]$ into n subintervals of equal width $\Delta x = x_i - x_{i-1}$. Then the **equipartition Trapezoidal Rule** T_n is the sum
>
> $$T_n = \frac{1}{2}\sum_{i=1}^{n}[f(x_{i-1}) + f(x_i)]\Delta x.$$

In practice we typically apply the Trapezoidal Rule in the form presented in (2.11). Another connection: $T_n = \frac{L_n + R_n}{2}$. That is, the Trapezoidal Rule is the average of the left- and right-hand Riemann sums.*

* To prove this, add the sums in (2.5) and (2.6) and then divide by 2.

EXAMPLE 2.5 Approximate $\int_1^2 \frac{1}{x^2}\, dx = 0.5$ with an $n = 5$ Trapezoidal Rule.

Solution Here $a = 1$, $b = 2$, and $n = 5$, so $\Delta x = \frac{2-1}{5} = \frac{1}{5} = 0.2$. Starting at $a = 1$ and then adding multiples of Δx until we get to $b = 2$ generates the equipartition

$$[1, 1.2, 1.4, 1.6, 1.8, 2].$$

The Trapezoidal Rule uses all these x-values as the function inputs. Thus, (2.11) becomes

$$T_5 = \frac{1}{2}\left[\frac{1}{1^2} + \frac{2}{(1.2)^2} + \frac{2}{(1.4)^2} + \frac{2}{(1.6)^2} + \frac{2}{(1.8)^2} + \frac{1}{2^2}\right](0.2) \approx 0.506,$$

where the highlighted 2s come from the 2s in the formula for the Trapezoidal Rule. ∎

EXAMPLE 2.6 Approximate $\int_0^3 (x^2 + 1)\, dx = 12$ with an $n = 3$ Trapezoidal Rule by using $T_3 = \frac{L_3 + R_3}{2}$. Compare the accuracy of this to our earlier results: $L_3 = 8$, $R_3 = 17$.

Solution We have $T_3 = \frac{8+17}{2} = 12.5$. This number matches what we calculated earlier and is much more accurate than L_3 or R_3. ∎

Related Exercises 5–6, A3(a).

This last example hints that the Trapezoidal Rule generates more accurate approximations than the left- and right-hand Riemann sums. And as you might suspect from glancing at figure 2.7(b), increasing the n-value used appears to generate even more accurate approximations.* Let's finally tackle these accuracy questions.

* Scan the QR code below to see an animation of how T_n changes as n changes.

2.4 How to Approximate Integrals to Any Desired Accuracy

Let's return to $\int_0^3 (x^2 + 1)\, dx = 12$ and our two Riemann sum approximations of it, $R_3 = 17$ and $L_3 = 8$, from figure 2.3(b)–(c). The errors in these approximations—which we'll denote by ER_3 and EL_3, respectively—are the differences between the value of the integral and the approximations R_3 and L_3:

$$ER_3 = \int_0^3 (x^2 + 1)\, dx - R_3 = 12 - 17 = -5$$

$$EL_3 = \int_0^3 (x^2 + 1)\, dx - L_3 = 12 - 8 = 4$$

EL_3 is positive because L_3 underestimates the area under the curve; ER_3 is negative because R_3 overestimates it. Figure 2.9(a) illustrates this: EL_3 is the sum of the areas of the darker gray regions, and $|ER_3| = |-5| = 5$ is the sum of the areas of the lighter gray regions. We see similar results if we look back at figure 2.4: L_n seems to always *underestimate* the area under the curve while R_n always seems to *overestimate* it. This turns out to be true in general for nondecreasing functions,[4] as the following theorem shows.

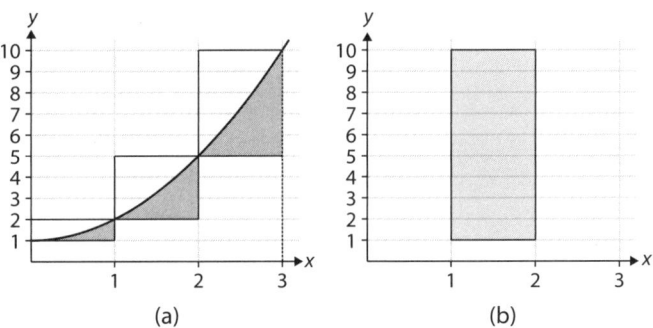

(a)   (b)

Figure 2.9: (a) The sum of the darker gray areas is how much L_3 underapproximates $\int_0^3 (x^2 + 1)\, dx$; the sum of the lighter gray areas is how much R_3 overapproximates it. (b) The gray-colored areas in (a) collapsed into one rectangle.

THEOREM 2.2: Let f be integrable on $[a, b]$,** and let L_n and R_n denote its associated left- and right-hand equipartition Riemann sums.

- If f is nondecreasing on $[a, b]$, then

$$L_n \leq \int_a^b f(x)\, dx \leq R_n. \tag{2.12}$$

** "Integrable" means that $\int_a^b f(x)\, dx$ is finite.

[4] A *nondecreasing* function's graph is *not* decreasing. Examples: $y = x$, $y = 2$. A *nonincreasing* function's graph is *not* increasing. Examples: $y = 5$, $y = -x$.

- If f is nonincreasing on $[a, b]$, then

$$R_n \leq \int_a^b f(x)\, dx \leq L_n. \qquad (2.13)$$

Let's now combine the insights of this theorem with our earlier calculations of EL_3 and ER_3 and the nice visual of figure 2.9(a). Here's something you might notice about that figure: If you combine the light and dark gray areas into one rectangle, you get the rectangle pictured in figure (b). That rectangle's base length is Δx, and its height is $10 - 1 = f(3) - f(0)$. Therefore, the area of the rectangle is $[f(3) - f(0)]\Delta x$. And because this area combines *both* the light and dark gray areas in figure (a), the shaded area in figure (b) is *larger than* the errors $|EL_3|$ and $|ER_3|$. In math speak: $|EL_3| < [f(3) - f(0)]\Delta x$ and $|ER_3| < [f(3) - f(0)]\Delta x$. Introducing now

$$EL_n = \int_a^b f(x)\, dx - L_n, \qquad ER_n = \int_a^b f(x)\, dx - R_n,$$

the errors in approximating a definite integral with L_n or R_n, respectively, our work suggests that perhaps $|EL_n| < [f(b) - f(a)]\Delta x$ and $|ER_n| < [f(b) - f(a)]\Delta x$. Recalling that $\Delta x = \frac{b-a}{n}$, we've anticipated the next theorem.*

> * Exercise D1 guides you through the theorem's proof. Remember: Exercises with a D prefix are derivations or explorations.

THEOREM 2.3: ERRORS FOR LEFT- AND RIGHT-HAND RIEMANN SUM APPROXIMATIONS. Let f be integrable on $[a, b]$, and let L_n and R_n denote its associated left- and right-hand equipartition Riemann sums. Suppose further that f is either nondecreasing on $[a, b]$ or nonincreasing on $[a, b]$. Then

$$|EL_n| \leq \frac{(b-a)|f(b) - f(a)|}{n}, \qquad |ER_n| \leq \frac{(b-a)|f(b) - f(a)|}{n}. \qquad (2.14)$$

This is the type of theorem we've been looking for. The inequalities in (2.14) tell us the *largest* errors that the left- and right-hand Riemann sum approximations generate. And since the right-hand side of (2.14) gets smaller as n increases, we can use the inequality to tell us how many terms n we need to add up in our left- or right-hand Riemann sums to reach a desired error tolerance. Here's a concrete example of how you'd do that.

EXAMPLE 2.7 (Video)   Use theorem 2.3 to determine the smallest n-value guaranteeing that a left-hand Riemann sum approximation to $\int_1^2 \frac{1}{x}\, dx$ is accurate to within 0.01.

Solution Here $a = 1$, $b = 2$, and $f(x) = \frac{1}{x}$, a decreasing function on $[1, 2]$. Furthermore, f is integrable on $[1, 2]$ since the corresponding area under the curve is finite. Theorem 2.3 therefore applies, which yields

$$|EL_n| \leq \frac{(2-1)|f(2) - f(1)|}{n} = \frac{|\frac{1}{2} - 1|}{n} = \frac{1}{2n}.$$

> ** The steps: Start with $\frac{1}{2n} \leq \frac{1}{100}$, take reciprocals to get $2n \geq 100$, and then divide by 2.

Setting $\frac{1}{2n} \leq 0.01 = \frac{1}{100}$ and solving for n yields $n \geq 50$.** ∎

The takeaway from this example: *Without knowing the value of $\int_1^2 \frac{1}{x}\, dx$, we are guaranteed* that L_{50} approximates that value within 0.01. I find that amazing.

Another takeaway from this example is the procedure we should follow to solve similar problems: First, ensure that f is integrable on $[a, b]$ and nonincreasing (or nondecreasing), so that theorem 2.3 applies; then, calculate the right-hand side of the relevant inequality in (2.14); finally, set that result less than or equal to the error tolerance (e.g., 0.01) and solve for n.

We saw in earlier examples that the Midpoint and Trapezoidal Rules seemed more accurate than the left- and right-hand Riemann sum approximations. A result similar to theorem 2.3 confirms this for the analogous errors EM_n and ET_n stemming from approximating $\int_a^b f(x)\, dx$ with the Midpoint or Trapezoidal Rules.[5] A more useful version is contained in the next theorem, half of which uses the fact that the Trapezoidal Rule overestimates the area under the graph of f when f is concave up (see figure 2.7(b)), and underestimates that area when f is concave down.

THEOREM 2.4:   ERRORS FOR MIDPOINT AND TRAPEZOIDAL RULE APPROXIMATIONS. Let f be integrable on $[a, b]$, and let M_n and T_n denote its associated equipartition Midpoint and Trapezoidal Rule approximations. Suppose further that f is either always concave up or always concave down on $[a, b]$, and that $|f''(x)| \leq U$ on $[a, b]$. Then

$$|EM_n| \leq U\frac{(b - a)^3}{24n^2}, \qquad |ET_n| \leq U\frac{(b - a)^3}{12n^2}. \qquad (2.15)$$

(Section A2.2 contains a result similar to this theorem but for the left- and right-hand Riemann sums.)

EXAMPLE 2.8 (Video)   Use theorem 2.4 to determine the smallest n-value guaranteeing that a Trapezoidal Rule approximation to $\int_1^2 \frac{1}{x}\, dx$ is accurate to within 0.01.

Solution We verified all but the concavity assumption in the previous example. Since $f(x) = x^{-1}$, $f'(x) = -x^{-2}$, and $f''(x) = 2x^{-3} > 0$, f is concave up for all x,[*] so the theorem applies. And since $f''(x) = \frac{2}{x^3}$ is a decreasing function on $[1, 2]$, its maximum value U occurs at $x = 1$ and is $\frac{2}{1^3} = 2$. Therefore, from (2.15),

> [*] Recall from Calculus 1 that $f''(x) > 0$ on (a, b) implies that f is concave up on (a, b).

$$|ET_n| \leq 2 \cdot \frac{(2 - 1)^3}{12n^2} = \frac{1}{6n^2}.$$

Setting $\frac{1}{6n^2} \leq 0.01 = \frac{1}{100}$ and taking reciprocals yields $6n^2 \geq 100$ or $n \geq \sqrt{100/6} \approx 4.08$. Conclusion: We'd need $n \geq 5$, much less than the $n \geq 50$ from example 2.7. ∎

Related Exercises 10–12.

This section's theorems give us methods to approximate a definite integral to a desired error *without* needing to know that integral's value beforehand. These theorems make assumptions—for example, that the integrand is nonincreasing on $[a, b]$—and so don't apply for every definite integral. But they're more than adequate

---

[5] In case you're interested, I've put the details in section A2.2.

for our purposes here—we can turn to them in the cases when we can't evaluate a definite integral exactly.

Stepping back now, notice that all the error bounds in the theorems above have powers of n in the denominator. This implies that as n increases those errors get smaller, finally confirming what we've been suspecting and seeing in figures like figure 2.4. We'd therefore expect that as $n \to \infty$ the corresponding Riemann sums (e.g., L_n, R_n, M_n) and Trapezoidal Rule T_n would yield the *exact* value of the definite integral. And indeed it does, under the right conditions. Exercise D2 explores this.

> **EXPLORATIONS**   Exercise D2 explores how to define the definite integral of a continuous function in terms of the limit of an equipartition Riemann sum. Note: This exercise uses the results of sections A2.1–A2.2 and the exercises suggested therein.

We now know how to approximate definite integrals to any desired accuracy. (Awesome!) So let's return to Evaluation Theorem 2.1. Recall that this gives us a way to evaluate a definite integral exactly. But it requires an antiderivative. In Calculus 1 you were taught a few techniques for finding antiderivatives. The most advanced of those was u-substitution (reviewed in appendix D). What if none of those techniques works for the integral you're trying to evaluate? That's where the rest of this chapter comes in. For the remainder of the chapter we'll work on developing additional integration techniques that hopefully apply in cases where u-substitution doesn't.

2.5   Integration by Parts

Let's start our tour of Calculus 2's new integration techniques by recalling the techniques you already learned in Calculus 1. Those fall into two categories: (1) using an integration formula (e.g., $\int x^n\, dx = \frac{x^{n+1}}{n+1} + C$, for $n \neq -1$), or (2) using u-substitution. (Both are reviewed in appendix D.) You may recall that u-substitution is the "anti-Chain Rule." That is, it's the integration technique that results from undoing the Chain Rule.* The technique we'll develop in this lesson—integration by parts—is the "anti-Product Rule."[6] Here's what I mean.

First, recall the Product Rule from Calculus 1:

$$[f(x)g(x)]' = f'(x)g(x) + f(x)g'(x). \quad \text{Example: } [x \sin x]' = \sin x + x \cos x$$

If we solve this for $f(x)g'(x)$,

$$f(x)g'(x) = [f(x)g(x)]' - f'(x)g(x), \quad \text{Continuing: } x \cos x = [x \sin x]' - \sin x$$

and then integrate both sides, we get a new integration technique.

* This is discussed in detail in *Calculus Simplified* [2], section 5.9.

---

[6] Recall that an indefinite integral, $\int f(x)\, dx$, produces an *antiderivative* of f, that is, a function F such that $F' = f$. So, integrating f is like undoing the (hidden) derivative on f, hence the "anti" in "antiderivative." It's in this sense that I'm carrying over the usage of "anti" to describe u-substitution and integration by parts. The former undoes the Chain Rule; the latter undoes the Product Rule.

> **THEOREM 2.5: INTEGRATION BY PARTS.** Let f and g be differentiable. Then
> $$\int f(x)g'(x)\,dx = f(x)g(x) - \int f'(x)g(x)\,dx. \qquad (2.16)$$

Continuing with our running example from above (the gray-colored equations), we could then write

$$\int x \cos x\,dx = x \sin x - \int \sin x\,dx,$$

which yields a final answer of $x \sin x + \cos x + C$.

Our success integrating $x \cos x$ hinged on knowing the derivative of $x \sin x$, manipulating that to obtain the integrand $x \cos x$, and then applying (2.16). That's too complicated. So let's systematize and simplify the process.

First, note that, being the "anti-Product Rule," this new integration technique works best on integrands that are products of two functions (like $x \cos x$). In practice, we often see the integration by parts formula written as

$$\int u\,dv = uv - \int v\,du, \quad \text{or} \quad \int_a^b u\,dv = [uv]_a^b - \int_a^b v\,du. \qquad (2.17)$$

We get to this by letting $u = f(x)$ and $dv = g'(x)\,dx$, from which we get $du = f'(x)\,dx$ and $v = g(x)$.* Substituting these into (2.16) yields (2.17). In this new representation of integration by parts, we choose u from the functions in the integrand, and the remainder of the integral (including dx) is dv. We then calculate du (the derivative of u multiplied by dx) and v (the integral of dv) and substitute the results into the right-hand side of (2.17). If the $\int v\,du$ integral is easier to evaluate than the integral we started with, then we've made progress.

* This is the one time a calculus professor will tell you that you don't need the "$+C$" when integrating indefinitely.

Like in u-substitution, choosing u is the most important step in applying integration by parts. What u should we choose, then? The acronym "ILATE" is a good guideline: choose u to be the function in the integrand that appears first in figure 2.10 (reading top to bottom). The rationale: This choice of u often yields a du that helps simplify the "$v\,du$" in the integral you'll be left with.

I	Inverse trigonometric functions	$\arctan x$, $\arcsin x$, ...
L	Logarithmic functions	$\ln x$, $\log x$, ...
A	Algebraic functions	$x, x^2, 3x^3 - 2x + 1, \ldots$
T	Trigonometric functions	$\sin x$, $\tan x$, $\cos x$, ...
E	Exponential functions	$e^x, 2^x, e^{-x}, \ldots$

Figure 2.10

EXAMPLE 2.9 (Video)

Evaluate the integral:

(a) $\displaystyle\int xe^x\,dx$ (b) $\displaystyle\int x \ln x\,dx$ (c) $\displaystyle\int x \cos x\,dx$

Solution

(a) Using "ILATE," the integrand has an "A" function (x) and an "E" function (e^x). "A" comes before "E" in ILATE, so let's try $u = x$, $dv = e^x\,dx$. This yields $du = dx$

and $v = e^x$,* so that (2.17) becomes

$$\int \underbrace{x}_{u} \; \underbrace{e^x \, dx}_{dv} = \underbrace{x}_{u} \; \underbrace{e^x}_{v} - \int \underbrace{e^x}_{v} \; \underbrace{dx}_{du} = xe^x - e^x + C.$$

(b) The integrand has an "A" function (x) and an "L" function ($\ln x$). "L" comes before "A" in ILATE, so let's try $u = \ln x$, $dv = x \, dx$. This yields $du = \frac{1}{x} \, dx$ and $v = \frac{x^2}{2}$, so that (2.17) becomes

$$\int \underbrace{\ln x}_{u} \cdot \underbrace{x \, dx}_{dv} = \underbrace{\ln x}_{u} \cdot \underbrace{\frac{x^2}{2}}_{v} - \int \underbrace{\frac{x^2}{2}}_{v} \cdot \underbrace{\frac{1}{x} \, dx}_{du}$$

$$= \frac{x^2}{2} \ln x - \frac{1}{2} \int x \, dx = \frac{x^2}{2} \ln x - \frac{x^2}{4} + C.$$

(c) The integrand has an "A" function (x) and a "T" function ($\cos x$). "A" comes before "T" in ILATE, so let's try $u = x$, $dv = \cos x \, dx$. This yields $du = dx$ and $v = \sin x$, so (2.17) becomes

$$\int \underbrace{x}_{u} \; \underbrace{\cos x \, dx}_{dv} = \underbrace{x}_{u} \; \underbrace{\sin x}_{v} - \int \underbrace{\sin x}_{v} \; \underbrace{dx}_{du} = x \sin x + \cos x + C. \quad \blacksquare$$

Related Exercises 13–18, A3(b)–(c), A4, D3–D4.

APPLICATIONS Applied exercise A4 uses integration by parts to explore *population density*, a measure of the concentration of a population within a prescribed radius about a city's center.

Using Integration by Parts More than Once

Sometimes we need to use integration by parts more than once. Often when that happens you'll end up obtaining an answer eventually. But sometimes, as the next example shows, you end up with a multiple of the integral you started with.

EXAMPLE 2.10 Evaluate $\int e^x \sin x \, dx$.

Solution The integrand has an "E" function (e^x) and a "T" function ($\sin x$), so let's choose $u = \sin x$, $dv = e^x \, dx$. Then, $du = \cos x \, dx$ and $v = e^x$. Integrating by parts:

$$\int e^x \sin x \, dx = e^x \sin x - \int e^x \cos x \, dx.$$

The new integrand, $e^x \cos x$, has another "E" function (e^x) and another "T" function ($\cos x$). Letting $u = \cos x$ and $dv = e^x \, dx$, we get $du = -\sin x \, dx$ and $v = e^x$. Integrating by parts again:

$$\int e^x \sin x \, dx = e^x \sin x - \left[e^x \cos x - \int e^x (-\sin x) \, dx \right].$$

This simplifies to

$$\int e^x \sin x \, dx = e^x \sin x - e^x \cos x - \int e^x \sin x \, dx.$$

This equation has the form $z = A - z$, where z is the integral we're trying to evaluate. And just like we'd solve $z = A - z$ by adding z to both sides and then dividing by 2, we'll do the same here:

$$2 \int e^x \sin x \, dx = e^x \sin x - e^x \cos x \quad \Rightarrow \quad \int e^x \sin x \, dx = \frac{e^x \sin x - e^x \cos x}{2} + C,$$

where I've added the "$+C$" since it's an indefinite integral.   ■

Related Exercise   D5.

When you don't end up grappling with one of these "get back the same integral" problems and you simply need to apply integration by parts multiple times, there are a couple of techniques that may help. One is what's called a *reduction formula*. Example: Consider $\int x^n e^{ax} \, dx$, where $n \geq 1$ is an integer. Here we'd need exactly n applications of integration by parts. Luckily, for this integral—and some others—we can derive a formula that reduces the integral to $\int x^m e^{ax} \, dx$, where $m < n$. (Exercise D6 asks you to derive that reduction formula.) Another technique to mention is the *tabular method*. Section A2.3 discusses this method, which uses tables of u and its derivatives, and v and its integrals, to speed up the application of multiple integration by parts rounds.

The "anti-Chain Rule" (u-substitution) and "anti-Product Rule" (integration by parts) round out the integration techniques developed to undo differentiation rules. (There is no "anti-Quotient Rule," for example.) So let's shift our focus to developing integration techniques for specific function families (e.g., trigonometric functions).

2.6   Trigonometric Integrals

Sometimes neither u-substitution nor integration by parts works—at least not right away. This happens quite often in *trigonometric integrals*: integrals whose integrands involve solely trigonometric functions. Take, for example, the integral

$$\int (\sin x + \cos x)^2 \, dx.$$

The choices $u = \cos x$ and $u = \sin x$, pretty much the only reasonable choices for u here, won't work. Integration by parts won't help either. But simplifying the integral will.

To start, recall that $\sin^2 x + \cos^2 x = 1$.[*] From this we get the handy conversions

$$\sin^2 x = 1 - \cos^2 x, \quad \cos^2 x = 1 - \sin^2 x,$$
$$\sec^2 x = \tan^2 x + 1, \quad \tan^2 x = \sec^2 x - 1. \tag{2.18}$$

[*] These and other trig. identities are reviewed in appendix B.

These identities can be used to replace $\sin^2 x$, $\cos^2 x$, $\sec^2 x$, and $\tan^2 x$ terms in an integrand with the right-hand sides above. In many cases this facilitates a successful follow-up u-substitution. Let's see this in action for the integral above.

EXAMPLE 2.11   Evaluate $\int (\sin x + \cos x)^2 \, dx.$

Solution Multiplying out (and recalling that $\sin^2 x = (\sin x)^2$):

$$(\sin x + \cos x)^2 = \sin^2 x + \cos^2 x + 2 \sin x \cos x = 1 + 2 \sin x \cos x,$$

where we've used (2.18) to replace $\sin^2 x + \cos^2 x$ with 1. Therefore,

$$\int (\sin x + \cos x)^2 \, dx = \int [1 + 2 \sin x \cos x] \, dx$$

$$= \int 1 \, dx + 2 \int \sin x \cos x \, dx$$

$$= x + 2 \int \sin x \cos x \, dx.$$

The u-substitution $u = \sin x, du = \cos x \, dx$ converts this last integral into $2 \int u \, du = 2u^2/2 + C = u^2 + C = \sin^2 x + C$. Thus,

$$\int (\sin x + \cos x)^2 \, dx = x + \sin^2 x + C.$$ ∎

In other cases, we want to break up the integrand and then use one of the identities from (2.18) to set ourselves up for a u-substitution. The next two examples illustrate that.

EXAMPLE 2.12 Evaluate $\int \cos^3 x \, dx$.

Solution Write $\cos^3 x = (\cos^2 x)(\cos x)$, and then use (2.18) to rewrite that as $(1 - \sin^2 x)(\cos x)$. From here,

$$\int (1 - \sin^2 x)(\cos x) \, dx,$$

we can then use the u-substitution $u = \sin x, du = \cos x \, dx$:

$$\int (1 - \sin^2 x)(\cos x) \, dx = \int (1 - u^2) \, du = u - \frac{u^3}{3} + C$$

$$= \sin x - \frac{\sin^3 x}{3} + C.$$ ∎

EXAMPLE 2.13 Evaluate $\int \tan^3 x \, dx$.

Solution Write $\tan^3 x = (\tan^2 x)(\tan x)$, and then use (2.18) to rewrite that as $(\sec^2 x - 1)(\tan x)$. The integral is now

$$\int (\sec^2 x - 1)(\tan x) \, dx = \int \tan x \sec^2 x \, dx - \int \tan x \, dx.$$

The rightmost integral was done in Calculus 1; it evaluates to $\ln |\sec x| + C$ (see appendix D). To evaluate the remaining integral we use the u-substitution $u = \tan x$, $du = \sec^2 x \, dx$:

$$\int \tan x \sec^2 x \, dx = \int u \, du = \frac{u^2}{2} + C = \frac{\tan^2 x}{2} + C.$$

Combining the results yields

$$\int \tan^3 x \, dx = \frac{\tan^2 x}{2} - \ln |\sec x| + C.$$ ∎

One takeaway from these examples is to simplify the integrand using trigonometric identities *before* trying to integrate. But how do you know which trigonometric identities to use? The answer depends on which integration technique you'll be using afterward. Often—like in the examples above—it's *u*-substitution, so you want to choose a trig. identity that will leave a trig. function in the integrand that can serve as part of your *du*. That's why we split $\cos^3 x$ into $(\cos^2 x)(\cos x)$ in example 2.13 (and similarly for $\tan^3 x$). That sole $\cos x$ combined with dx to give us the $du = \cos x \, dx$ we needed to complete the substitution.

Knowing what to transform and when will come with practice. Alternatively, there *are* some very specific rules that one could follow in certain cases. Section A2.4 discusses some of those rules in the case when the integrand is a product of powers of sine and cosine.

Related Exercises   21–23.

2.7 Trigonometric Substitution

Our next stop on our tour of new integration techniques is *trigonometric substitution*. This method takes an integral and attempts to convert it into a trigonometric integral. For example, consider the integral

$$\int \sqrt{1 - x^2} \, dx.$$

Neither *u*-substitution nor integration by parts will work here. The integral is also not a trigonometric integral (the integrand isn't a trig. function). But under the substitution $x = \sin\theta$ it can be converted into a trigonometric integral. Here's how.

First, let's address a question you probably have: Why the $x = \sin\theta$ substitution? One reason is that

$$\sqrt{1 - x^2} = \sqrt{1 - (\sin\theta)^2} = \sqrt{1 - \sin^2\theta} = \sqrt{\cos^2\theta} = \cos\theta,$$

where I've used one of the trig. identities in (2.18) after the highlighted equals sign.[7] This substitution therefore converts $\sqrt{1 - x^2}$ into $\cos\theta$, hinting that maybe if we complete the substitution we'll be left with a trigonometric integral we can evaluate using the previous section's methods. To complete the substitution we note that $x = \sin\theta \Rightarrow dx = \cos\theta \, d\theta$. The next example picks up the story here and finishes the evaluation.

EXAMPLE 2.14 (Video)   Evaluate $\int \sqrt{1 - x^2} \, dx$ via the substitution $x = \sin\theta$.

Solution   Substituting $x = \sin\theta$ and $dx = \cos\theta \, d\theta$ into $\int \sqrt{1 - x^2} \, dx$ yields

$$\int \sqrt{1 - x^2} \, dx = \int \cos\theta(\cos\theta \, d\theta) = \int \cos^2\theta \, d\theta.$$

---

[7] A quick note for what follows. In taking square roots of functions like $\cos\theta$ and $\tan\theta$ one needs to be careful, since those functions change sign on some domains. I'll assume hereafter that $0 < \theta < \pi/2$. This ensures, for example, that $\sqrt{\cos^2\theta} = \cos\theta$.

This is a trigonometric integral. Using the double-angle identity $\cos^2\theta = \frac{1+\cos(2\theta)}{2}$ (from appendix B) yields

$$\int \cos^2\theta\, d\theta = \frac{1}{2}\int [1 + \cos(2\theta)]\, d\theta$$

$$= \frac{1}{2}\left[\theta + \frac{\sin(2\theta)}{2}\right] + C$$

$$= \frac{1}{2}[\theta + \sin\theta\cos\theta] + C, \tag{2.19}$$

where I've used another double-angle identity: $\sin(2\theta) = 2\sin\theta\cos\theta$ (also from appendix B). We now have to convert back to xs. We know that $x = \sin\theta$, so $\theta = \arcsin(x)$ (the inverse sine function). And we've already calculated that $\cos\theta = \sqrt{1-x^2}$. Therefore,

$$\int \sqrt{1-x^2}\, dx = \frac{1}{2}\left[\underbrace{\arcsin(x)}_{\theta} + \underbrace{x}_{\sin\theta}\underbrace{\sqrt{1-x^2}}_{\cos\theta}\right] + C. \qquad \blacksquare$$

A few comments on our result and trigonometric substitution in general:

- In the example we were able to use the original trig. substitution and a trig. identity to help us convert back to xs. But that combo isn't always easy to work with. Luckily, there's a visual aid that can help: the right triangle defined by the trig. substitution. I discuss that in the video accompanying the example above. (It's easier to show this than to explain it in words.) I'll also describe these right triangles below.

- Because trig. substitution substitutes $x =$ trig. function (typically $\sin\theta$, $\tan\theta$, $\sec\theta$, or a multiple of one of these), it works best on integrands that involve a sum or difference of two squares. We'll see more of that in the examples below.

- If that sum or difference of two squares contains nonunit factors (e.g., $\sqrt{4-9x^2}$ versus $\sqrt{1-x^2}$) then we need to multiply the trig. function used in the substitution by a suitable number.* We'll talk soon about what that "suitable number" should be.

* For $\sqrt{4-9x^2}$ we'd substitute $x = \frac{2}{3}\sin\theta$, for example.

As you're finding out, trig. substitution can get quite complicated. (In my experience it's the most difficult integration technique for students.) So here are some explicit general guidelines that should help.**

** These guidelines are designed to remove the square roots you'll soon see.

General Guidelines

Assume that $a > 0$ and $b > 0$.

1. To convert $\sqrt{a^2 - b^2 x^2}$, use $x = \frac{a}{b}\sin\theta$:

$$\sqrt{a^2 - b^2 x^2} = \sqrt{a^2 - b^2\left(\frac{a^2}{b^2}\sin^2\theta\right)} = \sqrt{a^2(1 - \sin^2\theta)}$$

$$= \sqrt{a^2\cos^2\theta} = a\cos\theta.$$

Example: For $\sqrt{4-x^2}$, $a = 2$ and $b = 1$, so substituting $x = 2\sin\theta$ yields $\sqrt{4-x^2} = 2\cos\theta$.

2. To convert $\sqrt{a^2 + b^2 x^2}$, use $x = \frac{a}{b} \tan \theta$:

$$\sqrt{a^2 + b^2 x^2} = \sqrt{a^2 + b^2 \left(\frac{a^2}{b^2} \tan^2 \theta \right)} = \sqrt{a^2 (1 + \tan^2 \theta)}$$

$$= \sqrt{a^2 \sec^2 \theta} = a \sec \theta.$$

Example: For $\sqrt{4 + 9x^2}$, $a = 2$ and $b = 3$, so substituting $x = \frac{2}{3} \tan \theta$ yields $\sqrt{4 + 9x^2} = 2 \sec \theta$.

3. To convert $\sqrt{b^2 x^2 - a^2}$, use $x = \frac{a}{b} \sec \theta$. Then

$$\sqrt{b^2 x^2 - a^2} = \sqrt{b^2 \left(\frac{a^2}{b^2} \sec^2 \theta \right) - a^2} = \sqrt{a^2 (\sec^2 \theta - 1)}$$

$$= \sqrt{a^2 \tan^2 \theta} = a \tan \theta.$$

Example: For $\sqrt{9x^2 - 4}$, $a = 2$ and $b = 3$, so substituting $x = \frac{2}{3} \sec \theta$ yields $\sqrt{9x^2 - 4} = 2 \tan \theta$.

Once we've converted to θs and integrated, we can draw a right triangle containing the relationships implied by the substitutions to help us get back to the xs. Box 2.1 and the examples that follow illustrate this.

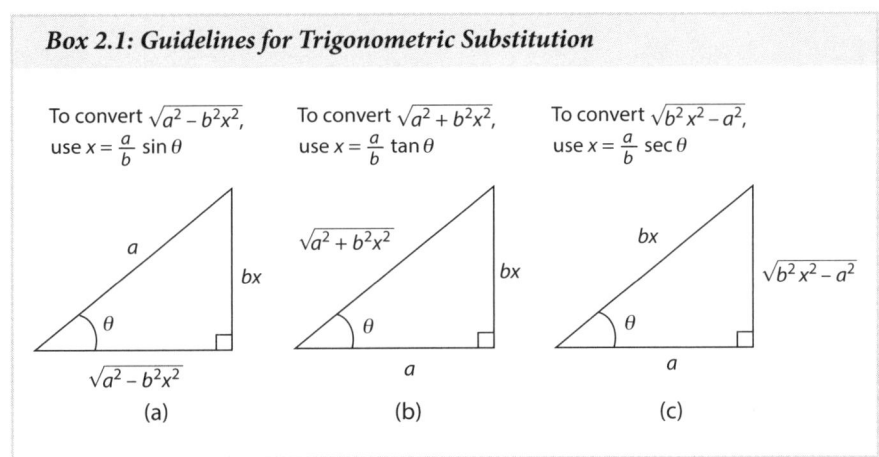

Box 2.1: Guidelines for Trigonometric Substitution

To convert $\sqrt{a^2 - b^2 x^2}$, use $x = \frac{a}{b} \sin \theta$

To convert $\sqrt{a^2 + b^2 x^2}$, use $x = \frac{a}{b} \tan \theta$

To convert $\sqrt{b^2 x^2 - a^2}$, use $x = \frac{a}{b} \sec \theta$

(a)   (b)   (c)

EXAMPLE 2.15 Evaluate $\displaystyle\int \frac{1}{\sqrt{x^2 - 25}}\, dx$.

Solution The denominator here fits into box 2.1(c) with $b = 1$ and $a = 5$. So, substituting $x = 5 \sec \theta$ and $dx = 5 \sec \theta \tan \theta\, d\theta$ into the integral yields

$$\int \frac{1}{\sqrt{x^2 - 25}}\, dx = \int \frac{5 \sec \theta \tan \theta}{\sqrt{25 \sec^2 \theta - 25}}\, d\theta$$

$$= \int \frac{5 \sec \theta \tan \theta}{\sqrt{25 (\sec^2 \theta - 1)}}\, d\theta \qquad \text{Factoring}$$

$$= \int \frac{5 \sec \theta \tan \theta}{5 \tan \theta}\, d\theta \quad \text{Since } \sqrt{25 (\sec^2 \theta - 1)} \qquad (2.20)$$

$$= \sqrt{25\tan^2\theta} = 5\tan\theta$$

$$= \int \sec\theta \, d\theta = \ln|\sec\theta + \tan\theta| + C.$$

From $x = 5\sec\theta$ we get $\sec\theta = \frac{x}{5}$, and from the triangle in box 2.1(c), $\tan\theta = \frac{\sqrt{b^2x^2 - a^2}}{a} = \frac{\sqrt{x^2 - 25}}{5}$, since $b = 1$ and $a = 5$ here. Thus,

$$\int \frac{1}{\sqrt{x^2 - 25}} \, dx = \ln\left| \frac{x}{5} + \frac{\sqrt{x^2 - 25}}{5} \right| + C. \qquad \blacksquare$$

Related Exercises 29–30.

EXAMPLE 2.16 Evaluate $\displaystyle\int \frac{\sqrt{1 - x^2}}{x^4} \, dx.$

Solution The numerator here fits into box 2.1(a) with $a = b = 1$. So, let's use $x = \sin\theta$, so that $dx = \cos\theta \, d\theta$. Substituting these into the integral yields

$$\int \frac{\sqrt{1 - x^2}}{x^4} \, dx = \int \frac{\sqrt{1 - \sin^2\theta}}{\sin^4\theta} (\cos\theta) \, d\theta$$

$$= \int \frac{\cos\theta}{\sin^4\theta} (\cos\theta) \, d\theta. \qquad \text{Since } \sqrt{1 - \sin^2\theta} = \sqrt{\cos^2\theta} = \cos\theta$$

Using now the fact that

$$\frac{\cos^2\theta}{\sin^4\theta} = \frac{\cos^2\theta}{\sin^2\theta} \cdot \frac{1}{\sin^2\theta} = \cot^2\theta \csc^2\theta,$$

the integral further simplifies to $\int \cot^2\theta \csc^2\theta \, d\theta$. We now employ the u-substitution $u = \cot\theta$, $du = -\csc^2\theta \, d\theta$:

$$\int \cot^2\theta \csc^2\theta \, d\theta = -\int u^2 \, du = -\frac{1}{3}u^3 + C = -\frac{1}{3}\cot^3\theta + C.$$

From the triangle in box 2.1(a), $\cot\theta = \frac{\sqrt{a^2 - b^2x^2}}{bx} = \frac{\sqrt{1 - x^2}}{x}$, since $a = b = 1$ here. Thus,

$$\int \frac{\sqrt{1 - x^2}}{x^4} \, dx = -\frac{1}{3}\left(\frac{\sqrt{1 - x^2}}{x} \right)^3 + C = -\frac{(1 - x^2)^{3/2}}{3x^3} + C. \qquad \blacksquare$$

Related Exercises 31–32, A5–A6.

APPLICATIONS Applied exercise A5 employs a trig. substitution to help calculate the electric field of a line of charge. Applied exercise A6 uses trig. substitution to study *laminar flow*, the flow rate of a fluid flowing through a cylindrical tube.

2.8 Partial Fraction Decomposition

Congratulations for making it through the previous section! We've got one last stop on our integration techniques tour: *integration by partial fraction decomposition*. To motivate that, consider the integral

$$\int \frac{1}{x^2 - 1} \, dx.$$

We could, *maybe*, try trig. substitution. But there's an easier way. You might recognize the difference of squares $x^2 - 1 = (x-1)(x+1)$ in the denominator. Might we then be able to claim that

$$\int \frac{1}{(x-1)(x+1)} \, dx = \int \frac{1}{x-1} \, dx + \int \frac{1}{x+1} \, dx?$$

Not quite, because

$$\frac{1}{(x-1)(x+1)} \neq \frac{1}{x-1} + \frac{1}{x+1}.$$

But maybe there's some combination of multiples of the two fractions on the right-hand side that *does* work out. To explore that, write

$$\frac{1}{(x-1)(x+1)} = \frac{A}{x-1} + \frac{B}{x+1}. \tag{2.21}$$

Our job now is to solve for the constants A and B. Multiplying the equation by $(x-1)(x+1)$ yields

$$1 = A(x+1) + B(x-1).$$

Substituting $x = 1$ yields $1 = 2A$, so $A = 1/2$; substituting $x = -1$ yields $1 = -2B$, so $B = -1/2$. Conclusion:

$$\frac{1}{(x-1)(x+1)} = \frac{1/2}{x-1} - \frac{1/2}{x+1} = \frac{1}{2}\left[\frac{1}{x-1} - \frac{1}{x+1}\right].$$

We've just *decomposed* $\frac{1}{x^2-1}$ into a sum of simpler *fractions*. That's why the method we're building up to is called the method of "partial fraction decomposition." Closing the loop, let's use what we've found to evaluate the original integral:

$$\int \frac{1}{x^2 - 1} \, dx = \frac{1}{2} \int \left[\frac{1}{x-1} - \frac{1}{x+1}\right] dx$$

$$= \frac{1}{2} [\ln|x-1| - \ln|x+1|] + C$$

$$= \frac{1}{2} \ln\left|\frac{x-1}{x+1}\right| + C, \tag{2.22}$$

where we've used the fact that $\ln a - \ln b = \ln \frac{a}{b}$.*

The decomposition we did in (2.21) can be generalized to the case when the denominator is a product of more than two distinct linear factors (e.g., $(x+1)$ $(x+7)(x-3)$). The next example illustrates this and adds the complication of the numerator not being equal to 1.

> * The Rules of Logarithms are reviewed in appendix B.

EXAMPLE 2.17 (Video)   Find constants A, B, and C such that

$$\frac{5x^2 - 12x - 12}{x(x-2)(x+2)} = \frac{A}{x} + \frac{B}{x-2} + \frac{C}{x+2}.$$

Then, use the resulting decomposition to evaluate

$$\int \frac{5x^2 - 12x - 12}{x(x-2)(x+2)} \, dx.$$

Solution Multiplying the first equation above by $x(x-2)(x+2)$ yields

$$5x^2 - 12x - 12 = A(x-2)(x+2) + Bx(x+2) + Cx(x-2).$$

When $x = 2$ this yields $20 - 24 - 12 = 8B \Rightarrow B = -2$. When $x = -2$ it yields $20 + 24 - 12 = 8C \Rightarrow C = 4$. When $x = 0$ it yields $-12 = -4A \Rightarrow A = 3$. Thus,

$$\frac{5x^2 - 12x - 12}{x(x-2)(x+2)} = \frac{3}{x} - \frac{2}{x-2} + \frac{4}{x+2}.$$

Therefore,

$$\int \frac{5x^2 - 12x - 12}{x(x-2)(x+2)} \, dx = \int \left[\frac{3}{x} - \frac{2}{x-2} + \frac{4}{x+2} \right] dx$$

$$= \int \frac{3}{x} \, dx - 2 \int \frac{1}{x-2} \, dx + 4 \int \frac{1}{x+2} \, dx$$

$$= 3 \ln |x| - 2 \ln |x - 2| + 4 \ln |x + 2| + C$$

$$= \ln |x^3| - \ln(x-2)^2 + \ln(x+2)^4 + C \quad \text{Using } \ln a^b = b \ln a$$

$$= \ln \left| \frac{x^3(x+2)^4}{(x-2)^2} \right| + C,$$

where we've used the Rules of Logarithms again. ∎

One takeaway from all this: *Partial fractions decomposition is intended for use on* rational *functions*—functions of the form $f(x) = \frac{p(x)}{q(x)}$, where p and q are both polynomials. As we've seen, when these polynomials get more complex, the decomposition gets more complex as well. The first complication that can arise is when f is "improper." This happens when the degree of p (the highest power of x in the polynomial) is greater than or equal to the degree of q.* When f is improper we can use **long division** to convert it into a proper fraction.[8] Long division often takes a *loooong* time. Since this is *Calculus 2 Simplified*, let's therefore deal only with proper fractions in this section.** The guidelines below then provide an explicit procedure for decomposing a proper rational function via the partial fractions technique.

* Examples of improper fractions are $\frac{x}{x+1}, \frac{x^3 - 2x}{x^2 + 4x - 7}$.

** Section A2.6 reviews long division and works through an example of partial fractions decomposition in this context. I recommend finishing this section before consulting that section.

Box 2.2: Guidelines for Partial Fraction Decomposition of a Proper Rational Function

Let $f(x) = \frac{p(x)}{q(x)}$ be a proper rational function. Factor q into factors of the form

$$(ax + b)^n \quad \text{and} \quad (ax^2 + bx + c)^m,$$

where $ax^2 + bx + c$ is irreducible (i.e., cannot be factored further).

1. For each *linear* factor—factor of the form $(ax + b)^n$—include the following sum in the partial fraction decomposition:

$$\frac{A_1}{(ax+b)} + \frac{A_2}{(ax+b)^2} + \cdots + \frac{A_n}{(ax+b)^n}$$

[8] In a proper fraction the degree of p is *less than* the degree of q.

2. For each *quadratic* factor—factor of the form $(ax^2 + bx + c)^m$—include the following sum in the partial fraction decomposition:

$$\frac{B_1x + C_1}{ax^2 + bx + c} + \frac{B_2x + C_2}{(ax^2 + bx + c)^2} + \cdots + \frac{B_mx + C_m}{(ax^2 + bx + c)^m}$$

Then, multiply your "$\frac{p(x)}{q(x)} =$ sum of fractions" equation by $q(x)$. Finally, substitute the x-values that make $q(x)$ zero (along with any other x-values, as needed) to get equations involving the As, Bs, and Cs, and solve these for those constants.

A couple of comments on these guidelines:

- *The numerators of the fractions in the decompositions are one degree less than the degree of the factor type*: constant numerators (0th-degree polynomials) A_1, A_2, etc., associated with a linear factor $ax + b$; linear numerators $B_1x + C_1$, $B_2x + C_2$, etc., associated with a quadratic factor $ax^2 + bx + c$. Examples:

$$\frac{1}{(x-1)(x+1)} = \frac{A_1}{x-1} + \frac{A_2}{x+1}, \qquad \frac{x}{x(x^2+1)} = \frac{A_1}{x} + \frac{B_1x + C_1}{x^2+1}.$$

- *The factor $(ax + b)^n$ of $q(x)$ contributes n terms to the decomposition; the factor $(ax^2 + bx + c)^m$ contributes m terms.* Example:

$$\frac{1}{(x+1)^3(x^2+7)^2} = \frac{A_1}{x+1} + \frac{A_2}{(x+1)^2} + \frac{A_3}{(x+1)^3} + \frac{B_1x + C_1}{x^2+7} + \frac{B_2x + C_2}{(x^2+7)^2}.^*$$

> * The factor $(x+1)^3$ contributes three terms to the decomposition; the factor $(x^2+7)^2$ contributes two terms.

EXAMPLE 2.18 Evaluate $\int \dfrac{3x - 4}{(x - 1)^2}\, dx.$

Solution The integrand is a proper rational function. And since $(x - 1)^2$ is a linear factor with power 2, guideline 1 in box 2.2 suggests the decomposition (I'll use A and B instead of A_1 and A_2)

$$\frac{3x - 4}{(x - 1)^2} = \frac{A}{x - 1} + \frac{B}{(x - 1)^2}.$$

Multiplying this equation by $(x - 1)^2$ yields

$$3x - 4 = A(x - 1) + B.$$

Now, when $x = 1$, this equation yields $-1 = B$. Another good x-value to use is $x = 0$.** When $x = 0$, we get $-4 = -A - 1 \Rightarrow A = 3$. Therefore,

$$\int \frac{3x - 4}{(x - 1)^2}\, dx = \int \frac{3}{x - 1}\, dx - \int \frac{1}{(x - 1)^2}\, dx = 3\ln|x - 1| + \frac{1}{x - 1} + C. \quad \blacksquare$$

> ** You can use any x-value, but the simpler the better.

Related Exercises 33–36, D7.

Thus far we've been able to solve for the constants fairly easily. One potential complication is when we're left with a system of linear equations involving those constants. The next example illustrates this.

EXAMPLE 2.19 Evaluate $\displaystyle\int \frac{x^2-1}{x(x^2+1)}\,dx$.

Solution Following the box 2.2 guidelines,

$$\frac{x^2-1}{x\,(x^2+1)} = \frac{A}{x} + \frac{Bx+C}{x^2+1}.$$

Multiplying this equation by $x\,(x^2+1)$ yields

$$x^2-1 = A(x^2+1)+(Bx+C)x.$$

Now, when $x=0$, this equation yields $-1=A$. Other good x-values to use are $x=\pm 1$. When $x=1$, we get $0=-2+B+C$. When $x=-1$, we get $0=-2+B-C$. So we're left with the system of equations $B+C=2$, $B-C=2$. Adding these equations yields $2B=4 \Rightarrow B=2$. (This isn't the only way to have solved this system of equations.)[9] Substituting this into any of the two equations and then solving for C yields $C=0$. Therefore,

$$\int \frac{x^2-1}{x^3+x}\,dx = -\int \frac{1}{x}\,dx + \int \frac{2x}{x^2+1}\,dx$$

$$= -\ln|x| + \int \frac{2x}{x^2+1}\,dx + C$$

$$= -\ln|x| + \ln\left|x^2+1\right| + C \qquad \text{Using } u=x^2+1,\, du=2x\,dx$$

$$= \ln\left|\frac{x^2+1}{x}\right| + C. \qquad\qquad\blacksquare$$

Related Exercises   37–38.

2.9   Parting Thoughts

We've now completed our study of the integration approximation methods and techniques normally covered in a Calculus 2 course. Woohoo! What we've learned considerably broadens our integration knowledge and toolbox from where it stood in Calculus 1. We'll use our new integration results frequently in chapter 3 when we return to the Geometry Big Question from chapter 1. So before you move on, I recommend trying exercises 39–48. Those exercises ask you to evaluate an integral, but since they aren't tied to a particular example in this chapter, you won't know ahead of time which integration technique to use. This will therefore help you practice the skill of determining which integration technique is best to use.

[9]We could also have solved one equation for, say, B and substituted the resulting expression into the other equation. Example: The first equation, $B+C=2$, solved for B yields $B=2-C$. Substituting that into the second equation, $B-C=2$, yields $(2-C)-C=2$. This simplifies to $2-2C=2$, or $-2C=0$, which yields $C=0$. Using this in $B=2-C$ yields $B=2-0=2$.

Chapter 2   Exercises

1–6: Calculate the indicated Riemann sum or trapezoidal approximation on the specified interval.

1. L_5 for $f(x) = 3x^2 - 1$, $[1, 2]$

2. R_4 for $f(x) = \sin x$, $\left[0, \frac{\pi}{2}\right]$

3. R_4 for $f(x) = \sqrt{x}$, $[0, 1]$

4. M_3 for $f(x) = \ln x$, $[1, 4]$

5. T_4 for $f(x) = e^x$, $[0, 1]$

6. T_5 for $f(x) = \dfrac{1}{\sqrt{x+1}}$, $[1, 3]$

7–9: Calculate R_n for the definite integral.

7. $\displaystyle\int_1^3 (5x + 3)\, dx$

8. $\displaystyle\int_0^2 (x^2 + x)\, dx$

9. $\displaystyle\int_0^1 5x^3\, dx$

10–12: Use theorem 2.3 or 2.4 to determine the smallest n-value guaranteeing that the indicated approximation to the specified integral is accurate to within 0.01.

10. R_n, $\displaystyle\int_e^{2e} \ln x\, dx$

11. M_n, $\displaystyle\int_0^1 e^{-x}\, dx$

12. T_n, $\displaystyle\int_1^4 \sqrt{x}\, dx$

13–38: Evaluate the integral.

13. $\displaystyle\int x^2 \ln x\, dx$ **14.** $\displaystyle\int 5x \sin x\, dx$

15. $\displaystyle\int \arcsin x\, dx$ **16.** $\displaystyle\int_1^e \frac{(\ln x)^2}{x^2}\, dx$

17. $\displaystyle\int 2x^3 e^{x^2}\, dx$ **18.** $\displaystyle\int_0^\pi x^2 \cos x\, dx$

19. $\displaystyle\int_0^1 x^4 e^{-x}\, dx$ **20.** $\displaystyle\int x^5 \sin x\, dx$

21. $\displaystyle\int_0^\pi 15 \sin^5 x\, dx$

22. $\displaystyle\int 15 \tan^2 x \sec^4 x\, dx$

23. $\displaystyle\int \frac{2 \cos^3 x}{1 - \sin x}\, dx$

24. $\displaystyle\int 15 \sin^3 x \cos^2 x\, dx$

25. $\displaystyle\int 35 \sin^4 x \cos^3 x\, dx$

26. $\displaystyle\int 35 \sin^2 x \cos^5 x\, dx$

27. $\displaystyle\int_0^{\pi/2} 32 \sin^4 x\, dx$

28. $\displaystyle\int_0^{\pi/2} 6 \cos^6 x\, dx$

29. $\displaystyle\int \frac{9}{x^2 \sqrt{x^2 - 9}}\, dx$

30. $\displaystyle\int \frac{3x^3}{\sqrt{x^2 - 81}}\, dx$

31. $\displaystyle\int \frac{2x^2}{\sqrt{9 - x^2}}\, dx$

32. $\displaystyle\int \frac{x}{(9 - x^2)^{3/2}}\, dx$

33. $\displaystyle\int \frac{x + 5}{x^2 - 3x + 2}\, dx$

34. $\displaystyle\int \frac{4x + 8}{x^3 + 6x^2 + 8x}\, dx$

35. $\displaystyle\int \frac{8x}{(x - 2)^2 (x + 2)}\, dx$

36. $\displaystyle\int \frac{36x + 72}{(x + 3)(x^2 - 6x + 9)}\, dx$

37. $\displaystyle\int \frac{x^2 + 5}{(2x + 2)(x^2 - 2x + 3)}\, dx$

38. $\displaystyle\int \frac{2x^3 - 4x^2 + 2x + 2}{(x^2 + 1)(x^2 + 4)}\, dx$

39–48: Evaluate the integral.

39. $\displaystyle\int_0^1 32x^2 e^{4x}\,dx$

40. $\displaystyle\int 80\cos^7 x \sin^3 x\,dx$

41. $\displaystyle\int \frac{18x+12}{x^2-1}\,dx$

42. $\displaystyle\int_0^1 \ln(2-x)\,dx$

43. $\displaystyle\int \frac{2}{x\sqrt{x^2-4}}\,dx$

44. $\displaystyle\int \frac{x^2-1}{(x+2)^3}\,dx$

45. $\displaystyle\int 2x\arctan x\,dx$

46. $\displaystyle\int 8\cos^2(2x)\,dx$

47. $\displaystyle\int \frac{4x}{x^4-1}\,dx$

48. $\displaystyle\int 32x^3 \ln^2 x\,dx$

A1. Cardiac output The **cardiac output** F of a person's heart is the volume (measured in liters) of blood the heart pumps per second. Cardiologists measure F by injecting a certain amount A (measured in milligrams) of dye into the right atrium of the heart and monitoring the concentration $c(t)$ (measured in mg/L) of dye in the aorta as the heart pumps. After some time T, all of the injected dye has flowed through the monitoring probe. Assuming F is constant, one can then show that

$$F = \frac{A}{\int_0^T c(t)\,dt}.$$

Suppose $c(t)$ is given in the figure below (here $T=4$ seconds). If $A=5$ mg, estimate F by using a left-hand Riemann sum with $n=4$. (You'll have to estimate the y-values $c(t)$ using the graph.)

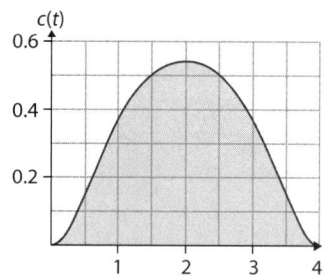

A2. Bell curves (normal distributions) The functions

$$n(x) = \frac{1}{\sigma\sqrt{2\pi}} e^{-(x-\mu)^2/(2\sigma^2)},$$

where $\sigma > 0$ and μ are constants, are called **normal density functions**, sometimes also called **Gaussian**

distributions. These functions are widely used in statistics to describe, for example, the distribution of human heights, students' exam scores, and even IQ scores. In such settings, if the values of a quantity X are normally distributed, then the probability that a measurement of X yields a result between a and b, denoted $P(a \le X \le b)$, is

$$P(a \le X \le b) = \int_a^b n(x)\,dx.$$

The graph of n is the familiar "bell-shaped" curve, shown below.

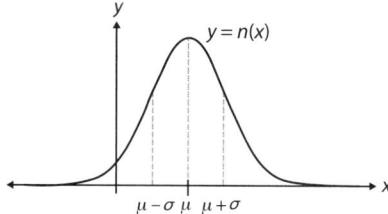

The parameter σ is the **standard deviation** and measures how spread out (large σ) or concentrated (small σ) n is around its **mean value** μ.

(a) Suppose that the exam scores X in a Calculus 2 course are normally distributed with mean 80 and standard deviation 5. Write down the integral representing the probability that a randomly chosen exam will have a score on it between 75 and 85.

(b) Write out the $n=5$ equipartition left-hand Riemann sum approximation to the integral obtained in part (a), and use it to estimate the probability described therein. (It turns out that $P(\mu - \sigma \le X \le \mu + \sigma) \approx 0.68$ for any normal distribution. Translation: If X is normally distributed, 68% of the values of X lie within one standard deviation of its mean.)

A3. Life expectancy A **period life table** approximates the mortality in a population by using death probabilities computed for short periods (less than 3 years) and applying those to a hypothetical cohort of 100,000 people. Such life tables provide a sense of how the population's mortality would change over time assuming the prevailing death rates continue. The 2020 period life table for the entire U.S. population is given below.

x	$s(x)$	x	$s(x)$
10	99,323	60	86,376
20	98,952	70	74,466
30	97,725	80	53,346
40	95,794	90	20,447
50	92,680	100	1,142

(*Source*: National Center for Health Statistics, [4].) Here x is measured in years, and $s(x)$, the **survival function**, records the number of those initial 100,000 individuals surviving to age x. Note that $s(0) = 100,000$ and $s(\omega) = 0$, where ω is the population's maximum life span.

(a) The **life expectancy at birth** $e_0 = \frac{1}{100,000} \int_0^\omega s(x)\, dx$. Use $\omega = 100$, the life table above, and the Trapezoidal Rule with $n = 10$ to estimate e_0.

(b) The **deaths distribution** $f(x)$, defined as $f(x) = -\frac{s'(x)}{100,000}$, describes how mortality is distributed across age in the population. Use the formula in (a) and integration by parts to show that $\int_0^\omega xf(x)\, dx = e_0$. This makes e_0 the mean value of the distribution f.

(c) The **variance of deaths** v is defined as $v = \int_0^\omega x^2 f(x)\, dx - e_0^2$. Use the definition of f from (b) and integration by parts to show that $v = \frac{2}{100,000} \int_0^\omega xs(x)\, dx - e_0^2$.

(d) The **standard deviation of deaths** d is defined as $d = \sqrt{v}$, where v is defined in (c) and the e_0 in v is defined in (a). Use $\omega = 100$, the life table above, and a right-hand Riemann sum with $n = 10$ to estimate v, and then take the square root of the result to yield an estimate of d. (Use the e_0-value in the answer to (a), in the Answers to Exercises at the end of the book.) A large d-value implies that mortality is dispersed across age in the population; a small d-value implies that it's concentrated in a specific age range. Therefore, d is one measure of **life span inequality**.

A4. Population density Let $p(r)$ denote the number of people (in thousands) per square mile living a distance r miles from a city center—the **population density**. The total population living within x miles of the city's center is then

$$P(x) = \int_0^x 2\pi r p(r)\, dr.$$

Calculate $P(x)$ if $p(r) = 100e^{-r}$.

A5. Electric field of a line of charge Imagine a rod of length 2ℓ lying on the x-axis so that its midpoint is at the origin. If a total charge Q is distributed uniformly on this rod, then the y-component of the electric field produced at a point $(0, z)$ on the y-axis ($z > 0$) is

$$E_y(z) = \frac{kQz}{2\ell} \int_{-\ell}^{\ell} \frac{1}{(z^2 + x^2)^{3/2}}\, dx,$$

where k is a physical constant. Use an appropriate trigonometric substitution to show that $E_y(z) = \frac{kQ}{z\sqrt{z^2+\ell^2}}$.

A6. Laminar flow In **laminar flow**, the horizontal flow rate Q (measured in cm^3/s) of a fluid flowing through a cylindrical tube of radius R is

$$Q = 2\pi \int_0^R rv(r)\, dr,$$

where $v(r)$ is the horizontal velocity (measured in cm/s) of the fluid at a distance r from the center of the tube. *Using a trig. substitution*, calculate Q for $v(r) = \frac{1}{(r^2+4)^{3/2}}$.

D1. This exercise proves the first inequality in (2.14) for the case of f nondecreasing on $[a, b]$. The proof of the second inequality is similar.

(a) Use (2.12) to show that

$$0 \leq EL_n \leq R_n - L_n.$$

(b) Use (2.5) and (2.6) to show that $R_n - L_n = \frac{(b-a)[f(b)-f(a)]}{n}$.

(c) Use (b) in (a) and explain how the result implies the first inequality in (2.14).

D2. It is a theorem that, if f is continuous on $[a, b]$, then f is integrable on $[a, b]$, and

$$\int_a^b f(x)\, dx = \lim_{n \to \infty} \sum_{i=1}^n f(x_i^*)\Delta x,$$

the infinite limit of an equipartition Riemann sum of f over $[a, b]$. (One can even use nonequal partitions in the Riemann sum here.) Return now to exercises 7–9 and calculate the infinite limit of the R_n obtained therein. Confirm that your answers agree with what you get from applying the Evaluation Theorem to the definite integrals.

D3. Evaluate $\int (\ln x)^2 \, dx$ by first using the substitution $x = e^{-z}$ and then using integration by parts on the resulting integral.

D4. Consider $\int x^3 \sin(x^2) \, dx$. What u would the ILATE guideline suggest here? Attempt the integration by parts with that u, and comment on the

obstacle you run into. What other u choice would work better?

D5. Let $a \neq 0$ and $b \neq 0$ be real numbers. Show that

$$\int e^{ax} \cos(bx) \, dx = \frac{e^{ax}(a \cos(bx) + b \sin(bx))}{a^2 + b^2} + C.$$

D6. Let $n \geq 1$ be an integer, and let $a \neq 0$ be a real number. Prove the *reduction formula*

$$\int x^n e^{ax} \, dx = \frac{1}{a}\left[x^n e^{ax} - \int nx^{n-1} e^{ax} \, dx \right].$$

D7. Let $a \neq 0$ be a real number. Prove that

$$\int \frac{1}{a^2 - x^2} \, dx = \frac{1}{2a} \ln \left| \frac{a+x}{a-x} \right| + C.$$

Chapter 2 Appendix

A2.1   Evaluating Riemann Sums Using Summation Formulas

THEOREM A2.1: SUMMATION FORMULAS. Let c be a real number. Then,

$$\sum_{i=1}^{n} c = cn, \quad \sum_{i=1}^{n} i = \frac{n(n+1)}{2}, \quad \sum_{i=1}^{n} i^2 = \frac{n(n+1)(2n+1)}{6}, \quad \sum_{i=1}^{n} i^3 = \left[\frac{n(n+1)}{2} \right]^2.$$

EXAMPLE A2.1   Calculate R_n for $f(x) = x^2 + 1$ over the interval $[0, 3]$. Then, verify that $R_{30} \approx 12.5$, as indicated in figure 2.4(d).

Solution Here $a = 0$, $b = 3$, and so $\Delta x = \frac{3-0}{n} = \frac{3}{n}$. From figure 2.5(a) we see that $x_i = a + i\Delta x = \frac{3i}{n}$, and so $f(x_i) = \left(\frac{3i}{n} \right)^2 + 1 = \frac{9i^2}{n^2} + 1$. From (2.5),

$$R_n = \sum_{i=1}^{n} f(x_i)\Delta x = \sum_{i=1}^{n} \left(\frac{9i^2}{n^2} + 1 \right) \cdot \frac{3}{n}$$

$$= \sum_{i=1}^{n} \frac{27i^2}{n^3} + \sum_{i=1}^{n} \frac{3}{n}$$

$$= \frac{27}{n^3} \sum_{i=1}^{n} i^2 + \frac{3}{n} \sum_{i=1}^{n} 1. \tag{A2.1}$$

Using now the first and third summation formulas in theorem A2.1,

$$R_n = \frac{27}{n^3} \left[\frac{n(n+1)(2n+1)}{6} \right] + \frac{3}{n}[n] = 3 + \frac{9(n+1)(2n+1)}{2n^2}. \tag{A2.2}$$

When $n = 30$, $R_{30} = 3 + \frac{9(31)(61)}{2(30)^2} = 12.455 \approx 12.5$.   ■

Related Exercises   Chapter 2: 7–9.

A2.2   Additional Error Theorems for Riemann Sums and the Trapezoidal Rule

THEOREM A2.2: ERRORS FOR LEFT- AND RIGHT-HAND RIEMANN SUM APPROXIMATIONS. Let f be integrable on $[a, b]$, and let L_n and R_n denote its associated left- and right-hand equipartition Riemann sums. Suppose further that f is either nondecreasing on $[a, b]$ or nonincreasing on $[a, b]$, and that $|f'(x)| \leq U$ on $[a, b]$. Then,

$$|EL_n| \leq U\frac{(b-a)^2}{n}, \qquad |ER_n| \leq U\frac{(b-a)^2}{n}.$$

THEOREM A2.3: ERRORS FOR MIDPOINT AND TRAPEZOIDAL RULE APPROXIMATIONS. Let f be integrable on $[a, b]$, and let M_n and T_n denote its associated equipartition Midpoint and Trapezoidal Rule approximations. Suppose further that f is either always concave up or always concave down on $[a, b]$. Then,

$$|EM_n| \leq \frac{(b-a)|T_n - M_n|}{n}, \qquad |ET_n| \leq \frac{(b-a)|T_n - M_n|}{n}.$$

A2.3   The Tabular Method for Integration by Parts

Suppose we want to show that

$$\int x^2 e^x \, dx = x^2 e^x - 2xe^x + 2e^x + C \tag{A2.3}$$

using integration by parts. It turns out you'd have to use integration by parts twice to do that. And if the integrand were $x^3 e^x$, you'd need to do it three times. Writing all that out takes time, but luckily there's a faster way called the **tabular method** for integration by parts. Figure A2.1 illustrates the process for the integral above.

We start with figure (a). First, choose your u and put it in the gray box in that table, and put the dv right below it. The u row is the "differentiate" row: Fill the row with the successive derivatives u', u'', etc. If eventually you get 0, then the tabular method will work out. The dv row is the "integrate" row: Fill the row with the successive integrals $\int v \, dx$, the integral of that, the integral of *that*, etc. Stop when

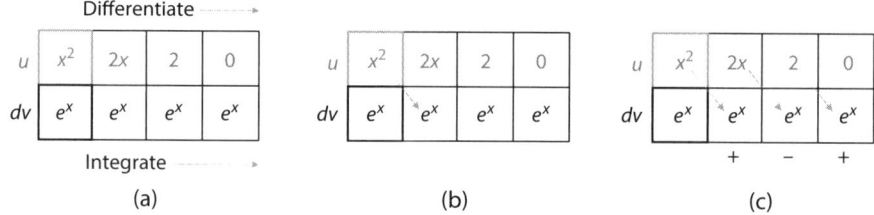

Figure A2.1: Illustrating the tabular method for integration by parts. Refer to the text for an explanation.

you get to the same column you got a 0 for in the first row. Figure (b) shows how this table encodes integration by parts. If we multiply diagonally across the gray-colored arrow, we get the "uv" term in the integration by parts formula, (2.17), bringing us to

$$x^2 e^x - \int 2x e^x \, dx.$$

Were we to apply integration by parts again here, we'd use $u = 2x$, $dv = e^x \, dx$, and get

$$x^2 e^x - \left(2x e^x - \int 2e^x \, dx \right) = x^2 e^x - 2x e^x + \int 2e^x \, dx.$$

Notice that the last gray-colored term here is the next diagonal product in the table, with the sign alternating from the previous term (negative from positive). This is illustrated in figure (c). Following the third arrow therein yields the last term:

$$x^2 e^x - 2x e^x + 2e^x + C.$$

(We stop here because the next product would involve the 0 in the table's last column.) To recap how to employ this method:

- **Step 1:** Write the u function in the gray box, and fill that row with its successive derivatives. Stop when you get a 0. (If you don't, abandon the method.)

- **Step 2:** Write the dv function under the u function, and fill that row with the integral of v and the result's successive integrals.

- **Step 3:** Multiply diagonally and put the alternating signs $+, -, +, \cdots$ in front of those products.

- **Step 4:** Add the results.

EXAMPLE A2.2 (Video)   Evaluate $\int x^3 \cos x \, dx$.

Solution   We have an "A" function (x^3) and a "T" function ($\cos x$) here, so let's use $u = x^3$ and $dv = \cos x \, dx$. Applying the tabular method yields the table in figure A2.2.

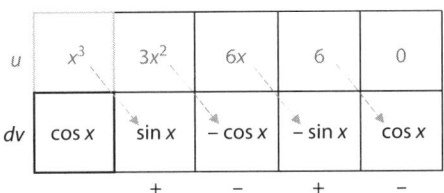

Figure A2.2

This yields

$$\int x^3 \cos x \, dx = x^3 \sin x - [3x^2(-\cos x)] + [6x(-\sin x)] - [6\cos x] + C,$$

which simplifies to $x(x^2 - 6)\sin x + 3(x^2 - 2)\cos x + C$. ■

Related Exercises   Chapter 2: 19–20.

A2.4   Integrands That Are Products of Powers of Sine and Cosine

The guidelines below summarize how to break up products of powers of sine and cosine functions to set ourselves up for a successful follow-up u-substitution. Note: Below, "$2k+1$" refers to an odd number and "$2k$" to an even number.

> **Box A2.1: Guidelines for Integrating Products of Powers of Sine and Cosine**
>
> 1. To integrate $\int (\cos^n x)(\sin^{2k+1} x)\, dx$:
>   - Pluck out a sine factor: $(\cos^n x)(\sin^{2k+1} x) = (\cos^n x)(\sin^{2k} x)(\sin x)$.
>   - Substitute $\sin^2 x = 1 - \cos^2 x$ to convert the even powers of sine into powers of cosine:
> $$(\cos^n x)(\sin^{2k} x)(\sin x) = (\cos^n x)(1 - \cos^2 x)^k(\sin x)$$
>   - Use $u = \cos x$, $du = -\sin x\, dx$: $(\cos^n x)(1 - \cos^2 x)^k(\sin x)\, dx = -u^n (1 - u^2)^k \, du$. Then multiply this out and integrate it.
>
> 2. To integrate $\int (\sin^n x)(\cos^{2k+1} x)\, dx$:
>   - Pluck out a cosine factor: $(\sin^n x)(\cos^{2k+1} x) = (\sin^n x)(\cos^{2k} x)(\cos x)$.
>   - Substitute $\cos^2 x = 1 - \sin^2 x$ to convert the even powers of cosine into powers of sine:

$$(\sin^n x)(\cos^{2k} x)(\cos x) = (\sin^n x)(1 - \sin^2 x)^k(\cos x)$$

- Use $u = \sin x$, $du = \cos x\, dx$: $(\sin^n x)(1 - \sin^2 x)^{2k}(\cos x) = u^n(1 - u^2)^k\, du$. Then multiply this out and integrate it.

EXAMPLE A2.3 (Video) Evaluate $\displaystyle\int \sin^3 x \cos^4 x\, dx$.

Solution Following the first guideline above,

$$\sin^3 x \cos^4 x = (\cos^4 x)(\sin^2 x)(\sin x) = (\cos^4 x)(1 - \cos^2 x)(\sin x).$$

Using now $u = \cos x$, $du = -\sin x\, dx$ yields

$$\int \sin^3 x \cos^4 x\, dx = -\int u^4(1 - u^2)\, du = -\int (u^4 - u^6)\, du$$

$$= \frac{u^7}{7} - \frac{u^5}{5} + C = \frac{\cos^7 x}{7} - \frac{\cos^5 x}{5} + C. \qquad \blacksquare$$

Related Exercises Chapter 2: 24–26.

In some cases we may also need to use additional trigonometric identities, like

$$\sin^2 x = \frac{1 - \cos(2x)}{2}, \qquad \cos^2 x = \frac{1 + \cos(2x)}{2}. \qquad \text{(A2.4)}$$

(These are reviewed in appendix B.) The example below illustrates such a case.

EXAMPLE A2.4 Evaluate $\displaystyle\int \cos^4 x\, dx$.

Solution Since $\cos^4 x = (\cos x)^4 = (\cos x)^2(\cos x)^2$, substituting the second equation in (A2.4) yields,

$$\int \cos^4 x\, dx = \int \left[\frac{1 + \cos(2x)}{2}\right]^2 dx = \int \left[\frac{1}{4} + \frac{\cos(2x)}{2} + \frac{\cos^2(2x)}{4}\right] dx.$$

We have three integrals to evaluate. The first one is the simplest: $\int \frac{1}{4}\, dx = \frac{x}{4} + C$. For the second we use the u-substitution $u = 2x$, $du = 2\, dx$, or $\frac{1}{2}\, du = dx$:

$$\int \frac{\cos(2x)}{2}\, dx = \frac{1}{2}\int \frac{\cos u}{2}\, du = \frac{1}{4}\sin u + C = \frac{1}{4}\sin(2x) + C.$$

For the remaining integral, we can use the second equation in (A2.4) again to transform the $\cos^2(2x)$ term:[10]

---

[10] Since $\cos^2 x = \frac{1 + \cos(2x)}{2}$ then $\cos^2(2x) = \frac{1 + \cos(4x)}{2}$. This follows from replacing "x" by "$2x$" in (A2.4).

$$\int \frac{\cos^2(2x)}{4}\, dx = \frac{1}{4} \int \left(\frac{1+\cos(4x)}{2} \right) dx = \frac{1}{8} \int 1\, dx + \frac{1}{8} \int \cos(4x)\, dx.$$

The first integral here evaluates to $\frac{x}{8} + C$. The u-substitution $u = 4x$, $du = 4\, dx$ evaluates the second integral to $\frac{\sin(4x)}{32} + C$. Adding all these results together finally yields

$$\int \cos^4 x\, dx = \frac{3x}{8} + \frac{\sin(2x)}{4} + \frac{\sin(4x)}{32} + C. \qquad \blacksquare$$

Related Exercises   Chapter 2: 27–28.

Believe it or not, *there are even more trig identities you might need to use* (e.g., for the specific case when the integrand is a product of sines and cosines with different arguments, like $\sin(2x)\cos(3x)$). The next section in this appendix covers those cases.

A2.5   Integrands That Are Sine-Cosine Products with Different Arguments

Let's look next at integrands of the form $\sin(mx)\sin(nx)$ and similar products of $\sin(mx)$ and/or $\cos(nx)$. Tackling these successfully relies on another set of trigonometric identities. The guidelines below provide the details.

> ### Box A2.2: Guidelines for Integrating Products of Sine and Cosine
>
> To integrate integrands of the form $\sin(mx)\sin(nx)$, $\sin(mx)\cos(nx)$, or $\cos(mx)\cos(nx)$, use one of the identities below to convert the integrand into a sum of trigonometric functions, and then do a u-substitution:
>
> - $\sin(mx)\sin(nx) = \dfrac{1}{2}\left[\cos((m-n)x) - \cos((m+n)x)\right]$
>
> - $\sin(mx)\cos(nx) = \dfrac{1}{2}\left[\sin((m-n)x) + \sin((m+n)x)\right]$
>
> - $\cos(mx)\cos(nx) = \dfrac{1}{2}\left[\cos((m-n)x) + \cos((m+n)x)\right]$

EXAMPLE A2.5 Evaluate $\int \sin(4x)\cos(3x)\,dx$.

Solution Using the second bullet point above,

$$\int \sin(4x)\cos(3x)\,dx = \frac{1}{2}\int [\sin((4-3)x) + \sin((4+3)x)]\,dx.$$

This simplifies to

$$\frac{1}{2}\int [\sin x + \sin(7x)]\,dx = -\frac{1}{2}\cos x - \frac{1}{14}\cos(7x) + C. \qquad \blacksquare$$

A2.6   A Brief Review of Long Division and Its Uses in Partial Fraction Decomposition

Long division is hard to explain in text (and hard to make sense of by reading). So, I invite you to watch the video for the example below. There I explain long division in the context of simplifying the rational function in the example.

EXAMPLE A2.6 (Video)   Use long division to show that

$$\frac{x^3 - x + 3}{x^2 + x - 2} = x - 1 + \frac{2x+1}{(x+2)(x-1)}.$$

Solution (Please watch the video for this example to see the solution.) $\blacksquare$

EXAMPLE A2.7 Evaluate $\int \frac{x^3 - x + 3}{x^2 + x - 2}\,dx$.

Solution Using our work in the previous example,

$$\int \frac{x^3 - x + 3}{x^2 + x - 2}\,dx = \int (x-1)\,dx + \int \frac{2x+1}{(x+2)(x-1)}\,dx$$

$$= \frac{x^2}{2} - x + \int \frac{2x+1}{(x+2)(x-1)}\,dx.$$

To calculate the second integral, let's use partial fractions. The factors $x+2$ and $x-1$ in the denominator contribute the terms $\frac{A}{x+2}$ and $\frac{B}{x-1}$ to the decomposition:

$$\frac{2x+1}{(x+2)(x-1)} = \frac{A}{x+2} + \frac{B}{x-1},$$

$$2x+1 = A(x-1) + B(x+2). \qquad \text{Multiplying this equation}$$
$$\text{by } (x+2)(x-1)$$

Now, when $x = -2$, this equation yields $-3 = -3A \Rightarrow A = 1$. When $x = 1$, it yields $3 = 3B \Rightarrow B = 1$. Therefore,

$$\int \frac{2x+1}{(x+2)(x-1)} \, dx = \int \left(\frac{1}{x+2} + \frac{1}{x-1} \right) dx$$

$$= \ln|x+2| + \ln|x-1| + C = \ln\left|x^2 + x - 2\right| + C.$$

Substituting this into our original result yields

$$\int \frac{x^3 - x + 3}{x^2 + x - 2} \, dx = \frac{x^2}{2} - x + \ln\left|x^2 + x - 2\right| + C. \qquad \blacksquare$$

3 Applications of Integration

 Chapter Preview. This chapter tackles the Geometry Big Question from chapter 1: Can we calculate the length of any curve, area of any surface, and volume of any solid? We'll begin by reviewing Leibniz's perspective on the definite integral, which we'll rely on often to derive the important formulas in this chapter. Then, we'll get to work tackling the Big Question. We'll start by developing formulas for the area between two curves, and then for the volumes of solids generated either by revolving a plane region about an axis or by gluing together known cross sections (e.g., as a loaf of bread is made by placing its slices side by side). We'll end with derivations of the formulas for the arc length of a curve and the surface area of a surface. That sounds like a lot, I know, but I'll scaffold the presentation to keep the adventure manageable.

3.1 A Quick Preview of What's to Come

We're going to derive lots of formulas involving definite integrals in this chapter. A traditional calculus textbook derives those formulas from Riemann sums. While this is more mathematically rigorous, this is *Calculus 2 **Simplified**,* not *Calculus 2 **Rigorized**.* I'll therefore opt for the more intuitive approach to deriving most of the important formulas in this chapter. And when it comes to integration, Leibniz's approach to integration is more intuitive than Riemann's. In this chapter I'll therefore use Leibniz's approach to derive most of the integral formulas we'll develop to answer our Geometry Big Question from chapter 1: Can calculate the length of any curve, area of any surface, and volume of any solid? You'll therefore see me drawing and discussing lots of "infinitesimal rectangles" (we discussed these in section 2.1).[1] In the next section we'll start the journey toward answering the Big Question by learning how to calculate the area between two curves. Then section 3.3 will mark our first use Leibniz's infinitesimal rectangles. Let's get started!

3.2 Area between Curves

In Calculus 1 you learned how to calculate the area between the graph of a (usually continuous) function $y = f(x)$ and the x-axis. What happens when we replace the x-axis with some other curve $y = g(x)$? How could we calculate the area *between* the two curves? Figure 3.1 illustrates such an area.

We *could* solve this problem using Riemann sums. (See section A3.1 for that approach.) We *could* also use Leibniz's infinitesimal rectangles. But let's go even *simpler.* Let's use what we already know: how to calculate the area between a curve

---

[1] For those curious to see the more rigorous (Riemannian) treatment of the formulas we'll develop in this chapter, I have included that in this chapter's appendix and will reference that as we go.

Area A between curves …   ... equals area under f(x) …   ... minus area under g(x).

A   = $\int_a^b f(x)\,dx$ − $\int_a^b g(x)\,dx$

Figure 3.1: The area A between the curves $y = f(x)$ and $y = g(x)$ is the difference between the area under f and the area under g.

and the x-axis. The rest of figure 3.1 illustrates this approach: The area *between* the graphs of f and g is the area under the graph of f minus the area under the graph of g. Using the difference property of integrals—the difference of two integrals is the integral of the difference of the integrands—then yields the following theorem.

THEOREM 3.1: AREA BETWEEN TWO FUNCTIONS OF X. Let f and g be continuous on $[a, b]$, with $f(x) \geq g(x)$ for all x in $[a, b]$. Then the area A between the graphs of f and g, and bounded by $x = a$ and $x = b$, is

$$A = \int_a^b [f(x) - g(x)]\,dx. \tag{3.1}$$

A couple of comments on this theorem:

- One way to remember the integrand in (3.1) is to think of it as "top minus bottom," since $f(x) \geq g(x)$ means that the graph of f is above the graph of g.*

- We can still use the theorem if the graph of f dips below the graph of g inside the interval $[a, b]$. In that case we need to split up the integral to ensure that the integrand always has the form "top minus bottom." Suppose, therefore, that $f(x) \geq g(x)$ on $[a, b]$ but $g(x) \geq f(x)$ on $[b, c]$. (Figure 3.2(b) illustrates an example of this swapping.) Then the area between the two graphs and bounded by $x = a$ and $x = c$ would be

$$\int_a^b [f(x) - g(x)]\,dx + \int_b^c [g(x) - f(x)]\,dx. \tag{3.2}$$

One can condense this result using absolute values.[2]

*We'll soon encounter a "right minus left" version of (3.1).

EXAMPLE 3.1 Set up *but do not evaluate* the definite integral yielding the area of the region between the graphs of $y = x^2 + 2x + 1$ and $y = 2x + 5$, and bounded by $x = -2$ and $x = 2$, shown shaded in figure 3.2(a).

---

[2]Indeed, (3.2) is equivalent to $\int_a^c |f(x) - g(x)|\,dx$.

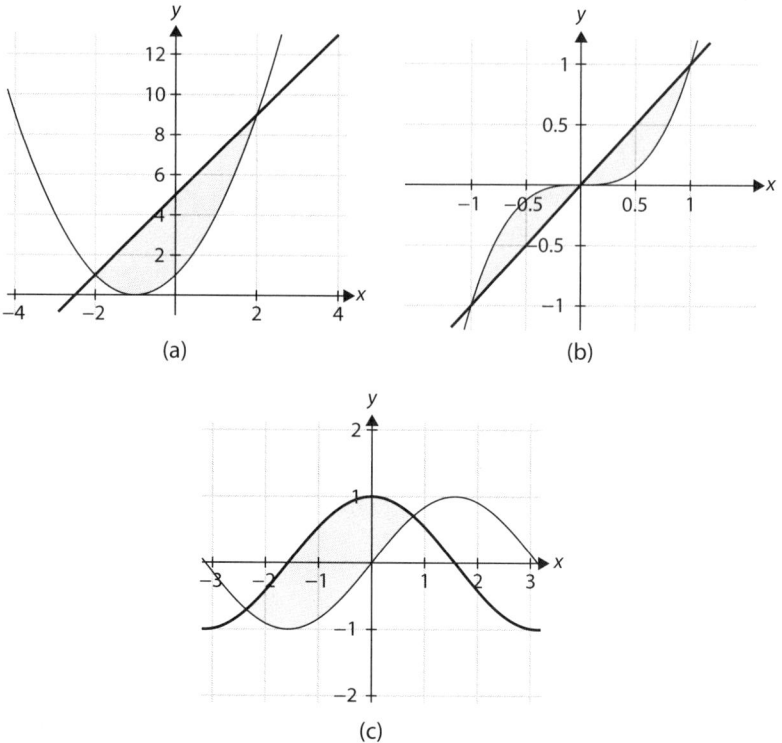

Figure 3.2: The plots and regions related to examples 3.1–3.3.

Solution Here the top graph is that of $y = 2x + 5$, so the shaded region's area is

$$\int_{-2}^{2} [(2x+5) - (x^2 + 2x + 1)] \, dx = \int_{-2}^{2} (4 - x^2) \, dx.$$ ■

EXAMPLE 3.2   Set up *but do not evaluate* the definite integral yielding the area of the region between the graphs of $y = x$ and $y = x^3$, shown shaded in figure 3.2(b).

Solution Here the graph on top changes when we cross $x = 0$. For $x < 0$ the top graph is that of $y = x^3$, so the area of the shaded region to the left of $x = 0$ is

$$\int_{-1}^{0} [x^3 - x] \, dx.$$

For $x > 0$, however, the top graph is that of $y = x$. The area of the shaded region to the right of $x = 0$ is therefore

$$\int_{0}^{1} [x - x^3] \, dx.$$

The total area of the shaded regions is thus

$$\int_{-1}^{0} (x^3 - x) \, dx + \int_{0}^{1} (x - x^3) \, dx.$$ ■

EXAMPLE 3.3   Find the area of the region between the graphs of $y = \cos x$ and $y = \sin x$, and bounded by $x = -3\pi/4$ and $x = \pi/4$, shown shaded in figure 3.2(c).

Solution Here the top graph is that of $y = \cos x$ (thicker black curve), so the shaded region's area is

$$\int_{-3\pi/4}^{\pi/4} (\cos x - \sin x)\, dx = [\sin x + \cos x]_{-3\pi/4}^{\pi/4}$$

$$= \left(\sin\frac{\pi}{4} + \cos\frac{\pi}{4}\right) - \left[\sin\left(-\frac{3\pi}{4}\right) + \cos\left(-\frac{3\pi}{4}\right)\right]$$

$$= \frac{\sqrt{2}}{2} + \frac{\sqrt{2}}{2} - \left[-\frac{\sqrt{2}}{2} - \frac{\sqrt{2}}{2}\right] = 2\sqrt{2}. \qquad \blacksquare$$

Related Exercise A1.

APPLICATIONS Applied exercise A1 investigates income inequality via the **Gini coefficient**, which is defined in terms of the area between two curves.

Two quick comments related to these examples:

- Sometimes you'll need to find the integration bounds yourself by finding where the graphs intersect. In example 3.2, for instance, we'd do this by solving

$$x = x^3 \quad \Longrightarrow \quad x^3 - x = 0 \quad \Longrightarrow \quad x(x^2 - 1) = 0,$$

which yields $x = 0$ and $x = \pm 1$. This generates the intervals $[-1, 0]$ and $[0, 1]$ we used in the example's solution.

- Implicit in any "find the area between the curves" question is that we're looking for a *finite* area. So, for example, if asked to "find the area between the curves $y = x^3$ and $y = x$," you should calculate the area of the shaded region in figure 3.2(b), which is the only finite such area. (For $x > 1$ and $x < -1$, the areas between the curves are infinite.)

Related Exercises 1–4.

Area between Two Functions of y

Sometimes we're asked to find the area between two functions of y. For example, what's the shaded area B in figure 3.3 of the region between the two curves $x = h(y)$ and $x = s(y)$?

Making sense of this question requires working with functions of y and their graphs. So let's talk about that first.

When we graph a function $y = f(x)$, we think in up and down terms. For example, to plot a point $y = f(3) = 9$, we'd go to $x = 3$ on the x-axis and then go *up* by 9 units. The x-axis is therefore the "foundation" of our plotting exercise. To graph

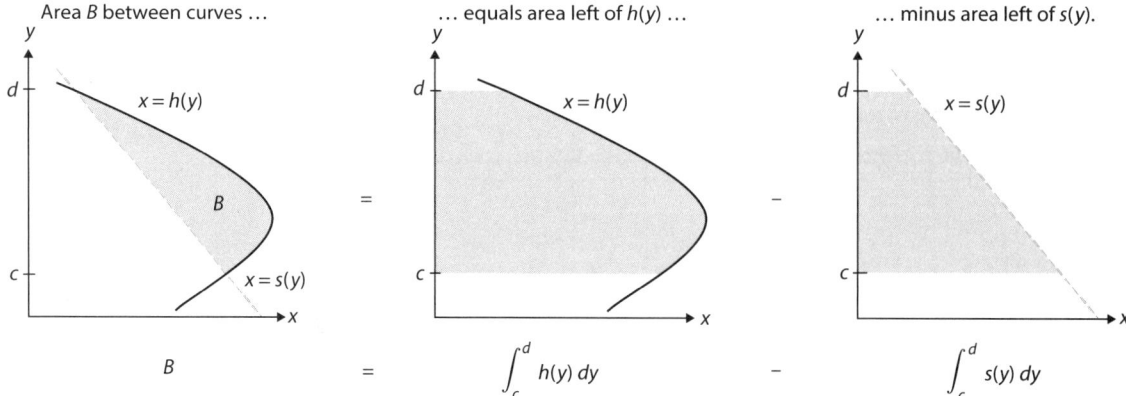

Figure 3.3: The area B between the curves $x = h(y)$ and $x = s(y)$ is the difference between the area left of h and the area left of g.

a function $x = g(y)$ we use the y-axis as our foundation, and we think in left and right terms. For example, to plot a point $x = g(4) = 5$, we'd go to $y = 4$ on the y-axis and then go *right* by 5 units. This shift from up/down thinking to right/left thinking can take some time to get used to if you're new to it, so I'll provide wordier solutions to the examples that follow to help you better understand graphing functions of y.

Okay, back to the question at hand: What's the area B of the shaded region in figure 3.3? In figure 3.1 we answered a similar question in the setting of functions of x. Compare the first plot in figure 3.1 to figure 3.3 and you might notice something: The plot in figure 3.3 is a clockwise rotation by $90°$ of the one in figure 3.1. I did that on purpose to show you that, just like in figure 3.1, where we expressed the area A as the difference of the two areas under f and g, we can do a similar thing with figure 3.3. Since we've swapped up/down thinking for right/left thinking now, though, to find B in figure 3.3 we find the area *left* of $x = h(y)$ and subtract the area *left* of $x = s(y)$. The rest of figure 3.3, analogues of figure 3.1, illustrate this and yield the following adaptation of theorem 3.1 to functions of y.

THEOREM 3.2: AREA BETWEEN TWO FUNCTIONS OF Y. Let $x = h(y)$ and $x = s(y)$ be continuous on $[c, d]$, with $h(y) \geq s(y)$ for all y in $[c, d]$. Then the area A between the graphs of h and s, and bounded by $y = c$ and $y = d$, is

$$A = \int_c^d [h(y) - s(y)]\, dy. \tag{3.3}$$

I remember the integrand above as "right minus left," because it's the difference between the rightmost graph that defines the region and the leftmost one. As before, that ordering may swap somewhere inside the interval $[c, d]$. In that case we do the same thing as before: We split up the integral (3.3) at the y-value where the ordering change occurs. (We'll see an example of this shortly.)

EXAMPLE 3.4 (Video)   Find the area of the region between the graphs of $x = 3 - y^2$ and $x = y + 1$, and bounded by $y = -2$ and $y = 1$, shown shaded in figure 3.4.

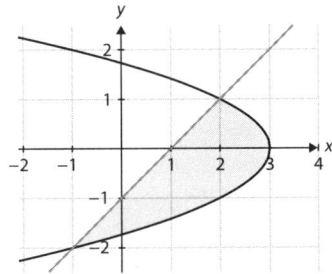

Figure 3.4

Solution Let's first talk about the graphs themselves. To plot $x = y + 1$ we start somewhere on the y-axis and then go left or right to get to $x = y + 1$. Example: If $y = 0$ (so that we start at the origin), then $x = y + 1 = 0 + 1 = 1$, taking us to the point $(1, 0)$. Another example: If $y = 1$ (so that we start at the point $(0, 1)$ on the y-axis), then $x = y + 1 = 1 + 1 = 2$, taking us to the point $(2, 1)$. I illustrate all this in the video accompanying this solution.

Alright, now back to calculating the shaded area. Since the graph of $x = 3 - y^2$ (black curve) is to the right of the graph of $x = y + 1$ (gray line) for the region of interest, (3.3) says that the area A of the shaded region is

$$A = \int_{-2}^{1} \left[\left(3 - y^2 \right) - (y + 1) \right] dy$$
$$= \int_{-2}^{1} \left(-y^2 - y + 2 \right) dy$$
$$= \left[-\frac{y^3}{3} - \frac{y^2}{2} + 2y \right]_{-2}^{1}$$
$$= \left(-\frac{1}{3} - \frac{1}{2} + 2 \right) - \left(\frac{8}{3} - 2 - 4 \right) = \frac{9}{2}. \qquad \blacksquare$$

EXAMPLE 3.5   Set up *but do not evaluate* the area of the region between the graphs of $x = y$ and $x = \sqrt[3]{y}$, and bounded by $y = -1$ and $y = 1$, shown shaded in figure 3.5.

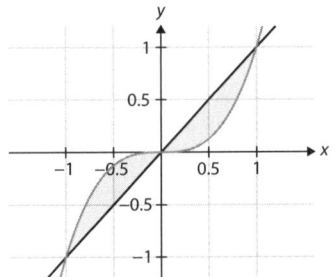

Figure 3.5

Solution The figure shows that for $x > 0$ the graph of $x = \sqrt[3]{y}$ (the gray curve) is to the right of the graph of $x = y$ (the black line). The area of the shaded region for $x > 0$ is therefore

$$\int_0^1 \left(\sqrt[3]{y} - y \right) dy.$$

For $x < 0$ the black line is to the right of the gray curve. So the area of the shaded region for $x < 0$ is

$$\int_{-1}^0 \left(y - \sqrt[3]{y} \right) dy.$$

The total area is thus

$$\int_{-1}^0 \left(y - \sqrt[3]{y} \right) dy + \int_0^1 \left(\sqrt[3]{y} - y \right) dy. \qquad \blacksquare$$

Related Exercises   5–8.

Additional Tips, Tricks, and Takeaways

You always have a choice of using either (3.1) or (3.3)—the "top minus bottom" or "right minus left," respectively, approach to find the area between two curves. An area is an area, after all, no matter how we calculate it. So which method is best? In many cases that depends on how hard (or not) it is to integrate the functions you're left with.

As an illustration, consider again example 3.4, but this time let's use the "top minus bottom" approach of (3.1). To do that we need to convert the functions of y in example 3.4 into functions of x. We do this by solving for y in each equation. That's easy for $x = y$: $y = x$. For $x = 3 - y^2$, however, we get $y^2 = 3 - x$ and then $y = \pm\sqrt{3 - x}$. The $y = \sqrt{3 - x}$ part is the portion of the black curve in figure 3.4 above the x-axis; the $y = -\sqrt{3 - x}$ part is the portion below the x-axis. (Things are getting more complicated already.) Staying with that figure, notice also that for $x < 2$ the gray line is the top graph, while for $x > 2$ the top graph is $y = \sqrt{3 - x}$. The same area A calculated in example 3.4 is therefore equivalent to

$$\int_{-1}^{2} [x - (-\sqrt{3 - x})]\, dx + \int_{2}^{3} [\sqrt{3 - x} - (-\sqrt{3 - x})]\, dx.$$

This integral is harder to evaluate than the one we got in the solution to example 3.4. Takeaway: Although one *can* choose either (3.1) or (3.3) to calculate the area between two curves, when given the choice, go with the one that's simplest.

3.3   Volumes by Cross Sections

We're now ready to tackle the volume portion of the Geometry Big Question from chapter 1: Can we calculate the length of any curve, area of any surface, and volume of any solid? We'll begin in this section by studying *volumes by cross sections*. This perspective on calculating volumes goes as follows. First, take your solid (a three-dimensional object) and put it on a table. Now draw a set of x- and y-axes on the table. Next, slice the solid perpendicular to the x-axis using a "mathematician's knife," an infinitesimally thick knife. The resulting slices will look like two-dimensonal shapes with infinitesimally small thickness.* To find the volume of the solid we calculate the volumes of those slices and then add up the results.[3] Let's now work this out for a concrete example.

Figure 3.6(d) shows a solid wedge-like object. Now, with your mathematician's knife in hand, slice the solid perpendicular to the (horizontal) t-axis at location $t = x$. What you'll get is the infinitesimally thick square pictured in figure (c). That square has area $A(x)$ and thickness dx, and so volume $A(x)\, dx$. Adding up the volumes of these "infinitesimal slices"** then yields the formula in the theorem below.

> **THEOREM 3.3:   VOLUMES OF SOLIDS WITH CROSS SECTIONS PERPENDICULAR TO THE *X*-AXIS.** Let S be a solid. If $A(x)$ denotes the cross-sectional areas of S perpendicular to the x-axis, where $a \leq x \leq b$, then

* The two-dimensonal shapes are called the *cross sections* of the solid.

** These are the analogues of Leibniz's "infinitesimal rectangles" in this setting.

[3] For you cooks out there, this is like slicing a tomato into very thin slices, then calculating the volumes of all those slices, then adding up those volumes to get the original tomato's volume.

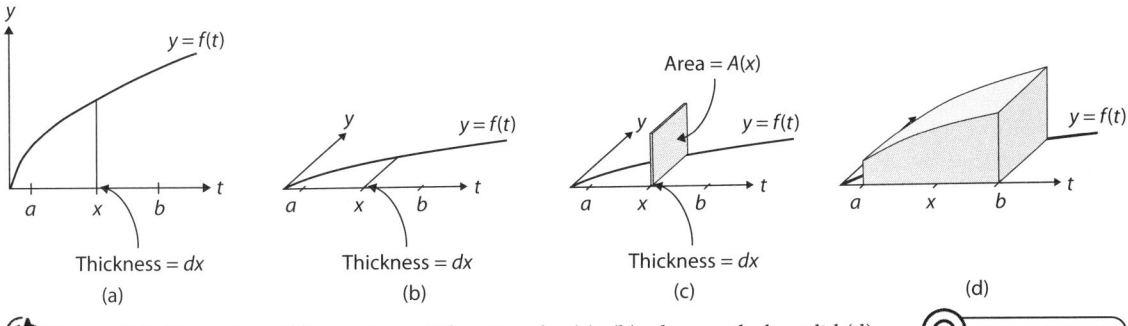

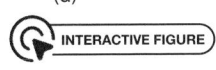

⟳**Figure 3.6:** Plots of $y = f(t)$ seen from different angles (a)–(b), along with the solid (d) whose cross sections perpendicular to the t-axis are squares of area $A(x)$ (c). Interact with this figure at sites.google.com/view/fernandezmath/apps/cs2.

the volume V of S is

$$V = \int_a^b A(x)\, dx. \tag{3.4}$$

EXAMPLE 3.6   The function graphed in figure 3.6(a) is $f(t) = \sqrt{t}$. If we use the area under this curve and bounded by $x = 1$ and $x = 2$ as the base of a solid, with cross sections perpendicular to the x-axis being squares, what is the volume of the resulting solid?

Solution   Figure 3.6(d) shows the resulting solid (here $a = 1$ and $b = 2$). The square of base area $A(x)$ in figure (c) is one of the cross sections of the solid. The base of that square is the height of the infinitesimal rectangle in figure (a), $y = \sqrt{x}$. And since the cross sections are squares, then $A(x) = [\sqrt{x}]^2$.* Substituting this into (3.4) yields

$$V = \int_1^2 [\sqrt{x}]^2\, dx = \int_1^2 x\, dx = \left[\frac{x^2}{2}\right]_1^2 = \frac{2^2 - 1^2}{2} = \frac{3}{2}. \qquad ■$$

* Recall that a square of side length s has area $A = s^2$.

Most of the time the base of the solid we're calculating the volume of is the area under the graph of a function $y = f(x)$ (as it was in the example above), the cross sections are perpendicular to the x-axis, and those cross sections are squares, semicircles, or equilateral triangles. In those cases we can determine $A(x)$ fairly easily. Figure 3.7 shows the results. To use those $A(x)$ formulas, you first need to figure out the semicircles' radii, or the squares' (or equilateral triangles') lengths. I've denoted those by $r(x)$ and $s(x)$, respectively, in the figure. The third column of the figure then tells us that $s(x) = f(x)$, while $r(x) = f(x)/2$.

EXAMPLE 3.7   Find the volume of the solid whose base is the area under the graph of $f(x) = \sqrt{x}$ and bounded by $x = 2$ and $x = 4$, and whose cross sections perpendicular to the x-axis are equilateral triangles.

Solution   The third row in figure 3.7 tells us that $A(x) = \frac{\sqrt{3}}{4}[\sqrt{x}]^2 = \frac{\sqrt{3}}{4}x$. Substituting this into (3.4) yields

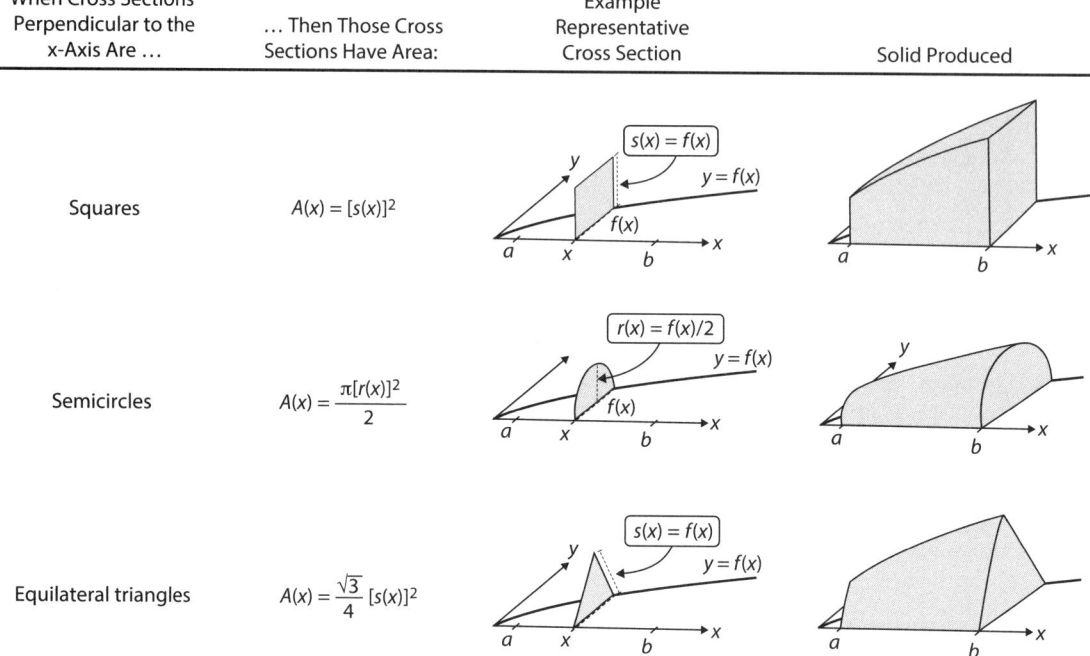

When Cross Sections Perpendicular to the x-Axis Are ...	... Then Those Cross Sections Have Area:	Example Representative Cross Section	Solid Produced
Squares	$A(x) = [s(x)]^2$		
Semicircles	$A(x) = \dfrac{\pi [r(x)]^2}{2}$		
Equilateral triangles	$A(x) = \dfrac{\sqrt{3}}{4}[s(x)]^2$		

Figure 3.7: A table of cross sections that are squares, semicircles, and equilateral triangles, along with their cross-sectional area formulas. Here $r(x)$ is the radius of the semicircular cross section and $s(x)$ is the side length of the square or equilateral triangle.

$$V = \frac{\sqrt{3}}{4} \int_2^4 x\,dx = \frac{\sqrt{3}}{4}\left[\frac{x^2}{2}\right]_2^4 = \frac{\sqrt{3}}{4} \cdot \frac{4^2 - 2^2}{2} = \frac{\sqrt{3}}{4} \cdot 6 = \frac{3\sqrt{3}}{2}.$$

The resulting solid resembles the last one in figure 3.7. ∎

Sometimes the base of the solid is an area *between* the graphs of two functions $f(x)$ and $g(x)$. In those cases the $s(x)$ and $r(x)$ in figure 3.7 become $s(x) = f(x) - g(x)$ and $r(x) = [f(x) - g(x)]/2$, respectively, assuming $f(x) \geq g(x)$ on the interval $[a, b]$ (i.e., the "top minus bottom" assumption from the previous section). Here are a few examples of this.

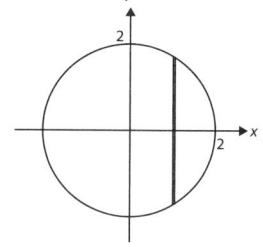

EXAMPLE 3.8 (Video) Find the volume of the solid whose base is bounded by the circle $x^2 + y^2 = 4$ and whose cross sections perpendicular to the x-axis are squares (see figure 3.8).

Solution Figure 3.8 (top) shows the graph of the circle along with an infinitesimal rectangle. The upper semicircle's equation is $f(x) = \sqrt{4 - x^2}$; the lower semicircle's is $g(x) = -\sqrt{4 - x^2}$. The height of the rectangle is therefore $f(x) - g(x)$ (top curve minus bottom curve). This simplifies to $2\sqrt{4 - x^2}$.

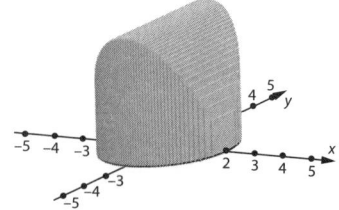

Figure 3.8

(This is our $s(x)$.) And since our cross sections are squares, they have area $A(x) = [2\sqrt{4 - x^2}]^2 = 4(4 - x^2)$. Substituting this into (3.4) yields

$$V = \int_{-2}^{2} 4(4 - x^2)\, dx = 4 \left[4x - \frac{x^3}{3} \right]_{-2}^{2} = \frac{128}{3}.$$

Figure 3.8 (bottom) illustrates the resulting solid. ■

Related Exercises   9(a)–14(a) ((a) only).

EXAMPLE 3.9   Redo the previous example for cross sections perpendicular to the x-axis that are (a) equilateral triangles and (b) semicircles.

Solution For (a), $s(x) = 2\sqrt{4 - x^2}$ again. But $A(x) = \frac{\sqrt{3}}{4}[s(x)]^2$ now (see the last row of figure 3.7). Thus, the volume V of the solid is

$$V = \frac{\sqrt{3}}{4} \int_{-2}^{2} [2\sqrt{4 - x^2}]^2\, dx = \sqrt{3} \int_{-2}^{2} (4 - x^2)\, dx = \frac{32\sqrt{3}}{3}.^*$$

For (b), $r(x) = s(x)/2 = \sqrt{4 - x^2}$ (half the $s(x)$ from part (a)). And with $A(x) = \pi[r(x)]^2/2$ now (see again figure 3.7), the volume V of the new solid is

$$V = \frac{\pi}{2} \int_{-2}^{2} [\sqrt{4 - x^2}]^2\, dx = \frac{\pi}{2} \int_{-2}^{2} (4 - x^2)\, dx = \frac{16\pi}{3}.^*$$ ■

* The integral in gray is equal to one-fourth the integral in the previous example (i.e., 32/3).

EXAMPLE 3.10 (Video)   Find the volume of the solid whose base is the region bounded by the graphs of $f(x) = 2 - x$, $g(x) = -2 + x$, and $x = 0$ and whose cross sections perpendicular to the x-axis are equilateral triangles (shown in figure 3.9).

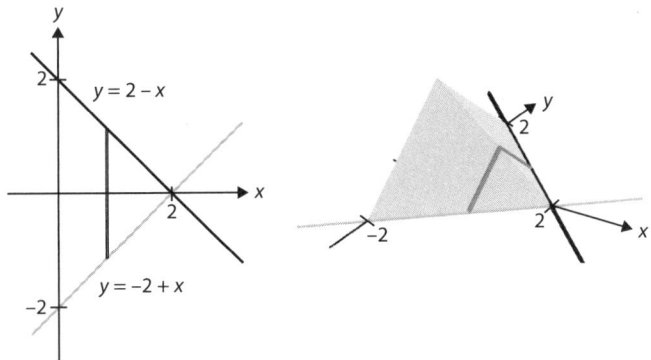

Figure 3.9

Solution Here $A(x) = \frac{\sqrt{3}}{4}[s(x)]^2$, with $s(x) = f(x) - g(x)$. And since $f(x) - g(x) = (2 - x) - (-2 + x) = 4 - 2x$, the volume of the solid is

$$V = \frac{\sqrt{3}}{4} \int_{0}^{2} (4 - 2x)^2\, dx$$

$$= \frac{\sqrt{3}}{4} \int_0^2 (16 - 16x + 4x^2)\, dx$$

$$= \frac{\sqrt{3}}{4} \left[16x - 8x^2 + \frac{4x^3}{3} \right]_0^2 = \frac{8\sqrt{3}}{3}. \qquad \blacksquare$$

Related Exercises　9(b)–14(b), 9(c)–14(c) ((b) and (c) only).

Cross Sections Parallel to the *x*-Axis

We've been slicing our solids perpendicular to the *x*-axis. We could also, however, slice them *parallel* to the *x*-axis. Why might we do this? Well, if the base of the solid is better expressed in terms of functions of *y*, then the integration will be easier if done with respect to *y*. Employing the same reasoning leading up to theorem 3.3, thinking this time in terms of slicing parallel to the *x*-axis, yields the following analogue.

> **THEOREM 3.4: VOLUMES OF SOLIDS WITH CROSS SECTIONS PARALLEL TO THE *X*-AXIS.** Let *S* be a solid. If $A(y)$ denotes the cross-sectional areas of *S* parallel to the *x*-axis, where $c \le y \le d$, then the volume *V* of *S* is
>
> $$V = \int_c^d A(y)\, dy. \tag{3.5}$$

The formulas for $A(y)$ when the cross sections are squares, semicircles, or equilateral triangles are the same as those in figure 3.7 but with *x* replaced by *y*. The meanings of $s(y)$ and $r(y)$ change, though, since now the infinitesimal rectangles are *horizontal*. (Figure 3.10 illustrates such a rectangle.) Now $s(y)$ is the *width* of those rectangles, while $r(y)$ is half their width. Furthermore, if the base of our solid is an area between the graphs of two functions $x = h(y)$ and $x = g(y)$, where $h(y) \ge g(y)$ on the interval $[c, d]$, then the widths of the rectangles are $s(y) = h(y) - g(y)$ ("right minus left," as in the previous section). The next example illustrates all this.

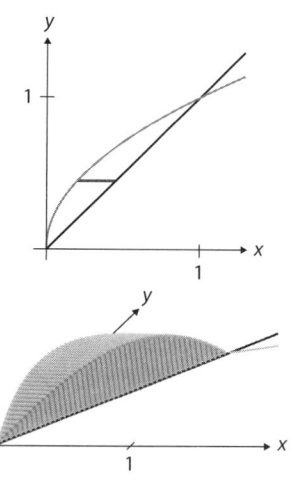

Figure 3.10

EXAMPLE 3.11 (Video)　Set up *but do not evaluate* the integrals yielding the volume of the solids whose base is bounded by the graphs of $x = \sqrt[3]{y}$ and $x = y$, between $y = 0$ and $y = 1$, and whose cross sections parallel to the *x*-axis are (a) squares, (b) semicircles, and (c) equilateral triangles.

Solution　Figure 3.10 plots the graphs and shows the infinitesimal rectangle.

For (a), $s(y) = y - \sqrt[3]{y}$ ("right curve (black) minus left curve (gray)") and $A(y) = [s(y)]^2 = (y - \sqrt[3]{y})^2$. Thus, the volume V of the solid is

$$V = \int_0^1 (y - \sqrt[3]{y})^2 \, dy.$$

Figure 3.10 (bottom) shows the resulting solid.

For (b), $r(y) = s(y)/2 = (y - \sqrt[3]{y})/2$ (half the $s(y)$ in part (a)). And with $A(y) = \pi[r(y)]^2/2$ now, the volume V of the new solid is

$$V = \frac{\pi}{2} \int_0^1 \left(\frac{y - \sqrt[3]{y}}{2} \right)^2 dy$$

$$= \frac{\pi}{8} \int_0^1 (y - \sqrt[3]{y})^2 \, dy.$$

For (c), $s(y) = y - \sqrt[3]{y}$, and with $A(y) = \frac{\sqrt{3}}{4}[s(y)]^2$, the volume V of the new solid is

$$V = \frac{\sqrt{3}}{4} \int_0^1 (y - \sqrt[3]{y})^2 \, dy. \qquad \blacksquare$$

Related Exercises   15–17.

I've taken here the Leibnizian approach to deriving equation (3.4), the main volume equation we've used in this section; check out section A3.2 if you'd like to see the Riemann sums approach.

Alright, we now know how to calculate the volumes of solids with known cross sections. Great! In the rest of this chapter these techniques will help us calculate the volumes of a surprising number of human-made solids. Indeed, look around wherever you are right now and focus on the human-made objects near you—the cups, tables, lamps, wheels, rings, etc. We can now calculate the volumes that many of those shapes contain or can hold (like in the case of a cup) using the methods we learned in this section. Some of those shapes have a special structure: they are "shapes of revolution," shapes one can obtain by revolving a curve one full revolution about an axis. A cup has that structure—it's the revolution of a line segment about a vertical axis. A wheel, many lamp shades, and many rings also have this structure. (Shapes of revolution are all around you!) In the next two sections we'll use our volumes by cross sections results to derive formulas for the volumes associated with these shapes of revolution.

3.4   Volumes of Revolution: The Disk Method

In the previous section we generated a solid by using a two-dimensional region as the solid's base and stipulating the geometry of the solid's cross sections (e.g., squares). Another way to generate a solid is to take that two-dimensional region and revolve it one full revolution about an axis in the plane. Solids produced in this way are called **solids of revolution**. Figures 3.11(a)–(b) illustrate the process for a function $y = f(x)$. (The function graphed is $f(x) = \sqrt{x}$, but let's pretend it's a generic function f.)

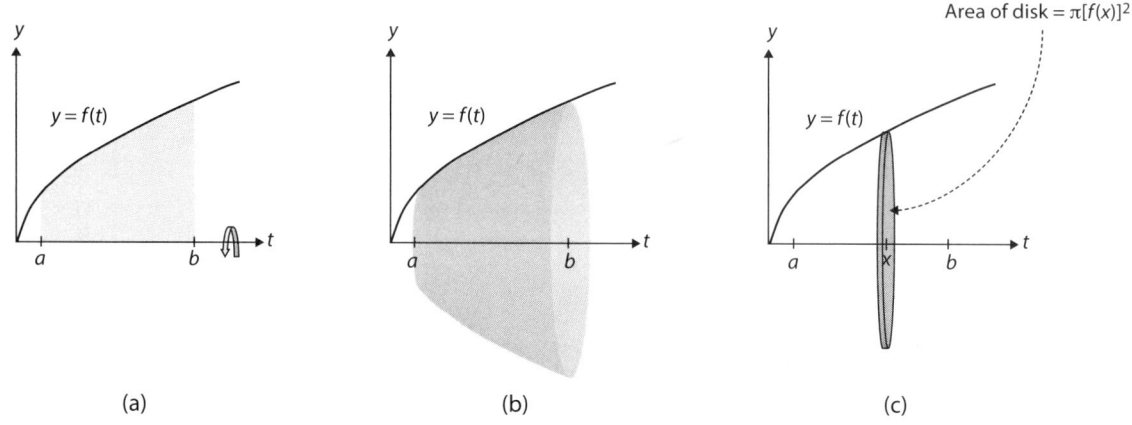

(a)   (b)   (c)

 INTERACTIVE FIGURE

Figure 3.11: Revolving the shaded region in (a) about the t-axis produces the solid in (b). That solid's cross sections perpendicular to the t-axis are disks, like the one in (c). Interact with this figure at sites.google.com/view/fernandezmath/apps/cs2.

I made a simple animation of this process to help you visualize how the solid is generated; scan the QR code on this page to see that animation. Looking now at the solid in figure (b), if we take out our mathematician's knife and slice that solid vertically at location $t = x$ on the t-axis, we obtain the infinitesimally thin cylinder in figure (c). That cylinder's cross sections perpendicular to the x-axis are disks. The radius r of the disk in figure (c) is the y-value of the function at $t = x$: $r = f(x)$. This yields the cross-sectional area $A(x) = \pi [f(x)]^2$ (from πr^2). Substituting this $A(x)$ into our formula for volumes by cross sections (3.4) from the previous section then yields the formula in the following theorem.

> **THEOREM 3.5:   THE DISK METHOD ABOUT THE X-AXIS.** Let $f(x)$ be continuous on $[a, b]$. Then the volume V of the solid obtained by revolving about the x-axis the plane region between the graph of $y = f(x)$ and the x-axis, bounded by $x = a$ and $x = b$, is
>
> $$V = \pi \int_a^b [f(x)]^2 \, dx. \qquad (3.6)$$

A couple of quick comments on this theorem.

- We call this method the *disk* method because the solids obtained in the theorem have cross sections that are disks.*

* See section A3.3 for the Riemann sums derivation of the disk method.

- If the solid you get after revolving the plane region about the x-axis *does not* have cross sections that are disks, *don't use the disk method*. What other cross sections are possible? We'll discuss one common occurrence in the next section.

EXAMPLE 3.12   Find the volume of the solid of revolution pictured in figure 3.12(b) generated by revolving the shaded plane region in figure (a) about the x-axis.

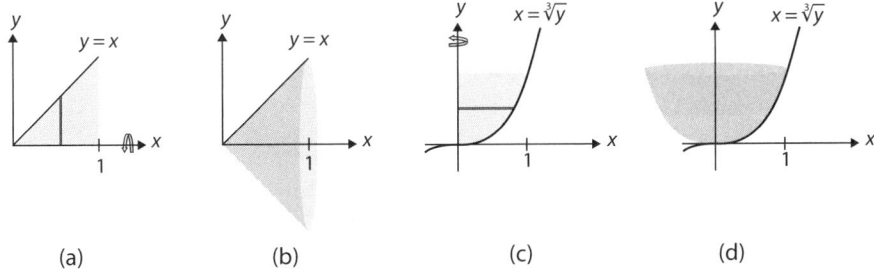

(a)   (b)   (c)   (d)

Figure 3.12: The plane regions and volumes of revolution associated with examples 3.12–3.13.

Solution Here $f(x) = x$, $a = 0$ and $b = 1$, so (3.6) yields

$$V = \int_0^1 \pi x^2 \, dx = \pi \left[\frac{x^3}{3} \right]_0^1 = \frac{\pi}{3}.$$ ■

Related Exercises   18(a), 19(a), 20–22, D1.

Thus far we've been revolving our regions about the x-axis to produce our solids of revolution. But we could also revolve our region about the y-axis, as illustrated in figures 3.12(c)–(d). Generalizing the disk method to this setting is fairly straight-forward (basically, replace the xs by ys and update the limits of integration). The result is the next theorem.

THEOREM 3.6: THE DISK METHOD ABOUT THE Y-AXIS. Let $h(y)$ be continuous on $[c, d]$. Then the volume V of the solid obtained by revolving about the y-axis the plane region between the graph of $x = h(y)$ and the y-axis, bounded by $y = c$ and $y = d$, is

$$V = \pi \int_c^d [h(y)]^2 \, dy. \tag{3.7}$$

EXAMPLE 3.13   Find the volume of the solid of revolution pictured in figure 3.12(d) generated by revolving the shaded plane region in figure (c) about the y-axis.

Solution Here $h(y) = \sqrt[3]{y} = y^{1/3}$, $c = 0$, and $d = 1$, so (3.7) yields

$$V = \int_0^1 \pi y^{2/3} \, dy = \pi \left[\frac{3y^{5/3}}{5} \right]_0^1 = \frac{3\pi}{5}.$$ ■

Related Exercises   18(b), 19(b), 23.

Additional Tips, Tricks, and Takeaways

- The disk method is deceptively simple to use: Just grab your function and the correct bounds, and then integrate. The (hidden) deception arises from the requirement that the solid's cross sections be disks. I'm stressing this again because a common mistake is to use the disk method when the cross sections *aren't* disks. To help you ensure the cross sections are disks, I recommend drawing an infinitesimal rectangle in your plane region, as I've done in figures 3.12(a) and (c). *If you move the rectangle around the plane region and it doesn't always touch the axis of revolution*, then the cross sections won't all be disks. In that case you shouldn't use the disk method.*

* Flip forward to the top right plot in figure 3.17 to see an example of this.

• Notice that solids produced by revolving a plane region about the *x*-axis have *vertical* infinitesimal rectangles, while those produced by revolving about the *y*-axis have *horizontal* rectangles.

These two tips combined should help you avoid the most common pitfalls in applying the disk method. Next up: What happens when the cross sections are not disks.

3.5   Volumes of Revolution: The Washer Method

The solid in figure 3.12(d) looks like a bowl filled in with concrete. The bowls in your cupboard, however, are hollow—they look more like what's pictured in figure 3.13(b). The cross sections of that solid that are perpendicular to the *t*-axis *aren't* disks, so we can't use the disk method to calculate that solid's volume. But a simple tweak to that method saves the day.

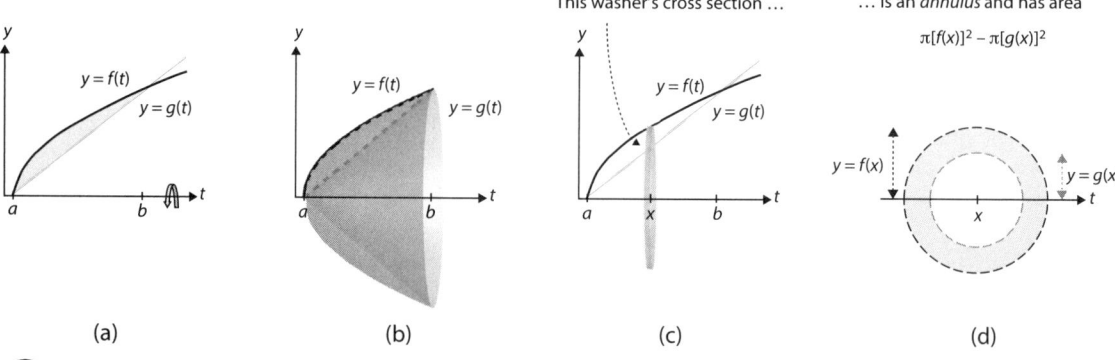

(a)   (b)   (c)   (d)

(INTERACTIVE FIGURE)

Figure 3.13: Revolving the shaded region in (a) about the *t*-axis produces the solid in (b). Slicing that solid at location *x* on the *t*-axis produces the thin washer in (c). That washer's cross sections are annuli with outer radius *f*(*x*) and inner radius *g*(*x*) (d). Interact with this figure at sites.google.com/view/fernandezmath/apps/cs2.

Let's first begin by noticing that the new "hole-y" solid pictured in figure 3.13(b) results from revolving the shaded plane region in figure (a) about the *t*-axis. Slice the resulting solid (figure (b)) perpendicular to the *t*-axis with our mathematician's knife and you end up with washers, like the one pictured in figure (c). The cross sections of those washers are *not* disks. They are *annuli*.** Figure (d) shows the annular cross section of the washer in figure (c). Since every annulus is made up of two disks—a larger one and a smaller one that defines the hole—we have two radii that characterize the annulus: the larger radius, *R*, and the smaller radius, *r*.

** An *annulus* is a disk with a smaller concentric disk removed.

Figure (d) shows that the annular cross section at location $t = x$ in figure (c) has $R = f(x)$ and $r = g(x)$. The *area* of the shaded region in figure (d) is therefore the area of the larger disk ($\pi R^2 = \pi [f(x)]^2$) minus the area of the smaller disk ($\pi r^2 = \pi [g(x)]^2$). This yields $\pi [f(x)]^2 - \pi [g(x)]^2 = \pi \left([f(x)]^2 - [g(x)]^2 \right)$. Substituting this into our volumes by cross sections formula (3.4), then yields the formula in the following theorem.*

* See section A3.4 for the Riemann sums derivation of the same result.

> **THEOREM 3.7: THE WASHER METHOD ABOUT THE *X*-AXIS.** Let $f(x)$ and $g(x)$ be continuous on $[a, b]$, with $f(x) \geq g(x)$. Then the volume V of the solid obtained by revolving about the x-axis the plane region bounded by the graphs of $y = f(x)$ and $y = g(x)$, $x = a$ and $x = b$, is
>
> $$V = \pi \int_a^b \left([f(x)]^2 - [g(x)]^2 \right) \, dx. \tag{3.8}$$

If you look closely at the formula above, you might notice two things: (1) It looks like a modification of the disk method, and (2) if $g(x) = 0$, then it *is* the disk method. The latter makes sense: If $g(x) = 0$, then the annulus in figure 3.13(d) is no longer an annulus—it's a disk. Hence, we're back to the disk method. To explain (1), have a look at figure 3.14.

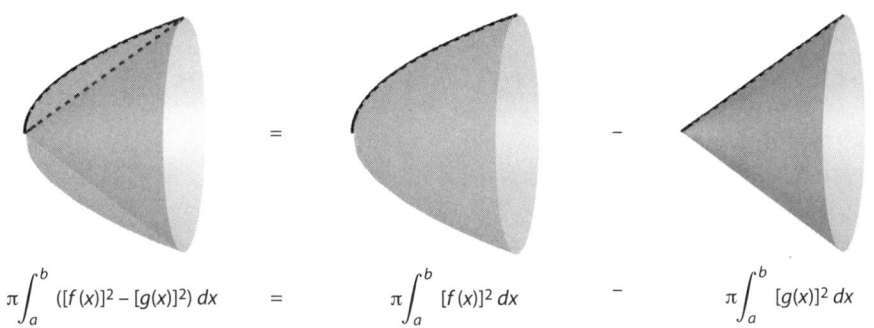

$$\pi \int_a^b ([f(x)]^2 - [g(x)]^2) \, dx \quad = \quad \pi \int_a^b [f(x)]^2 \, dx \quad - \quad \pi \int_a^b [g(x)]^2 \, dx$$

Figure 3.14: An illustration of how the washer method works: It takes a volume and removes a (smaller) volume from it.

This approach expresses (3.8) as the difference of two *disk method* applications: the first for $y = f(x)$ and the second for $y = g(x)$.

EXAMPLE 3.14   Find the volume of the solid of revolution generated by revolving the region bounded by the graphs of $y = \sqrt{x}$ and $y = x^2$ about the x-axis (shown in figure 3.15).

Solution Here $f(x) = \sqrt{x}, g(x) = x^2, a = 0$, and $b = 1$. Thus, (3.8) yields

$$V = \pi \int_0^1 ([\sqrt{x}]^2 - [x^2]^2) \, dx$$

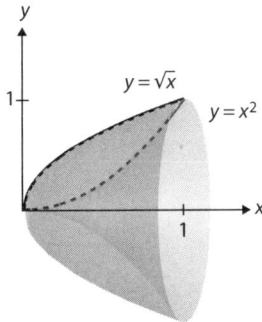

Figure 3.15

$$= \pi \int_0^1 (x - x^4)\, dx$$

$$= \pi \left[\frac{x^2}{2} - \frac{x^5}{5} \right]_0^1$$

$$= \frac{3\pi}{10}.$$

■

Related Exercises   24–26, A2.

APPLICATIONS   Applied exercise A2 investigates the volume of a "mathematical doughnut" using the methods taught in this section.

Just like we adapted the disk method for revolutions about the y-axis, let's likewise adapt the washer method. The next theorem describes the result.

THEOREM 3.8: THE WASHER METHOD ABOUT THE Y-AXIS. Let $h(y)$ and $s(y)$ be continuous on $[c, d]$, with $h(y) \geq s(y)$. Then the volume V of the solid obtained by revolving about the y-axis the plane region bounded by the graphs of $x = h(y)$ and $x = s(y)$, $y = c$ and $y = d$, is

$$V = \pi \int_c^d \left([h(y)]^2 - [s(y)]^2 \right) dy. \tag{3.9}$$

EXAMPLE 3.15   Set up *but do not evaluate* the integral yielding the volume of the solid of revolution generated by revolving the region bounded by the graphs of $x = y^2/4$ and $x = \sqrt[3]{2y}$ about the y-axis (shown in figure 3.16).

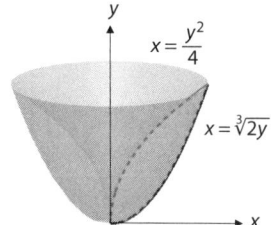

Figure 3.16

Solution   Here $h(y) = \sqrt[3]{2y}$ and $s(y) = y^2/4$. To figure out c and d (the limits of integration we'll need), we look for where these functions' graphs intersect:

$$\sqrt[3]{2y} = y^2/4 \quad \Rightarrow \quad 2y = \frac{y^6}{64} \quad \Rightarrow \quad \frac{y^6}{64} - 2y = 0.$$

Factoring: $y \left(\frac{y^5}{64} - 2 \right) = 0$. Solving yields $y = 0$ and $y^5/64 = 2$, or $y^5 = 128$. Taking the fifth root yields $y = \sqrt[5]{128} = 2\sqrt[5]{4}$. Thus, (3.9) yields

$$V = \pi \int_0^{2\sqrt[5]{4}} \left([(2y)^{1/3}]^2 - \left[\frac{y^2}{4} \right]^2 \right) dy$$

$$= \pi \int_0^{2\sqrt[5]{4}} \left[(2y)^{2/3} - \frac{y^4}{16} \right] dy.$$

■

Related Exercises   27–29.

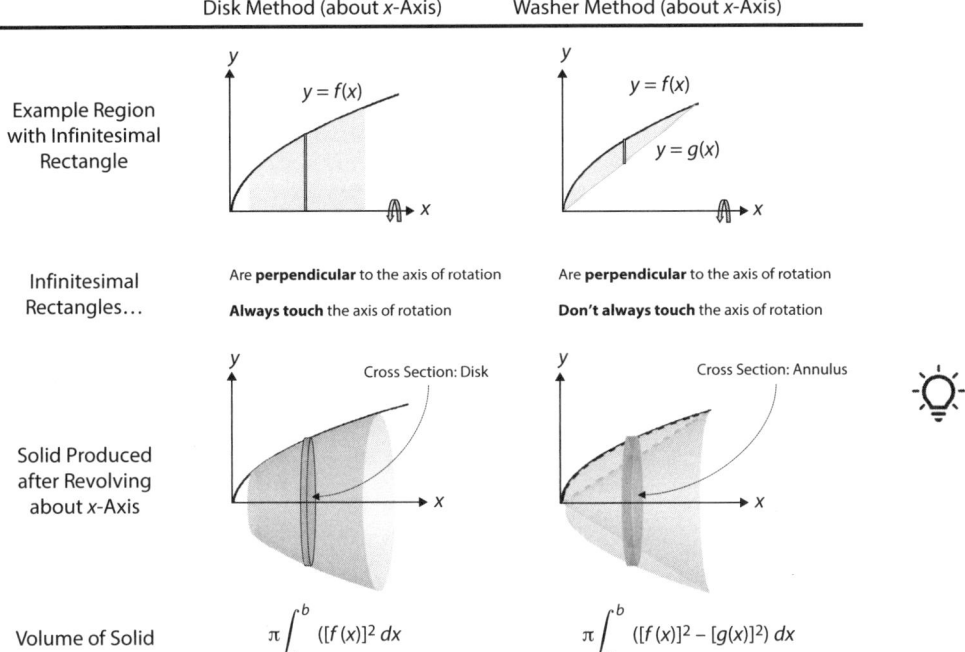

Figure 3.17: A comparison of the disk and washer methods when revolving about the x-axis.

Additional Tips, Tricks, and Takeaways

Figure 3.17 compares and contrasts the disk and washer methods for revolutions about the x-axis.

I encourage you to study this figure closely. Not only does it recap what we've done in this and the previous section, but it also comes in handy when deciding when to use the disk versus the washer method. One way to determine this: Draw an infinitesimal rectangle in the plane region to be revolved and imagine it moving around the region. If that infinitesimal rectangle doesn't *always* touch the axis of rotation, then the solid of revolution produced will have a cavity in it (like the one shown in figure 3.17, lower right). That will indicate nondisk cross sections and therefore tell you *not* to use the disk method (and to use the washer method instead). For revolutions about the y-axis, make these modifications to figure 3.17: the infinitesimal rectangles are *horizontal* (as we've previously discussed), and the volume integrals and integrands are in terms of y.

3.6   Volumes of Revolution: The Shell Method

Cross sections, disks, washers—we've covered a lot! Yet all these methods are based on one assumption: that the solid of revolution has cross sections whose areas $A(x)$ (or $A(y)$) we can calculate. What if we can't? This might happen because the solid isn't a solid of revolution, or because it is but we can't calculate the $A(x)$ (or $A(y)$) equation, for some reason. A good example of this is the function $f(t) = 1 - 0.5t +$

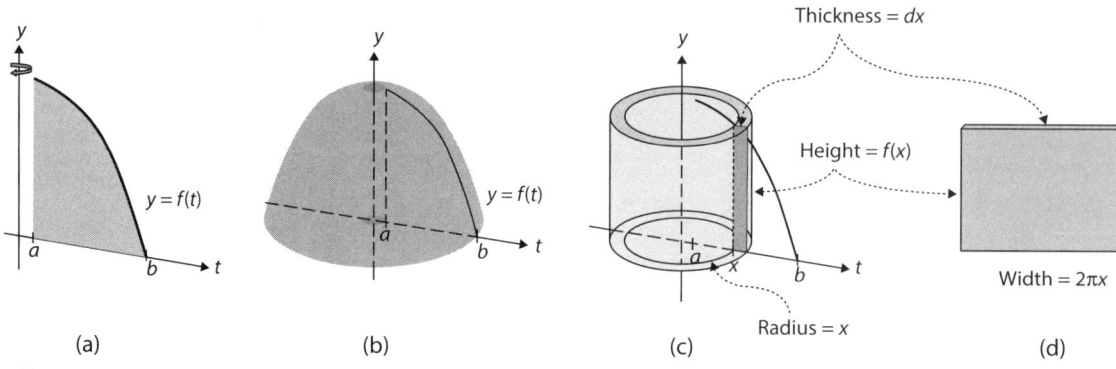

(a)   (b)   (c)   (d)

Figure 3.18: Revolving the shaded region in (a) about the y-axis produces the cored "half apple" in (b). Using a "mathematician's corer" to slice that solid at location x on the t-axis produces the inner cylinder of the shell pictured in (c). Unwrapping that shell produces the thin box in (d). Interact with this figure at sites.google.com/view/fernandezmath/apps/cs2.

$t^2 - 2t^3$. Figure 3.18(a) plots this function between $t = a$ and $t = b$ and shades the region below its graph and bounded by those t-values.

Revolving that region about the y-axis produces the cored "half apple" pictured in figure (b).[4] The cross sections parallel to the t-axis of that "half apple" are annuli. (Great!) But using the washer method for revolutions about the y-axis would require solving $y = 1 - 0.5t + t^2 - 2t^3$ for t, which is too difficult. (Ugh.) Luckily, there's another method that avoids this issue and lets us stick with $f(t)$: the **shell method**.* In the disk and washer methods we used a "mathematician's knife" to slice our solid into infinitesimally thin disks and washers (and then added up the results to get the volume). In the shell method we use a "mathematician's corer" to slice the solid into infinitesimally thin shells of different radii. Figure 3.18(c) illustrates one such shell; I've thickened it to help you better visualize its shell nature. (I also encourage you to scan the QR code on this page to see an animation of this new slicing approach.) If we now cut that shell vertically along the rectangle in figure (c) and unwrap the shell, we obtain the thin box shown in figure (d). The circumference of the shell becomes the width of the box; the height and thickness of the shell become the box's height and thickness. From figure (d) we can then calculate the volume of the shell: $2\pi x f(x)\,dx$. We now follow Leibniz again and add up these infinitesimal shell volumes (i.e., integrate) to get the volume of our original "half apple" (figure (b)). This yields the formula in the following theorem.**

* "Shell" here is in the sense of an egg shell (a thin coating of something) and not a seashell.

SCAN ME

** See section A3.5 for the Riemann sums derivation of the same result.

THEOREM 3.9:   THE SHELL METHOD, ABOUT THE Y-AXIS. Let $y = f(x) \geq 0$ be continuous on $[a, b]$, where $a \geq 0$. Then the volume V of the solid of revolution obtained by revolving about the y-axis the region bounded by the graph of f, $x = a$ and $x = b$, is

$$V = \int_a^b 2\pi x f(x)\,dx. \tag{3.10}$$

---

[4] "Coring" an apple refers to removing the apple's core using a cylindrical tool called an "apple corer." You hold the tool above the apple's stem, push down hard, and then pull up slowly. When done successfully this removes a cylindrical portion of the apple that contains the core.

EXAMPLE 3.16 Find the volume of the solid obtained by revolving the region under the graph of $f(x) = 1 - x + 2x^2 - 2x^3$ and bounded by $x = 0$ and $x = 1$ about the y-axis.

Solution From (3.10),

$$V = 2\pi \int_0^1 x\left(1 - x + 2x^2 - 2x^3\right) dx$$

$$= 2\pi \int_0^1 \left(x - x^2 + 2x^3 - 2x^4\right) dx$$

$$= 2\pi \left(\frac{x^2}{2} - \frac{x^3}{3} + \frac{x^4}{2} - \frac{2x^5}{5}\right)\Bigg|_0^1 = \frac{8\pi}{15}.$$

(The resulting solid looks like the one in figure 3.18(b) but with a pointier tip at $a = 0$.) ■

Related Exercises   30–32, A3.

APPLICATIONS Applied exercise A3 explores the volume of a "mathematical bundt cake" using the methods taught in this section.

We can also use the shell method to find the volume of solids of revolution generated by revolving regions about the x-axis.* The result is the following analogue of the previous theorem.

* Rotating figure 3.18 clockwise 90° gives you a picture of the setup for this.

> **THEOREM 3.10: THE SHELL METHOD, ABOUT THE X-AXIS.** Let $x = h(y) \geq 0$ be continuous on $[c, d]$, where $c \geq 0$. Then the volume V of the solid of revolution obtained by revolving about the x-axis the region bounded by the graph of h, $y = c$ and $y = d$, is
>
> $$V = \int_c^d 2\pi y h(y)\, dy. \tag{3.11}$$

EXAMPLE 3.17 Find the volume of the solid obtained by revolving the region under the graph of $f(x) = 4 - x^2$ bounded by $x = 0$ and $x = 2$ about the x-axis (shown in figure 3.19).

Solution Here $c = 0$, $d = 4$, and $x = h(y) = \sqrt{4 - y}$ (we get this by solving $y = 4 - x^2$ for x and choosing $\sqrt{4 - y}$, not $-\sqrt{4 - y}$, since $x \geq 0$ in our region). Substituting this into (3.11) then yields

$$V = 2\pi \int_0^4 y\sqrt{4 - y}\, dy.$$

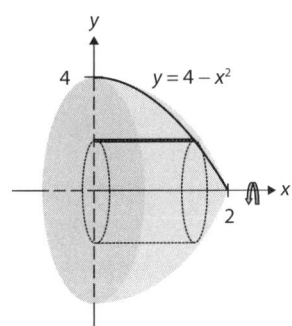

Figure 3.19

To evaluate this integral we employ the u-substitution $u = 4 - y$, $du = -dy$. When $y = 0$, $u = 4 - 0 = 4$. When $y = 4$, $u = 4 - 4 = 0$. And since $u = 4 - y \Rightarrow y = 4 - u$,

$$V = -2\pi \int_4^0 (4 - u)\sqrt{u}\, du \qquad \text{Substituting } u = 4 - y,\ dy = -du$$

$$= 2\pi \int_0^4 (4 - u)\sqrt{u}\, du \qquad \text{Using } \int_b^a f(u)\, du = -\int_a^b f(u)\, du$$

$$= 2\pi \int_0^4 \left(4u^{1/2} - u^{3/2}\right)\, du \qquad \text{Simplifying}$$

$$= 2\pi \left[\frac{8u^{3/2}}{3} - \frac{2}{5}u^{5/2}\right]_0^4 = \frac{256\pi}{15}. \qquad \text{Evaluation Theorem} \quad \blacksquare$$

Related Exercises 33–35.

Additional Tips, Tricks, and Takeaways

We've now got the disk method, the washer method, and the shell method. How do we know which method to use when? Here's one part of that answer: If you're given $y = f(x)$ and asked to revolve about the y-axis, the shell method is usually easiest, since as we saw above you can keep using $y = f(x)$ and not have to solve for x in that equation. However, if you're given $y = f(x)$ and asked to revolve about the x-axis, the disk/washer method is usually easiest, for the same reason—you'll be able to use the function as given. Deciding between the disk or washer method goes back to figure 3.17 and the question of whether the infinitesimal rectangles always (or don't always) touch the axis of rotation as you scan the region that's being revolved. Figure 3.20 does a similar comparison between the washer method and the shell method for solids obtained by revolving a plane region about the x-axis.

 I encourage you to study this figure closely, too. While doing so you'll notice that the formula in the shell method column is a generalization of theorem 3.10 to the case when the plane region to be revolved is between two curves. I'll discuss that in a minute. But first, let me make a few other comments on the figure.

- Notice that in the shell method the infinitesimal rectangle is *parallel* to the axis of revolution (as in figure 3.19), whereas in the washer (and disk) method it's *perpendicular* to it.

- Note that there's no mention of "cross sections" in the shell method column in figure 3.20. That's because of the different way we assemble the solid in the shell method (refer to our chat about that at the start of this section, and revisit the animation linked to the QR code there).

- Notice the squared functions in the integrand in the volume formula for the washer method. The shell method volume integrand has no squared functions. Notice, too, the π in the disk and washer volume formulas. The shell method formula has 2π. The reason for all these differences traces back to the geometries behind these methods—the washer method traces back to disks (area: πr^2), while the shell method traces back to the surface area of shells (area: $2\pi rh$). Remembering these geometric foundations of each method should help you remember which volume formula goes with which method.

Alright, now back to what I mentioned earlier about the formula for the shell method in the figure: It's a generalization of theorem 3.10 to the case when the plane region to be revolved is between two curves. This is not the only generalization you might

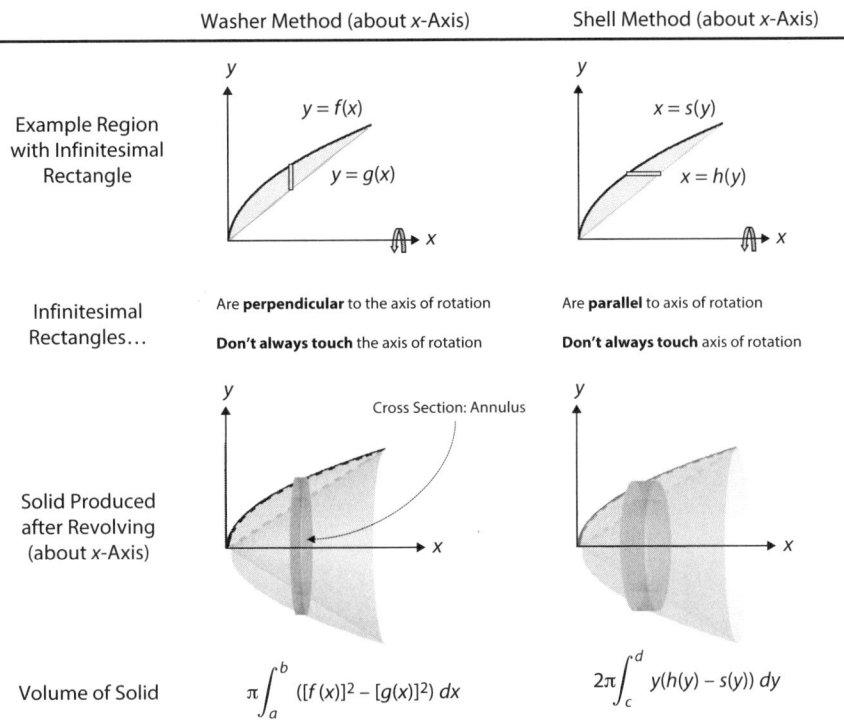

	Washer Method (about x-Axis)	Shell Method (about x-Axis)
Example Region with Infinitesimal Rectangle	$y = f(x)$ $y = g(x)$	$x = s(y)$ $x = h(y)$
Infinitesimal Rectangles…	Are **perpendicular** to the axis of rotation **Don't always touch** the axis of rotation	Are **parallel** to axis of rotation **Don't always touch** axis of rotation
Solid Produced after Revolving (about x-Axis)	Cross Section: Annulus	
Volume of Solid	$\pi \displaystyle\int_a^b ([f(x)]^2 - [g(x)]^2)\, dx$	$2\pi \displaystyle\int_c^d y(h(y) - s(y))\, dy$

Figure 3.20: A comparison of the washer and shell methods when revolving about the x-axis.

run into. For any of the three methods we've studied—disk, washer, or shell—we may also want to revolve the region about a *noncoordinate* axis (examples: $y = 1$, $x = 3$). If you'd like to learn that content, I've put the details in section A3.6.

We've now learned multiple methods for calculating the volume of a solid. So let's move on to the remaining parts of the Geometry Big Question from chapter 1: calculating the length of a curve and the area of a surface. We'll tackle those next, in that order. Thus far in this chapter we've taken Leibniz's approach to developing the integral formulas. In the next section, which tackles the curve length problem, we'll take Riemann's approach. Thus far I've been relegating that approach to this chapter's appendix, but I do want you to see it at least once. We'll also use Riemann's approach in the subsequent section, which tackles the surface area problem, to give you more exposure to it and practice with it.

3.7   Calculating the Length of a Curve

Grab a piece of string and mold it into the shape of a curve. What's that curve's length? It's a simple question that eluded mathematicians for centuries. If the curve were a *line* we could easily calculate its length. And therein lies the insight: Can we use what we know (how to calculate the length of a line segment) to figure out what we don't (how to calculate the length of a curve)? The first row of figure 1.5 (chapter 1) sketched out that approach. Therein we imagined *approximating* the length of the curve using an increasing number of line segments, and then taking the limit

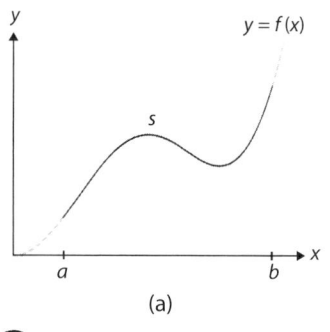
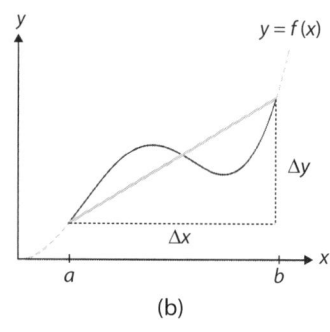
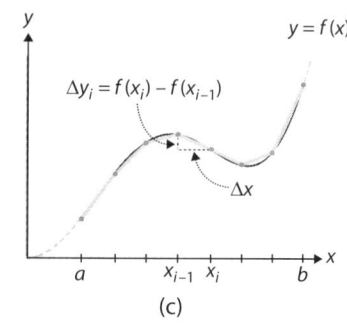

(a)   (b)   (c)

Figure 3.21: (a) The arc length s of the part of the graph of $y = f(x)$ between $x = a$ and $x = b$. (b) The approximation of s by the line segment connecting the endpoints of the curve. (c) The approximation of s by 7 line segments. Interact with this figure at sites.google.com/view/fernandezmath/apps/cs2.

as the number of line segments used tends to infinity to get the **arc length**—the length of the curve. Let's now fill in the details by following Riemann's approach to this problem,[5] using figure 3.21(a) for reference.

The black portion of the curve in figure (a) measures some arc length s. Riemann's approach to calculating s starts by approximating it with a quantity we do know how to calculate; in this case that's the length of a line segment. Let's start off with a *very* coarse approximation to s:

$$s \approx \text{length of gray line segment in figure 3.21(b)}.$$

That gray line segment is the hypotenuse of the dashed right triangle in the figure, so by the Pythagorean Theorem it's length is

$$\sqrt{(\Delta x)^2 + (\Delta y)^2} = \sqrt{\left\{ 1 + \left(\frac{\Delta y}{\Delta x} \right)^2 \right\} (\Delta x)^2} = \sqrt{1 + \left(\frac{\Delta y}{\Delta x} \right)^2} \, \Delta x. \quad (3.12)$$

Success! But like I said, this is a very coarse approximation to the arc length s. To improve the approximation let's partition the interval $[a, b]$ into n subintervals of equal width Δx.* Then, draw the line segments corresponding to each subinterval. Figure 3.21(c) illustrates this in the case of $n = 7$ subintervals. The length of the line segment corresponding to the interval $[x_{i-1}, x_i]$ pictured in the figure is just the right-hand side of (3.12) with Δy replaced by Δy_i:

$$\sqrt{1 + \left(\frac{\Delta y_i}{\Delta x} \right)^2} \, \Delta x, \quad (3.13)$$

where $\Delta y_i = f(x_i) - f(x_{i-1})$. Since $x_i = x_{i-1} + \Delta x$ (recall figure 2.5), we can rewrite (3.13) as

$$\sqrt{1 + \left[\frac{f(x_{i-1} + \Delta x) - f(x_{i-1})}{\Delta x} \right]^2} \, \Delta x. \quad (3.14)$$

You may recognize the gray-colored term here from your Calculus 1 knowledge: It's the "difference quotient," the quantity we take the limit of (as $\Delta x \to 0$) to get the

* This forms an *equipartition* of $[a, b]$, as we discussed in section 2.2.

[5]In case you've read through section A3.1, we're about to follow the approach outlined in box A3.1 therein.

derivative, $f'(x)$. We'll use that fact in a minute. First, let's remember that (3.14) measures the length of the ith line segment. There are n such line segments being used to approximate the arc length s, so the total length of those line segments is the sum of lengths (3.14):

$$\sum_{i=1}^{n} \sqrt{1 + \left[\frac{f(x_{i-1} + \Delta x) - f(x_{i-1})}{\Delta x}\right]^2} \, \Delta x.$$

This is a Riemann sum.* As we increase the number n of line segments, that Riemann sum approximation of s gets better. (Scan the QR code on this page to see an animation of that.) When we take the limit as $n \to \infty$, that approximation yields the exact arc length s and results in the theorem below.[6]

* We discussed these in section 2.2.

SCAN ME

THEOREM 3.11: ARC LENGTH OF A FUNCTION OF X. Let $y = f(x)$ be a function that's differentiable on $[a, b]$, with f' continuous on that interval as well. Then the **arc length** s of f between $x = a$ and $x = b$ is

$$s = \int_a^b \sqrt{1 + [f'(x)]^2} \, dx. \qquad (3.15)$$

EXAMPLE 3.18   Find the arc length of the portion of the graph of $f(x) = x$ between $x = 0$ and $x = 1$.

Solution Here $f'(x) = 1$, which is a continuous function. Since $\sqrt{1 + [f'(x)]^2} = \sqrt{2}$, using this in (3.15) yields

$$s = \int_0^1 \sqrt{2} \, dx = \left[\sqrt{2} x\right]_0^1 = \sqrt{2}.$$

Note: This matches what we'd get by using the Pythagorean Theorem to calculate the distance between the points $(0, 0)$ and $(1, 1)$.   ∎

EXAMPLE 3.19   Find the arc length of the portion of the graph of $f(x) = \frac{2}{3}(x^2 + 1)^{3/2}$ between $x = 0$ and $x = 2$.

Solution Here $f'(x) = (x^2 + 1)^{1/2}(2x) = 2x\sqrt{1 + x^2}$, which is a continuous function. Since $[f'(x)]^2 = \left(2x\sqrt{1 + x^2}\right)^2 = 4x^2(x^2 + 1)$, then

$$1 + [f'(x)]^2 = 1 + 4x^2(x^2 + 1) = 4x^4 + 4x^2 + 1 = \left(2x^2 + 1\right)^2.$$

Using this in (3.15) yields

$$s = \int_0^2 \sqrt{(2x^2 + 1)^2} \, dx$$

$$= \int_0^2 \left(2x^2 + 1\right) dx \qquad \text{Since } \sqrt{(2x^2 + 1)^2} = 2x^2 + 1$$

[6]The $f'(x)$ in (3.15) comes from the $\Delta x \to 0$ limit of the gray-colored term in (3.14). That $\Delta x \to 0$ follows from $\Delta x = \frac{b-a}{n}$ and us taking $n \to \infty$ to get (3.15).

$$= \left[\frac{2x^3}{3} + x \right]_0^2 = \frac{22}{3}. \qquad \text{Evaluation Theorem} \qquad ■$$

EXAMPLE 3.20   Find the arc length of the portion of the graph of $f(x) = x^2 - \frac{1}{8}$ $\ln x$ between $x = 1$ and $x = 2$.

Solution   Here $f'(x) = 2x - \frac{1}{8x}$, which is continuous on the interval $[1, 2]$ of interest, and

$$[f'(x)]^2 = \left(2x - \frac{1}{8x} \right)^2$$

$$= 4x^2 - \frac{1}{2} + \frac{1}{64x^2},$$

$$1 + [f'(x)]^2 = 1 + 4x^2 - \frac{1}{2} + \frac{1}{64x^2}$$

$$= 4x^2 + \frac{1}{2} + \frac{1}{64x^2} = \left(2x + \frac{1}{8x} \right)^2.$$

Using this in (3.15) yields

$$s = \int_1^2 \left(2x + \frac{1}{8x} \right) dx = \left[x^2 + \frac{1}{8} \ln x \right]_1^2$$

$$= \left(4 + \frac{1}{8} \ln 2 \right) - 1 = 3 + \frac{\ln 2}{8}. \qquad ■$$

Related Exercises   36–39, A4.

APPLICATIONS   Applied exercise A4 explores the length of a *catenary*, the shape a rope or cable hangs in when supported only at its ends (e.g., the shape of an overhead power line).

One thing you may have noticed about these last two examples is that $1 + [f'(x)]^2$ ended up being a perfect square. This happens often in the exercises you'll run into in calculus textbooks, since it makes simplifying $\sqrt{1 + [f'(x)]^2}$ easier.

We've now derived formulas that help us calculate the volumes of solids and the lengths of curves. We're two-thirds done tackling the Geometry Big Question from chapter 1. What remains is to develop a method to calculate the area of any surface. That's what's up next. We took a Riemann sums approach to deriving this section's main integral formula; we'll do the same in this last section of this chapter to ensure we see that approach at least once more.

3.8   Calculating the (Lateral) Area of a Surface

Let's first narrow down our objective: In Calculus 2 we restrict our attention to calculating the lateral surface area of a **surface of revolution**, the surface generated by revolving a graph one full revolution about an axis. The "lateral" area of that surface excludes the areas of the top and bottom of the resulting three-dimensional

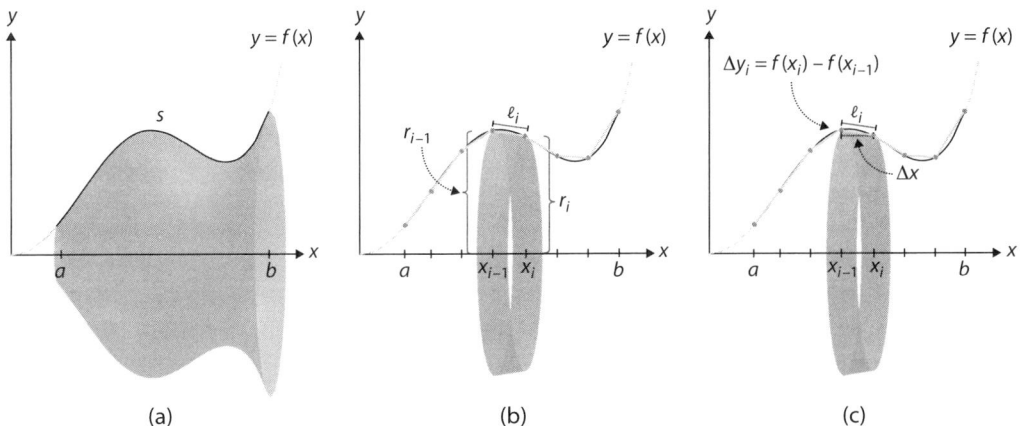

Figure 3.22: The revolution of the curve of length s in (a) around the x-axis produces the surface pictured in (a). We approximate the lateral surface area of that surface using the ribbon-like tubes pictured in (b)–(c).

object. Sometimes an animation helps, and so I invite you to scan the QR code on this page to see both how we generate a surface of revolution and the lateral surface area that process generates. I bet you're already familiar with these concepts. For example, revolving the graph of a horizontal line about the x-axis produces a cylinder; revolving the graph of the upper semicircle $f(x) = \sqrt{1 - x^2}$ about the x-axis produces a sphere. In both these cases we can use formulas from geometry to calculate the resulting surface areas. But if we revolve the graph of some general function $y = f(x)$ about the x-axis, how can we calculate the resulting lateral surface area? To answer that, let's once more use a Riemann sums approach—let's find an approximation to that surface area using a finite number n of surfaces we *do* know the areas for, and then take the limit as $n \to \infty$ to get our integral formula.

We'll begin with figure 3.22(a), where we see the graph of $y = f(x)$ (gray) and a portion of that graph (black) that measures an arc length s. When we revolve that arc length about the x-axis we get the surface of revolution pictured. To calculate that surface's lateral surface area via the Riemann sums approach, we first equipartition $[a, b]$ into n subintervals of equal width Δx. Figure 3.22(b) illustrates this. Next, we approximate the arc length s by n line segments, just like we did in the previous section (recall figure 3.21(c)). The line segment ℓ_i corresponding to the interval $[x_{i-1}, x_i]$ is shown in figure 3.22(b), along with the surface of revolution generated by revolving that line segment a full revolution about the x-axis. That ribbon-like tube is the lateral surface area of a **frustum**.* Exercise D2 helps you show that the lateral surface area of a frustum with larger radius R, smaller radius r, and "slant length" ℓ is

$$2\pi \left(\frac{R + r}{2} \right) \ell.$$

* A frustum is the shape obtained by taking a cone and chopping off its pointy tip.

Applying this to the surface pictured in figure 3.22(b) yields the lateral surface area

$$2\pi \left(\frac{r_{i-1} + r_i}{2} \right) \ell_i = 2\pi \left(\frac{f(x_{i-1}) + f(x_i)}{2} \right) \ell_i,$$

since r_{i-1} and r_i are the y-values at x_{i-1} and x_i, respectively, of the function $y = f(x)$. Substituting in expression (3.14) for ℓ_i from the previous section yields

$$2\pi \left(\frac{f(x_{i-1}) + f(x_i)}{2} \right) \sqrt{1 + \left[\frac{f(x_{i-1} + \Delta x) - f(x_{i-1})}{\Delta x} \right]^2} \, \Delta x.$$

This is the lateral surface area of the ribbon-like tube pictured in figure 3.22(c). The sum of the lateral surface areas of n such tubes,

$$\sum_{i=1}^{n} 2\pi \left(\frac{f(x_{i-1}) + f(x_i)}{2} \right) \sqrt{1 + \left[\frac{f(x_{i-1} + \Delta x) - f(x_{i-1})}{\Delta x} \right]^2} \, \Delta x,$$

is the Riemann sum we're after. As $n \to \infty$ the gray-colored term above tends to $\sqrt{1 + [f'(x)]^2} \, dx$ (as we saw in the previous section). And since $f(x_{i-1})$ and $f(x_i)$ get closer together as $n \to \infty$, the parenthetical term in the sum tends to $\frac{f(x) + f(x)}{2} = \frac{2f(x)}{2} = f(x)$. This all leads to the following theorem.

> **THEOREM 3.12: THE AREA OF A SURFACE OF REVOLUTION.** Let $y = f(x)$ be differentiable on $[a, b]$, with f' continuous on that same interval. Then the lateral surface area S of the surface of revolution generated by revolving the graph of f between $x = a$ and $x = b$ about the x-axis is
>
> $$S = 2\pi \int_a^b f(x) \sqrt{1 + [f'(x)]^2} \, dx. \tag{3.16}$$

EXAMPLE 3.21 Calculate the surface area of the surface of revolution generated by revolving $y = x^3$, $0 \le x \le 1$, about the x-axis (shown in figure 3.23).

Solution Here $f'(x) = 3x^2$, so $\sqrt{1 + [f'(x)]^2} = \sqrt{1 + 9x^4}$. Substituting this, and $f(x) = x^3$, into (3.16) then yields

$$S = 2\pi \int_0^1 x^3 \sqrt{1 + 9x^4} \, dx.$$

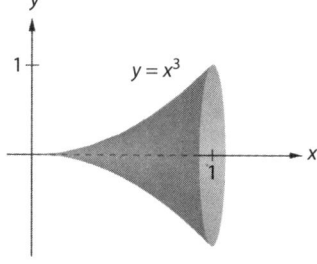

Figure 3.23

We now use the u-substitution $u = 1 + 9x^4$, $du = 36x^3 \, dx$. When $x = 0$, $u = 1$; when $x = 1$, $u = 10$. Thus,

$$S = 2\pi \int_1^{10} \frac{\sqrt{u}}{36} \, du = \frac{\pi}{18} \left[\frac{2}{3} u^{3/2} \right]_1^{10}$$

$$= \frac{\pi}{27} \left(10^{3/2} - 1 \right). \qquad \blacksquare$$

EXAMPLE 3.22 Calculate the surface area of the surface of revolution generated by revolving $y = \sqrt{9 - x^2}$, $-1 \le x \le 1$, about the x-axis (shown in figure 3.24).

Solution The black-colored curve in the figure is the portion of the (gray-colored) semicircle we're revolving to obtain the ring shape in the figure. With $f(x) = \sqrt{9 - x^2}$,

$$f'(x) = \frac{1}{2}\left(9 - x^2\right)^{-1/2}(-2x) = \frac{-x}{\sqrt{9 - x^2}}.$$

This function is defined and continuous on $[-1, 1]$, so theorem 3.12 applies. Let's next calculate $\sqrt{1 + [f'(x)]^2}$:

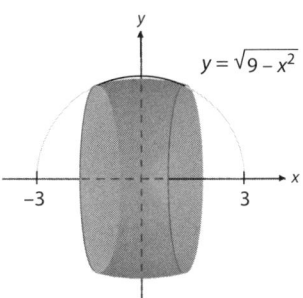

$$\sqrt{1 + [f'(x)]^2} = \sqrt{1 + \frac{x^2}{9 - x^2}} \qquad \text{Substituting } f'(x)$$

$$= \sqrt{\frac{9 - x^2 + x^2}{9 - x^2}} \qquad \text{Common denominator}$$

Figure 3.24

$$= \frac{3}{\sqrt{9 - x^2}}. \qquad \text{Simplifying}$$

Substituting this, and $f(x) = \sqrt{9 - x^2}$, into (3.16) then yields

$$S = 2\pi \int_{-1}^{1} \sqrt{9 - x^2}\left(\frac{3}{\sqrt{9 - x^2}}\right) dx \qquad \blacksquare$$

$$= 6\pi \int_{-1}^{1} 1\, dx = 6\pi(2) = 12\pi.$$

Related Exercises 40–43.

3.9 Parting Thoughts

This completes our Geometry Big Question work. We certainly haven't answered that question in full generality. But we've made enough progress on it that we can move on. If you're heading to chapter 4 next, you're in for a detailed discussion of the Infinite Sum Question and the Approximation Question. If you're heading to chapter 5 next, you'll find there a discussion of the interrelationships among integration, volumes of revolution, and sequences and series. That chapter is intended to be both an opportunity to review what you've learned and a capstone chapter that'll help you make new connections among these concepts. See you in the next chapter (whichever one that is)!

Chapter 3 Exercises

1–8: Sketch the region between the graphs of the functions and bounded by the indicated interval, and then find its area.

1. $y = \dfrac{x^2}{2}, y = \dfrac{x^3}{2}$

2. $y = \dfrac{x^2}{2}, y = x$

3. $y = 2x - 6, y = \dfrac{\ln x^2}{x^2}, 1 \le x \le e$

4. $y = \dfrac{2}{x^2 - 1}, y = 2x, 2 \le x \le 3$

5. $y = x^3, y = -x, 0 \le y \le 2$

6. $x = 6y^2 + 12, x = 6y + 48$

7. $x = 4ye^{-y}, x = 2y$

8. $x = y\sqrt{2 - y^2}, x = y$

9–14: Find the volume of the solid whose base is the indicated region and whose cross sections perpendicular to the x-axis are (a) squares, (b) semicircles, and (c) equilateral triangles.

9. The region bounded by $x^2 + y^2 = 9$

10. The region between the graphs of $y = x$ and $y = x^2$, for $0 \le x \le 1$

11. The region between the graphs of $y = \sqrt[3]{x}$ and $y = x^2$, for $0 \le x \le 1$

12. The region between the graph of $y = \sin x$ and the x-axis, for $0 \le x \le \pi$

13. The region between the graph of $y = xe^x$ and the x-axis, for $0 \le x \le 1$

14. The region between the graph of $y = \dfrac{\ln x}{\sqrt{x}}$ and the x-axis, for $1 \le x \le e$

15–17: Find the volume of the solid whose base is the indicated region and whose cross sections parallel to the x-axis are (a) squares, (b) semicircles, and (c) equilateral triangles.

15. The region between the graphs of $x = y$ and $x = \sqrt{y}$, for $0 \le y \le 1$

16. The region between the graphs of $x = y$ and $x = y^2$, for $0 \le y \le 1$

17. The region between the graphs of $x = y^2$ and $x = \sqrt{y}$, for $0 \le y \le 1$

18–29: Find the volume of the solid of revolution generated by revolving the indicated plane region about the specified axis.

18. The region between the graph of $y = \sqrt{4 - x^2}$ and the x-axis, $0 \le x \le 1$, revolved about (a) the x-axis and (b) the y-axis

19. The region between the graph of $y = (3 - x)^{1/3}$ and the x-axis, $0 \le x \le 3$, revolved about (a) the x-axis and (b) the y-axis

20. The region between the graph of $y = \sec x$ and the x-axis, $-\dfrac{\pi}{4} \le x \le \dfrac{\pi}{4}$, revolved about the x-axis

21. The region between the graph of $y = \sqrt{\ln x}$ and the x-axis, $e \le x \le e^2$, revolved about the x-axis

22. The region between the graph of $y = \cos x$ and the x-axis, $0 \le x \le \dfrac{\pi}{2}$, revolved about the x-axis

23. The region between the graph of $x = ye^{y/2}$ and the y-axis, $0 \le y \le 1$, revolved about the y-axis

24. The region between the graphs of $y = 1 - x$ and $y = x$, for $0 \le x \le \dfrac{1}{2}$, revolved about the x-axis

25. The region between the graphs of $y = 2$ and $y = e^x$, for $0 \le x \le \ln 2$, revolved about the x-axis

26. The region between the graphs of $y = \sin x$ and $y = \dfrac{1}{2}$, for $\dfrac{\pi}{6} \le x \le \dfrac{\pi}{2}$, revolved about the x-axis

27. The region between the graphs of $x = 2 - y$ and $x = 1$, for $1 \le x \le 2$, revolved about the y-axis

28. The region between the graphs of $x = 2 - y^2$ and $x = 1$, revolved about the y-axis

29. The region between the graphs of $x = \sin y$ and $x = \cos y$, for $0 \le y \le \dfrac{\pi}{4}$, revolved about the y-axis

30–32: Use the shell method to find the volume of the solid of revolution generated by revolving the indicated plane region about the y-axis.

30. The region between the graph of $y = \sqrt{4 - x^2}$ and the x-axis, $0 \le x \le 2$

31. The region between the graph of $y = 3\sqrt{1 + x^2}$ and the x-axis, $0 \le x \le 4$

32. The region between the graph of $y = \sin x$ and the x-axis, $0 \le x \le \pi$

33–35: Use the shell method to find the volume of the solid of revolution generated by revolving the indicated plane region about the x-axis.

33. The region between the graph of $x = 1 - y$ and the y-axis, $0 \le y \le 1$

34. The region between the graph of $x = 1 - y^2$ and the y-axis, $0 \le y \le 1$

35. The region between the graph of $x = (1 - y^2)^{2/3}$ and the y-axis, $0 \le y \le 1$

36–39: Find the arc length of the portion of the graph of the function over the indicated interval.

36. $y = 5x + 2$, $[0, 1]$

37. $y = \frac{4}{3}x^{3/2} + 3$, $[0, 1]$

38. $y = \frac{\ln[\cos(2x)]}{2}$, $\left[0, \frac{\pi}{6}\right]$

39. $y = x^{2/3} + 4$, $[0, 2^{3/2}]$

40–43: Calculate the surface area of the surface of revolution generated by revolving the function on the indicated interval about the x-axis.

40. $y = 5x$, $[0, 3]$

41. $y = \sqrt{x+1}$, $[1, 3]$

42. $y = 2x^3$, $[0, 1]$

43. $y = \sqrt{9 - x^2}$, $[-1, 1]$

44–49: Use theorem A3.13 to find the volume of the solid of revolution generated by revolving the indicated plane region about the specified line.

44. The region between the graphs of $y = 2 + x^2$ and $y = 1$, $-1 \le x \le 1$, about (a) the line $y = 1$ and (b) the line $y = -1$

45. The region between the graphs of $y = \sin x$ and $y = 2$, $0 \le x \le \pi$, about (a) the x-axis and (b) the line $y = 3$

46. The region between the graphs of $y = \sqrt{x}$ and $y = x$, about the line $y = 2$

47. The region between the graphs of $x = y^2$ and $x = 1$, $-1 \le y \le 1$, about the line $x = 2$

48. The region between the graphs of $x = \sqrt{1 - y}$ and $x = 0$, $0 \le y \le 1$, about the line $x = 2$

49. The region between the graph of $x = \sin y$ and $x = 1$, $0 \le y \le \frac{\pi}{2}$, about the line $x = 2$

50–53: Use the shell method to find the volume of the solid of revolution generated by revolving the indicated plane region about the specified line.

50. The region between the graphs of $y = \sqrt{x}$ and $y = x^3$, about the y-axis

51. The region between the graphs of $y = 1 - x^2$ and $y = x^2$, $0 \le x \le \frac{1}{\sqrt{2}}$, about the line $x = 1$

52. The region between the graphs of $x = y^2$ and $x = y$, $0 \le y \le 1$, about the line $y = 2$

53. The region between the graphs of $x = 1 - y$ and $x = y$, $0 \le y \le 1/2$, about the line $y = 1$

A1. Measuring income inequality In many countries income is distributed unevenly among the country's wage earners; for example, in 2013 the bottom 99% of wage earners in the United States received only about 80% of the nation's pretax income. Economists quantify this income distribution disparity using a *Lorenz curve L(x)*, defined to be the percentage of the nation's income earned by the bottom x% of households (here both x and $L(x)$ are in decimal form). Given a country's Lorenz curve, its **Gini coefficient** G, defined as

$$G = 2 \int_0^1 [x - L(x)]\, dx,$$

can be used to measure the degree of income inequality in that country. The range for G is $0 \le G \le 1$, with higher values indicating greater income inequality. Calculate the Gini coefficient of a country with Lorenz curve $L(x) = x^2$.

A2. The volume of a mathematical doughnut The figure below shows a "mathematical doughnut." I generated it by revolving the region between the curves $y = 2 + \sqrt{1 - x^2}$ and $y = 2 - \sqrt{1 - x^2}$, $-1 \le x \le 1$, about the x-axis.

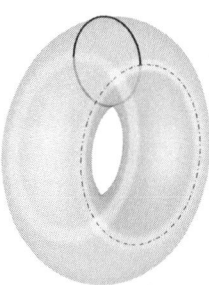

Set up *but do not evaluate* the integral yielding the volume of the doughnut.

A3. Volume of a mathematical bundt cake

The figure below shows a "mathematical bundt cake." I generated it by revolving the curve $y = \sin(x^2)$, $0 \leq x \leq \sqrt{\pi}$, about the y-axis.

(a) Would it be easier to use the washer method or the shell method to find the volume of this cake?

(b) Find the volume of the cake using the easiest method.

A4. Length of a catenary

You've likely seen catenary-like curves all around the city or town you live in. A **catenary** is the shape a rope or cable hangs in when supported only at its ends. Overhead power lines, cables in some suspension bridges, and even (to an extent) the Gateway Arch in St. Louis follow catenary-like curves. The equation of a catenary is $y = a\cosh(x/a)$, where $a > 0$. Here $\cosh(x) = \frac{e^x + e^{-x}}{2}$ is the **hyperbolic cosine** function. Calculate the arc length of the simple catenary $f(x) = \cosh x$ for $-1 \leq x \leq 1$.

D1. This exercise uses the disk method to derive the familiar formulas for the volume of a cylinder, cone, and sphere.

(a) Let $f(x) = r$, where $r > 0$. Use the interval $[0, h]$ (here $h > 0$) and the disk method to show that the volume of the cylinder produced by revolving f about the x-axis is $V = \pi r^2 h$.

(b) Let $f(x) = rx/h$, where $r > 0$ and $h > 0$. Use the interval $[0, h]$ and the disk method to show that the volume of the cone produced by revolving f about the x-axis is $V = \frac{1}{3}\pi r^2 h$.

(c) Let $f(x) = \sqrt{r^2 - x^2}$, where $r > 0$. Use the interval $[-r, r]$ and the disk method to show that the volume of the sphere produced by revolving f about the x-axis is $V = \frac{4}{3}\pi r^3$.

D2. This exercise derives the formula for the lateral surface area of a frustum. To do so, we'll use the fact that the lateral surface area S of a cone of radius r and slant length ℓ (the length from the circular base of the cone to its tip) is $S = \pi r \ell$.

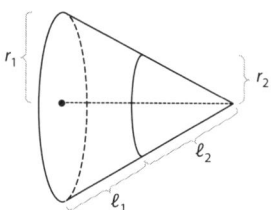

(a) Consider the larger cone of radius r_1 and slant length $\ell_1 + \ell_2$ pictured. Let S_1 denote its lateral surface area. We can view S_1 as $S_1 = S + S_2$, where S_2 is the lateral surface area of the smaller cone of radius r_2 and slant length ℓ_2, and S is the lateral surface area of the frustum obtained by removing that smaller cone from the larger one. Show that $S = \pi[(r_1 - r_2)\ell_2 + r_1\ell_1]$.

(b) Using similar triangles,

$$\frac{\ell_2}{r_2} = \frac{\ell_1 + \ell_2}{r_1}.$$

Use this to show that

$$S = \pi(r_2\ell_1 + r_1\ell_1) = 2\pi r\ell_1,$$

where $r = \frac{r_1 + r_2}{2}$.

Chapter 3 Appendix

A3.1 Area between Two Curves: The Riemann Sums Approach

To employ a Riemann sums approach to deriving a definite integral—whether it measures the area between two curves or some other quantity—we'll run through the four-step procedure in the box below.

Box A3.1: Steps for Deriving an Integral Formula via Riemann Sums

1. Equipartition the interval $[a, b]$ into n subintervals of equal width Δx. (See section 2.2 for a refresher on equipartitions.)

2. Then, focus on the ith subinterval and come up with a formula for the mathematical object of interest (e.g., area) associated with just that subinterval. That formula should have the form

   "(Function of i) $\cdot \Delta x$"   (Example: $f(x_i)\Delta x$.)

   that is, some quantity that depends on i multiplied by Δx.

3. Now, add up those n quantities to get a Riemann sum:

   $$\sum_{i=1}^{n} \text{"(Function of } i) \cdot \Delta x\text{"}$$

4. Finally, take the limit as $n \to \infty$ of that Riemann sum to get the definite integral formula:

   $$\lim_{n\to\infty} \sum_{i=1}^{n} \text{"(Function of } i) \cdot \Delta x\text{"} = \int_{a}^{b} \text{"(Function of } x)\text{"} \, dx$$

The fact that this limit becomes a definite integral (when the integrand is continuous) is something we discussed connection with exercise D2 in chapter 2. Let me point out, too, that this last step—taking the infinite limit—is the last step in the Calculus 2 workflow from figure 1.3, and that the Riemann sum in step 3 is the "Y_n" in that workflow.

  Alright, let's test drive this four-step procedure now and use it to help us verify the integral formula (3.1) for the area between the graphs of two functions of x, the area A in figure A3.1(a). The first step in the procedure is to equipartition $[a, b]$. Figure A3.1(b) illustrates this. In the next step, we focus on the ith subinterval and, in this case, try to quantify the area between the graphs of f and g. The rectangle shown in the figure at location x_i^* on the x-axis approximates that area. That rectangle's width is Δx. It's top is at the y-value $f(x_i^*)$, and its bottom at the y-value $g(x_i^*)$, so its height is $f(x_i^*) - g(x_i^*)$. That makes its area $[f(x_i^*) - g(x_i^*)]\Delta x$. And *there's* the "(Function of i) $\cdot \Delta x$" expression we're looking for. For our third step we add up

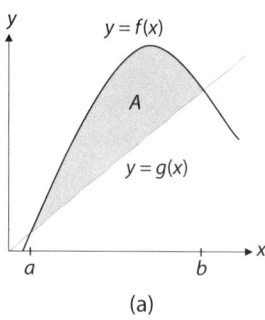
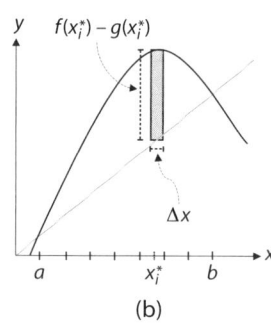
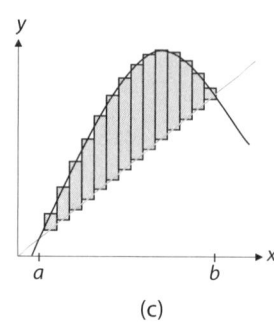

(a)   (b)   (c)

Figure A3.1: Plots of two functions $y = f(x)$ and $y = g(x)$ and (a) the area A between them bounded by $x = a$ and $x = b$, (b) the *very* coarse of approximation A by the area of a single rectangle of width Δx and height $f(x_i^*) - g(x_i^*)$, and (c) the more accurate approximation of A via an equipartition Riemann sum with 14 subdivisions. Interact with this figure at sites.google.com/view/fernandezmath/apps/cs2 and scan the QR code to see an animation.

the areas of many of these types of rectangles to get the Riemann sum

$$\sum_{i=1}^{n} [f(x_i^*) - g(x_i^*)]\Delta x.$$

(Figure A3.1(c) illustrates this sum for $n = 14$ rectangles.) The last step takes the infinite limit of that Riemann sum (scan the QR code next to the figure to see an animation of that limiting process) and yields the integral formula (3.1).

A3.2   Riemann Sums Approach to Volumes by Cross Sections

Let's return to figure 3.6 but modify it to take a Riemann sums approach to deriving equation (3.4). Figure A3.2(d) shows the same wedge-like solid we studied in section 3.3, whose cross sections perpendicular to the x-axis are squares. Let's now follow the four-step procedure outlined in box A3.1 of the previous section. The first step in the procedure is to equipartition $[a, b]$. Figure A3.2(a) illustrates this. In the next step, we focus on the ith subinterval and, in this case, try to quantify the volume of a square cross-sectional slice of thickness Δx (figure (c)). The base of that slice is the rectangle shown in figure (a) at location x_i^* on the x-axis. That rectangle's width is Δx, and its height is $f(x_i^*)$. So the volume of the slice in figure (c) is $[f(x_i^*)]^2\Delta x$. *There's* the "(Function of i) $\cdot \Delta x$" expression we're looking for. For our third step we add up the volumes of n of these types of rectangles to get the Riemann sum

$$\sum_{i=1}^{n} [f(x_i^*)]^2\Delta x.$$

The last step takes the infinite limit of this Riemann sum (scan the QR code on the left to see an animation of that limiting process) and yields the integral formula

$$V = \int_{a}^{b} [f(x)]^2 \, dx.$$

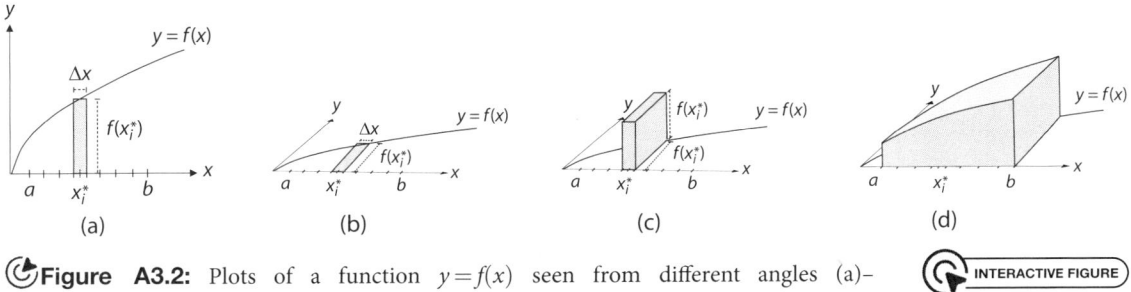

⟳Figure A3.2: Plots of a function $y = f(x)$ seen from different angles (a)–(b), along with the solid (d) whose cross sections perpendicular to the x-axis are squares with base length equal to $y = f(x)$ (c). Interact with this figure at sites.google.com/view/fernandezmath/apps/cs2.

⟳ INTERACTIVE FIGURE

The integrand here is the area of each cross section of the solid. Those cross sections were squares. If we replaced them with some other geometric shape with area $A(x)$, this would replace the integrand with $A(x)$, yielding (3.4).

A3.3   Riemann Sums Approach to the Disk Method

Let's return to figure 3.11 and modify it to take a Riemann sums approach to deriving (3.6). We'll consider $f(x) = \sqrt{x}$ to keep things concrete and then generalize our results. The first step in our four-step procedure from box A3.1 of section A3.1 is to equipartition $[a, b]$. Figure A3.3(c) illustrates this.

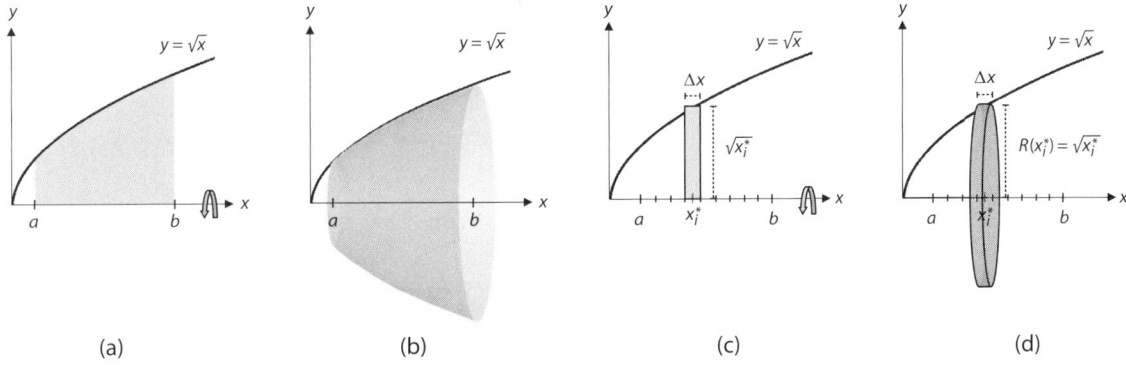

Figure A3.3: Revolving the shaded region in (a) about the x-axis produces the solid in (b). Revolving the rectangle in (c) about the x-axis produces the "pancake" in (d). Interact with this figure at sites.google.com/view/fernandezmath/apps/cs2.

In the next step, we focus on the ith subinterval and, in this case, try to quantify the volume of the solid of revolution pictured in figure (d). That solid is obtained by revolving the rectangle in figure (c) about the x-axis. That rectangle's width is Δx, and its height is $\sqrt{x_i^*}$. So the volume of the "pancake" in figure (d) is $\pi[\sqrt{x_i^*}]^2 \Delta x$ (this is the area of a disk of radius R, πR^2, multiplied by the thickness Δx of the pancake). *There's* the "(Function of i) · Δx" expression we need. For our third step we add up the volumes of n of these types of pancakes to get the Riemann sum

$$\sum_{i=1}^{n} \pi[\sqrt{x_i^*}]^2 \Delta x.$$

The last step takes the infinite limit of this Riemann sum (scan the QR code on the left to see an animation of that limiting process) and yields the integral formula

$$V = \pi \int_a^b [\sqrt{x}]^2 \, dx.$$

When we replace $\sqrt{x}$ here by a general function $f(x)$, we obtain the general formula for V given in (3.6).

A3.4   Riemann Sums Approach to the Washer Method

Let's return to figure 3.13 and modify it to take a Riemann sums approach to deriving (3.8). We'll again consider $f(x) = \sqrt{x}$ to keep things concrete and then generalize our results. The first step in our four-step procedure from box A3.1 of section A3.1 is to equipartition $[a, b]$. Figure A3.4(c) illustrates this.

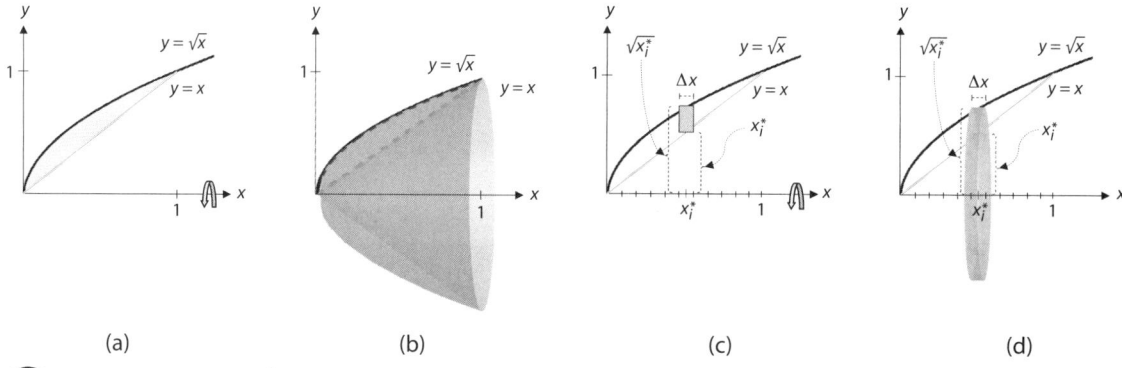

(a)   (b)   (c)   (d)

INTERACTIVE FIGURE

Figure A3.4: Revolving the shaded region in (a) about the x-axis produces the solid in (b). Revolving the rectangle in (c) produces the washer in (d). Interact with this figure at sites.google.com/view/fernandezmath/apps/cs2.

In the next step, we focus on the ith subinterval and, in this case, try to quantify the volume of the solid of revolution pictured in figure (d). That solid is obtained by revolving the rectangle in figure (c) about the x-axis. That rectangle's width is Δx, and its top is at a distance $\sqrt{x_i^*}$ from the x-axis while its bottom is at a distance x_i^* from the axis. So the volume of the washer in figure (d) is $\pi[\sqrt{x_i^*}]^2 \Delta x$ minus $\pi[x_i^*]^2 \Delta x$, that is, $\pi\left([\sqrt{x_i^*}]^2 - [x_i^*]^2\right)\Delta x$. We've obtained the "(Function of i) · Δx" expression we need. For our third step we add up the volumes of n of these types of washers to get the Riemann sum

$$\sum_{i=1}^{n} \pi \left([\sqrt{x_i^*}]^2 - [x_i^*]^2\right) \Delta x.$$

The last step takes the infinite limit of this Riemann sum and yields the integral formula

$$V = \pi \int_a^b \left([\sqrt{x}]^2 - [x]^2\right) dx.$$

When we replace $\sqrt{x}$ by a general function $f(x)$, and the "x" in the second brackets by another function $g(x)$, we obtain the general formula for V given in (3.8). (This assumes that $f(x) \geq g(x)$ on $[a, b]$.)

A3.5   Riemann Sums Approach to the Shell Method

Let's return to figure 3.18 and modify it to take a Riemann sums approach to deriving (3.10). The first step in our four-step procedure from box A3.1 of section A3.1 is to equipartition $[a, b]$. Figure A3.5(c) illustrates this.

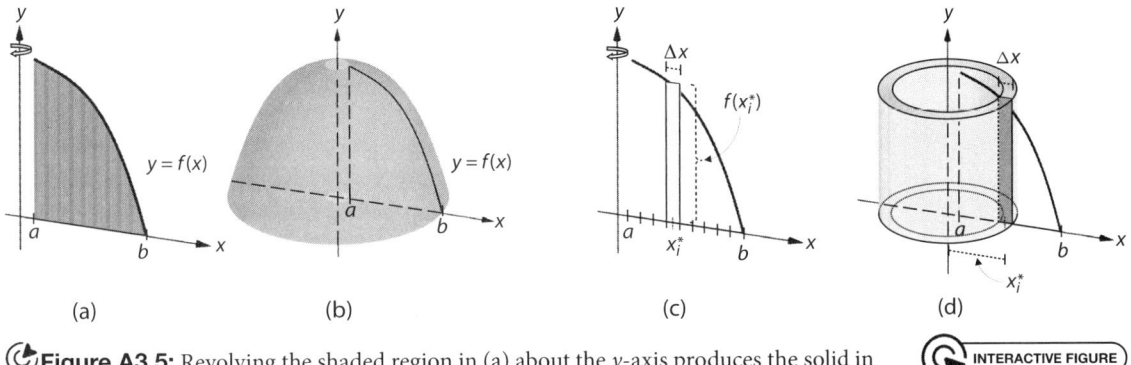

(a)   (b)   (c)   (d)

⟳Figure A3.5: Revolving the shaded region in (a) about the y-axis produces the solid in (b). Revolving the rectangle in (c) about the y-axis produces the shell in (d). Interact with this figure at sites.google.com/view/fernandezmath/apps/cs2.

INTERACTIVE FIGURE

In the next step, we focus on the ith subinterval and, in this case, try to quantify the volume of the shell pictured in figure (d). That solid is obtained by revolving the rectangle at location x_i^* on the x-axis in figure (c) about the y-axis. That rectangle's width is Δx and its height is $f(x_i^*)$. When it's revolved about the y-axis we get the shell in figure (d). The distance x_i^* from the origin, when revolved about the y-axis, produces a circumference $2\pi x_i^*$. And since the shell's height is $f(x_i^*)$ and thickness is Δx, its volume is $2\pi x_i^* f(x_i^*)\Delta x$. This is the "(Function of i) $\cdot \Delta x$" expression we need. For our third step we add up the volumes of n of these types of shells to get the Riemann sum

$$\sum_{i=1}^{n} 2\pi x_i^* f(x_i^*)\Delta x.$$

The last step takes the infinite limit of this Riemann sum and yields the integral formula in (3.10).

A3.6   Volumes of Revolution: Noncoordinate Axes of Revolution

In our volumes of revolution work we've always assumed that the axis of revolution is either the x-axis or the y-axis. But what if we'd like to use a different horizontal or vertical line as the axis of revolution? Figure A3.6 provides a few handy illustrations we can use (complete with infinitesimal rectangles) to generalize the washer and shell methods to handle these noncoordinate axes of revolution. (Please read the figure column by column—each plot at the top is intended to go with the plot directly beneath it.)

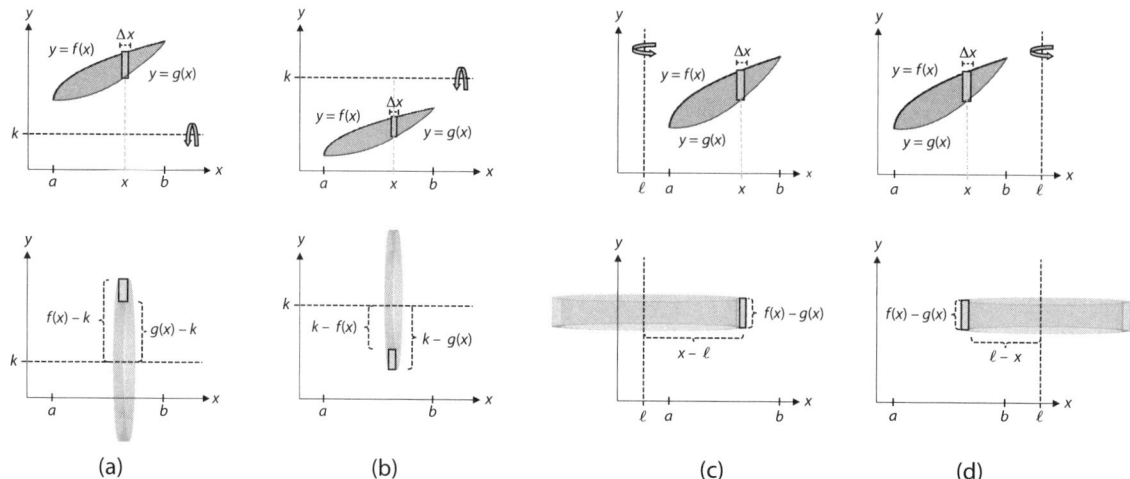

Figure A3.6: (a)–(b): Two plane regions (top) and the thin washers generated (bottom) when the regions are revolved about a horizontal line $y = k$. (c)–(d): Two plane regions (top) and the thin shells generated (bottom) when the regions are revolved about a vertical line $x = \ell$.

The Generalized Washer Method

Consider first figure A3.6(a). When the shaded region in the top plot is revolved about the line $y = k$ (which we'll assume is below the region) we get the washer of thickness Δx pictured in the bottom plot. The volume of that cylindrical washer is

$$\pi \left([f(x) - k]^2 - [g(x) - k]^2 \right) \Delta x.$$

(Recall that the volume of a cylindrical washer is $\pi(R^2 - r^2)t$, with R the larger radius, r the smaller one, and t the thickness.) If the plane region is below the line $y = k$ (figure A3.6(b), top), then $f(x) - k$ and $g(x) - k$ are replaced with $k - f(x)$ and $k - g(x)$, respectively. These insights lead to the following generalization of the washer method.

THEOREM A3.13: THE WASHER METHOD, NONCOORDINATE AXIS OF REVOLUTION. Let f and g be continuous on $[a, b]$, with $f(x) \geq g(x)$, and let h and s be continuous on $[c, d]$, with $h(y) \geq s(y)$.

- The volume V of the solid obtained by revolving the plane region bounded by the graphs of $y = f(x)$ and $y = g(x)$, $x = a$ and $x = b$, about the line $y = k$, is

$$V = \pi \int_a^b \left([f(x) - k]^2 - [g(x) - k]^2 \right) dx, \quad \text{if } g(x) \geq k \qquad (A3.1)$$

$$V = \pi \int_a^b \left([k - g(x)]^2 - [k - f(x)]^2 \right) dx, \quad \text{if } f(x) \leq k. \qquad (A3.2)$$

- The volume V of the solid obtained by revolving the plane region bounded by the graphs of $x = h(y)$ and $x = s(y)$, $y = c$ and $y = d$, about the line $x = \ell$, is

$$V = \pi \int_c^d \left([h(y) - \ell]^2 - [s(y) - \ell]^2 \right) dy, \quad \text{if } s(y) \geq \ell \qquad \text{(A3.3)}$$

$$V = \pi \int_c^d \left([\ell - s(y)]^2 - [\ell - h(y)]^2 \right) dy, \quad \text{if } h(y) \leq \ell. \qquad \text{(A3.4)}$$

A few comments on this theorem:

- When $k = \ell = 0$, the volume equations in this theorem reduce to the ones in the original washer method (e.g., theorem 3.7). If in addition $g(x) = 0$ and $s(y) = 0$, then they reduce further to the equations in the disk method (e.g., theorem 3.5). Finally, when $g(x) = k$ or $s(y) = \ell$ (equivalently, when the infinitesimal rectangle always touches the axis of revolution), then the second term in all the integrands above is zero, and the result is a generalization of the disk method to handle noncoordinate axes of revolution. (The next example illustrates this.) Takeaway: The theorem above is the most general disk/washer method we've studied yet.

- The conditions "$g(x) \geq k$" and "$f(x) \leq k$" in (A3.1)–(A3.2) translate visually to: "the region to be revolved is above the axis of revolution" (as in figure A3.6(a), top) and "the region to be revolved is below the axis of revolution" (as in figure A3.6(b), top), respectively.

- The conditions "$s(y) \geq \ell$" and "$h(y) \leq \ell$" in (A3.3)–(A3.4) translate visually to "the region to be revolved is to the right of the axis of revolution" and "the region to be revolved is to the left of the axis of revolution," respectively.

EXAMPLE A3.1 (Video)   Find the volume of the solid of revolution obtained by revolving the region between the graphs of $f(x) = 2 - x^4$ and $g(x) = 1$, and bounded by $x = -1$ and $x = 1$, about the line $y = 1$ (shown figure A3.7).

Solution Here $k = 1$, $a = -1$, and $b = 1$. Now, since the region of interest is above the axis of revolution ($y = 1$), we are therefore in the case "$g(x) \geq k$," and so we'll use (A3.1):

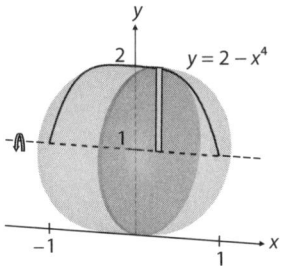

Figure A3.7

$$V = \pi \int_{-1}^1 \left([\underbrace{\{2 - x^4\}}_{f(x)} - \underbrace{1}_{k}]^2 - [\underbrace{1}_{g(x)} - \underbrace{1}_{k}]^2 \right) dx$$

$$= \pi \int_{-1}^1 [1 - x^4]^2 \, dx$$

$$= \pi \int_{-1}^1 (1 - 2x^4 + x^8) \, dx$$

$$= \pi \left[x - \frac{2x^5}{5} + \frac{x^9}{9} \right]_{-1}^{1}$$

$$= \frac{64\pi}{45}.$$ ∎

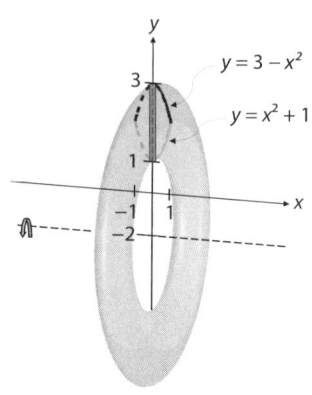

Figure A3.8

EXAMPLE A3.2 Find the volume of the solid of revolution generated by revolving the region bounded by the graphs of $y = x^2 + 1$ and $y = 3 - x^2$ about the line $y = -2$ (shown figure A3.8).

Solution Here $k = -2$, $a = -1$, and $b = 1$. The region of interest is above the axis of revolution ($y = -2$), so we'll use (A3.1) again:

$$V = \pi \int_{-1}^{1} \left([\underbrace{\{3 - x^2\}}_{f(x)} - (\underbrace{-2}_{k})]^2 - [\underbrace{\{x^2 + 1\}}_{g(x)} - (\underbrace{-2}_{k})]^2 \right) dx$$

$$= \pi \int_{-1}^{1} \left([5 - x^2]^2 - [x^2 + 3]^2 \right) dx$$

$$= -16\pi \int_{-1}^{1} (x^2 - 1) \, dx = -16\pi \left[\frac{x^3}{3} - x \right]_{-1}^{1}$$

$$= \frac{64\pi}{3}.$$ ∎

Related Exercises Chapter 3: 44–46.

EXAMPLE A3.3 Set up *but do not evaluate* the integral representing the volume of the solid of revolution generated by revolving the region bounded by the graphs of $x = y^2/4$ and $x = \sqrt[3]{2y}$ about the line $x = -1$.

Solution The region here is the same one pictured in figure 3.16. But since we're revolving about $x = -1$ and not $x = 0$, the solid obtained in this example will look different than the one in the figure. (Our solid is the same height but has a wider bowl shape.) Since the region of interest is to the right of the axis of revolution, we're in the "$s(y) \geq \ell$" case, so we'll use (A3.3). Referring again to figure 3.16, the infinitesimal rectangle indicates that $h(y) = \sqrt[3]{2y}$, $s(y) = y^2/4$, so that (A3.3) becomes[7]

$$V = \pi \int_{0}^{2\sqrt[5]{4}} \left([\underbrace{\{\sqrt[3]{2y}\}}_{h(y)} - (\underbrace{-1}_{\ell})]^2 - \left[\underbrace{\left\{ \frac{y^2}{4} \right\}}_{s(y)} - (\underbrace{-1}_{\ell}) \right]^2 \right) dy$$

[7]We found the limits of integration in example 3.15.

$$= \pi \int_0^{2\sqrt[5]{4}} \left([\sqrt[3]{2y} + 1]^2 - \left[\frac{y^2 + 4}{4}\right]^2 \right) dy. \qquad \blacksquare$$

Related Exercises Chapter 3: 47–49.

The Generalized Shell Method

Consider now figure A3.6(c). When the shaded region in the top plot is revolved about the line $x = \ell$ (which we'll assume is to the left of the region), we get the shell of thickness Δx pictured in the bottom plot. The volume of that shell is

$$2\pi (x - \ell)[f(x) - g(x)]\Delta x.$$

(Recall that the volume of a shell is $2\pi r h t$, with radius r, height h, and shell thickness t.) If the axis of revolution $x = \ell$ is to the right of the plane region (figure A3.6(d), top), then $x - \ell$ is replaced with $\ell - x$. These insights lead to the following generalization of the shell method.

> **THEOREM A3.14: THE SHELL METHOD, NONCOORDINATE AXIS OF REVOLUTION.** Let f and g be continuous on $[a, b]$, where $a \geq 0$, with $f(x) \geq g(x) \geq 0$. Let $h(y)$ and $s(y)$ be continuous on $[c, d]$, where $c \geq 0$, with $h(y) \geq s(y) \geq 0$.
>
> - The volume V of the solid obtained by revolving the plane region bounded by the graphs of $y = f(x)$ and $y = g(x)$, $x = a$ and $x = b$, about the line $x = \ell$, is
>
> $$V = 2\pi \int_a^b (x - \ell)[f(x) - g(x)] \, dx, \quad \text{if } \ell \leq a \qquad (A3.5)$$
>
> $$V = 2\pi \int_a^b (\ell - x)[f(x) - g(x)] \, dx, \quad \text{if } \ell \geq b. \qquad (A3.6)$$
>
> - The volume V of the solid obtained by revolving the plane region bounded by the graphs of $x = h(y)$ and $x = s(y)$, $y = c$ and $y = d$, about the line $y = k$, is
>
> $$V = 2\pi \int_c^d (y - k)[h(y) - s(y)] \, dy, \quad \text{if } k \leq c \qquad (A3.7)$$
>
> $$V = 2\pi \int_c^d (k - y)[h(y) - s(y)] \, dy, \quad \text{if } k \geq d. \qquad (A3.8)$$

Like before, the conditions in this theorem can be interpreted visually. For example, "$\ell \leq a$" is equivalent to "the region to be revolved is to the right of the axis of revolution," as in the top plot in figure A3.6(c); "$\ell \geq b$" is equivalent to "the region to be revolved is to the left of the axis of revolution," as in the top plot in figure A3.6(d). Note also that when $k = \ell = 0$ and $g(x) = s(y) = 0$, the formulas in this theorem reduce to the original shell method formulas from section 3.6. Takeaway: The theorem above is the most general shell method we've studied yet.

EXAMPLE A3.4 Set up *but do not evalu-*
ate the integrals representing the volumes
below.

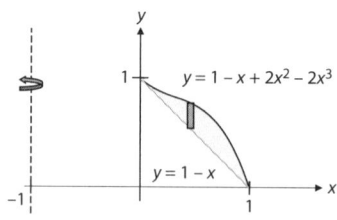

(a) The volume of the solid obtained by
revolving the region bounded by the
graphs of $f(x) = 1 - x + 2x^2 - 2x^3$ and
$g(x) = 1 - x$, $x = 0$ and $x = 1$, about the
line $x = -1$

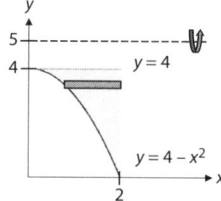

(b) The volume of the solid obtained by
revolving the region bounded by the
graphs $s(y) = \sqrt{4 - y}$, $h(y) = 2$, and $y = 4$, about the line $y = 5$

Figure A3.9

Solution

(a) Here $a = 0$, $b = 1$, and $\ell = -1$. Thus, $\ell < a$ and so we'll use (A3.5). Figure A3.9
(top) shows the region to be revolved. The height of the infinitesimal rectangle
is

$$f(x) - g(x) = (1 - x + 2x^2 - 2x^3) - (1 - x)$$
$$= 2x^2 - 2x^3,$$

so (A3.5) becomes

$$V = 2\pi \int_0^1 (x - \underbrace{(-1)}_{\ell}) \underbrace{[2x^2 - 2x^3]}_{f(x) - g(x)} dx = 4\pi \int_0^1 (x + 1)(x^2 - x^3)\, dx.$$

(b) Here $c = 0$, $d = 4$, and $k = 5$. Since $k > d$, we'll use (A3.8). Figure A3.9 (bottom)
shows the region to be revolved. The width of the infinitesimal rectangle is

$$h(y) - s(y) = 2 - \sqrt{4 - y},$$

so (A3.8) becomes

$$V = 2\pi \int_0^4 (\underbrace{5}_{k} - y) \underbrace{[2 - \sqrt{4 - y}]}_{h(y) - s(y)}\, dy. \qquad \blacksquare$$

Related Exercises   Chapter 3: 50–53.

<table>
<tr><td>

4

</td><td>

Sequences and Series

</td></tr>
</table>

Chapter Preview. Does an infinite sum have a sum, and if so, what's the sum? With-out knowing the exact value of a function, can we accurately approximate it? This chapter tackles these Big Questions from chapter 1. We'll start by studying infinite sequences—infinite lists of numbers—because as we'll discover, infinite sums are built from infinite sequences. Then we'll study infinite series—sums of terms in an infinite list of numbers (an infinite sequence)—and follow the Calculus 2 workflow (see the sec-ond row in figure 1.5) to investigate the sum of an infinite series. When that sum is finite we'll say the series *converges*. We'll then develop a plethora of series conver-gence tests. But only in a few cases will we be able to calculate the sum of convergent series. In section 4.7 we'll have an epiphany: We'll realize that the two Big Questions above—the Infinite Sum Question and the Approximation Question—are connected! We'll then spend the rest of the chapter expounding on the Approximation Question (see the third row in figure 1.5) to develop polynomial approximations to differentiable functions. Amazingly, we'll eventually learn how to use "infinite-degree" polynomial rep-resentations of functions to sum infinite series! Along the way I'll give you lots of tips, tricks, and takeaways to help you master all this content. I'll assume you're comfortable with the content in appendixes A–B (precalculus content) and in appendix C (calculus up to differentiation), so skim those first if you haven't already. Alright, let's start the adventure!

4.1   Introduction to Sequences

Imagine you're heading to the grocery store with a shopping list of items to buy: bread, milk, tortillas, and beans. That list is a *sequence*, a sequence of foods. If for some reason you needed to buy the items in that exact order—and not, for ex-ample, the milk before the bread—then the sequence is an *ordered* sequence. And since there are only a finite number of foods on your list, it's a *finite* ordered se-quence. Were we to track the number of food items you buy instead (e.g., 2 loaves of bread, 1 gallon of milk, etc.), then we'd be talking about a finite ordered sequence of *numbers*. This brings us to the following definition.

> **DEFINITION 4.1: SEQUENCES.** A **sequence** is an ordered list of numbers. If the list is finite, the sequence is a **finite sequence**. If the list is infinite, the sequence is an **infinite sequence**.

Three quick examples:

$$S_1 = (2, 4, 6, 8), \quad S_2 = (1, 1.5, 1.75, 1.825, \dots),$$

$$S_3 = \left(1, \frac{1}{2}, \frac{1}{3}, \dots\right). \tag{4.1}$$

Sequence S_1 is a finite sequence; sequences S_2 and S_3 are infinite sequences.* We'll refer to the actual numbers in a sequence as the **values** of the sequence, and to the position in the sequence as the **term** of the sequence. Example: 1.5 is the *value* of the second *term* in sequence S_2.

In this book we'll put the sequence's values in parentheses and keep track of their location in the sequence via a nonnegative integer n called the **index**. We'll denote the corresponding term in the sequence by a_n. Examples: In S_1, $a_3 = 6$ (the third term has value 6); in S_2, $a_3 = 1.75$ (the third term has value 1.75). We'll refer to an entire sequence like this:

$$(a_n)_{n=k}^m, \quad \text{or} \quad (a_n), \quad \text{or} \quad (a_n)_{n=k}^\infty,$$

where k is the **starting index** and m the **ending index**. We use the first notation for finite sequences, the second for infinite sequences with $k = 1$, and the third for infinite sequences with $k \neq 1$. (We'll work through examples of these shortly.) In this notation, a general infinite sequence (a_n), with starting index $k = 1$, looks like

$$(a_1, a_2, a_3, a_4, \ldots).$$

We've already met one infinite sequence (in chapter 1): the sequence $(d_1, d_2, d_3, \ldots)$ of distances traveled by Zeno during his walk. Equation (1.1) gave us a formula for the values d_n of this sequence, and equation (1.4) helped us determine what number those values approach as $n \to \infty$. Can we generalize those results? That is, for a general infinite sequence (a_n), what's the pattern its values follow? And do those values approach some number as $n \to \infty$? The first question is a static mindset one: "Give me the formula for a_n." The second question is a dynamics mindset one, and is ultimately a question about whether the sequence **converges**, which involves investigating the limit as $n \to \infty$ of a_n. We'll tackle the first question first. (We'll discuss the convergence of sequences in the next section.) But first, a quick note: The convergence question only makes sense if we're dealing with an infinite sequence. For this reason, *in calculus we focus on studying infinite sequences*, and so, *henceforth, when we say "sequence" we'll understand it to mean "infinite sequence," unless otherwise specified.*

Quantifying Patterns in a Sequence's Values

Most sequences' values a_n follow one of two types of patterns: a **functional relationship** to the index n, or a **recurrence relation**. The former is when $a_n = f(n)$, that is, when the values are the outputs of some function n.** Example: If a bank account pays 5% yearly interest, then \$1 deposited into the account grows to $a_n = (1.05)^n$ dollars after n years. The latter—the recurrence relation pattern—is where a_n is related to prior terms via some equation. Examples: $a_n = 1 + a_{n-1}$, and the Fibonacci sequence from chapter 1 (which we'll explore in more detail soon). Let's work through a few examples of these cases.

EXAMPLE 4.1   Returning to the sequence S_1 from (4.1), verify that

$$S_1 = (2n)_{n=1}^4.$$

Solution We're being told that $a_n = 2n$ for $n = 1, \ldots, 4$. This generates the terms

$$a_1 = 2, \qquad a_2 = 4, \qquad a_3 = 6, \qquad a_4 = 8,$$

which are indeed all the terms—in the correct order and with the correct values—in the sequence S_1 from (4.1).   ∎

This example gave us the a_n formulas. But how do we find them ourselves? Here's a handy procedure for that.

Box 4.1: How to Find the a_n Formula for a Sequence

To find the formula that a sequence's values follow:

1. Write down a_1, a_2, a_3, etc., next to each of the sequence's values.

2. Determine how the index value n of a_n can be transformed to produce the sequence's values.

One tip for that second step: Explore that transformation via five arithmetical operations—addition, subtraction, multiplication, division, or exponentiation. One last tip: Once you're done with all this, verify that your a_n formula indeed generates the sequence you're working with.

As an example, let's apply this procedure to the sequence S_1:

$$a_1 \leftrightarrow 2, \qquad a_2 \leftrightarrow 4, \qquad a_3 \leftrightarrow 6, \qquad a_4 \leftrightarrow 8.$$

Our job now is to determine how the index value (the bolded subscripts) transforms to produce each sequence value. For this sequence this means, How does 1 turn into 2? How does 2 turn into 4? And so on. Using the multiplication lens gets us to $a_n = 2n$.* And as we verified in Example 4.1, this functional relationship indeed generates the sequence S_1.

* When $a_n = b + mn$ the sequence is called an **arithmetic sequence**.

EXAMPLE 4.2 (Video)   Assuming that the patterns below continue, determine a_n and the appropriate starting index for each sequence:

(a) $\left(1, \dfrac{1}{2}, \dfrac{1}{3}, \dfrac{1}{4}, \ldots \right)$ (b) $\left(1, \dfrac{1}{2}, \dfrac{1}{4}, \dfrac{1}{8}, \ldots \right)$ (c) $\left(0, -\dfrac{1}{5}, \dfrac{2}{25}, -\dfrac{3}{125}, \ldots \right)$

(d) $\left(\dfrac{1}{2}, \dfrac{2}{3}, \dfrac{3}{4}, \ldots \right)$

Solution

(a) Here $a_1 = 1$, $a_2 = \frac{1}{2}$, $a_3 = \frac{1}{3}$, etc. The subscripts are the denominators in the fractions on right-hand sides of the equations, and the numerators are always 1. Thus, $a_n = \frac{1}{n}$, $n \geq 1$.

(b) Here the denominators in the fractions on right-hand sides of the equations are powers of 2 (recall that $1 = 2^0$), so that

$$a_1 = \frac{1}{2^0}, \qquad a_2 = \frac{1}{2^1}, \qquad a_3 = \frac{1}{2^2}, \qquad a_4 = \frac{1}{2^3}, \qquad \ldots$$

Therefore, $a_n = \frac{1}{2^{n-1}}$, $n \geq 1$.

(c) Here the denominators are powers of 5:

$$a_1 = \frac{0}{5^0}, \qquad a_2 = -\frac{1}{5^1}, \qquad a_3 = \frac{2}{5^2}, \qquad a_4 = -\frac{3}{5^3}, \qquad \ldots$$

Notice next that the numerators are one less than the index value, and that every other term is negative. We can account for that alternation in sign by including a factor of $(-1)^{n+1}$, which alternates between $+1$ and -1 as n increments. Altogether, this yields

$$a_n = (-1)^{n+1} \left[\frac{n-1}{5^{n-1}} \right], \quad n \geq 1.$$

(d) If we choose a starting index value of $n = 1$, then the first fraction in the sequence is $\frac{n}{n+1}$. This pattern works for the other terms too. Thus, $a_n = \frac{n}{n+1}$, $n \geq 1$. ∎

A few comments about the results of this example:

- The sequence in part (b) is an example of a **geometric sequence**. In these sequences, $a_n = ar^n$ for some nonzero a- and r-values. When the starting index is $n = 0$, the terms in a geometric sequence are . . . .

$$a_0 = a, \qquad a_1 = ar, \qquad a_2 = ar^2, \qquad a_3 = ar^3, \qquad \ldots$$

Takeaway: *The value of each term in a geometric sequence is r times the value of its preceding term.*

To find a and r for a given geometric sequence we set the value of the first term in the sequence equal to a, and the second equal to ar, and solve for the a- and r-values. For the sequence in (b) of the example above, this gives $a = 1$, $ar = \frac{1}{2}$, which yields $a = 1$ and $r = 1/2$. Thus, we can express the sequence as

$$a_n = 1 \cdot \left(\frac{1}{2} \right)^n = \frac{1}{2^n}, \quad n \geq 0.$$

Note that this is equivalent to the solution in the example, $a_n = \frac{1}{2^{n-1}}$, $n \geq 1$, since both generate the same sequence. Takeaway: *The same sequence can be expressed using different a_n and starting index combinations.*

- The values of the sequence in part (c) start with zero and then alternate sign after that. That alternating portion of the sequence is an example of an **alternating sequence**. In these sequences,

$$a_n = (-1)^n b_n \qquad \text{or} \qquad a_n = (-1)^{n+1} b_n$$

for some other sequence b_n whose values are all positive. For example,

$$b_n = \frac{n-1}{5^{n-1}}$$

in the solution to part (c) above.

Related Exercises   1–6, A1(a)–A4(a), A1(b)–A4(b), A5(a), A6(a)–(c), A7.*

APPLICATIONS Applied exercises A1–A5 explore the sequences hidden in basketball, population growth, and economics. Applied exercise A6 explains how Archimedes used sequences to derive the formula for the area of a circle. Applied exercise A7 explains how the chromatic scale—the foundation of Western music—is based on a particular infinite sequence.

Let's now return to the second most common pattern type for sequences: recurrence. In such sequences there's an equation—the recurrence relation—that describes how a sequence's terms interrelate. We've already discussed (in chapter 1) perhaps the most famous example of such a sequence: the Fibonacci sequence.

EXAMPLE 4.3   The **Fibonacci sequence** (F_n) is defined by

$$F_1 = 1, \quad F_2 = 1, \quad F_n = F_{n-1} + F_{n-2} \quad \text{for } n \geq 3.$$

Calculate the values of the first six terms in this sequence.

Solution We're given the first two terms, so let's find the next four. The recurrence relation $F_n = F_{n-1} + F_{n-2}$ says that the value of the nth term, F_n, is the sum of the values of the preceding two terms. So,

$$F_3 = F_2 + F_1 = 1 + 1 = 2, \qquad F_5 = F_4 + F_3 = 3 + 2 = 5,$$
$$F_4 = F_3 + F_2 = 2 + 1 = 3, \qquad F_6 = F_5 + F_4 = 5 + 3 = 8.$$

Thus, the first six terms in the sequence are $(1, 1, 2, 3, 5, 8, \dots)$. ■

Related Exercises   7–8, A1(c)–A4(c), A5(b), D1(a)–(d).*

* Remember: Exercises with a D prefix are derivations or explorations.

EXPLORATIONS Exercise D1 explores the Fibonacci sequence in more depth. It shows how the sequence's values are connected to the quadratic equation $x^2 = x + 1$, how this leads to an explicit formula for the nth Fibonacci number that involves the golden ratio, and how that can be used to show that the infinite limit of the ratio of successive Fibonacci numbers equals the golden ratio (we saw this briefly in chapter 1).

Factorials

Some sequences' a_n formulas contain a new type of object called a *factorial*.

DEFINITION 4.2: FACTORIALS. The **factorial** of a nonnegative integer n, denoted by $n!$, is defined as the product of n with all positive integers less than n:

$$n! = n(n-1)(n-2)(n-3) \cdots 2 \cdot 1.$$

We also define $0! = 1$. (Note: $1! = 1$ too.)

Examples: $2! = 2 \cdot 1 = 2$, $3! = 3 \cdot 2 \cdot 1 = 6$, and $4! = 4 \cdot 3 \cdot 2 \cdot 1 = 24$. Notice here how factorials nest: $3! = 3 \cdot (2 \cdot 1) = 3 \cdot 2!$, and $4! = 4 \cdot (3 \cdot 2 \cdot 1) = 4 \cdot 3!$. Indeed, this is a general property of factorials:

$$n! = n(n-1)! \quad \text{or, equivalently,} \quad (n+1)! = (n+1)n! \qquad (4.2)$$

* This nesting property is also a recurrence relation: If $a_n = na_{n-1}$ and $a_0 = 1$, then $a_n = n!$.

This follows from the fact that $n! = n\,(n-1)(n-2)\cdots 2\cdot 1 = n\,(n-1)!\,$.*

Now that you know what factorials are and how they nest, I recommend trying out some of the exercises suggested below to practice working with them.

Related Exercises   9–12.

4.2   Convergence of Sequences

Let's now turn to the second question regarding infinite sequences: Do their values approach some number as $n \to \infty$? To explore that, let's first turn on our dynamics mindset and find a way to animate sequences—to see them as *dynamic* objects, not static lists of numbers.

Visualizing Sequences

One way to visualize a sequence (a_n) is to graph the points (n, a_n) for each index n in the sequence. In other words, we plot the points in which the x-values are the index numbers n and the y-values are the values a_n of the sequence. Figure 4.1 does this for the sequence $(1/n)$. In (a) I've plotted just the first point, $(1, 1)$; and in (b) the first two points, in (c) the first three points, and in (d) the first 10. This new perspective helps us *see* the patterns in a sequences' values, and by comparing these graphs in order (from (a) to (d)) we come to understand this particular sequence's "personality" (every mathematical object has a personality): The sequence's values start at 1 and then get progressively smaller, appearing to approach zero as n increases.

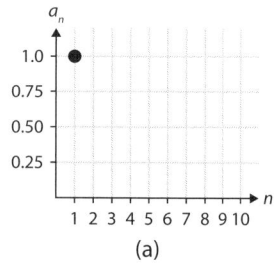

(a)

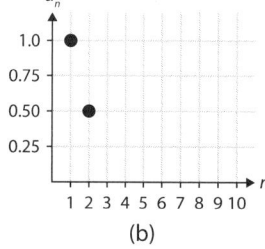

(b)

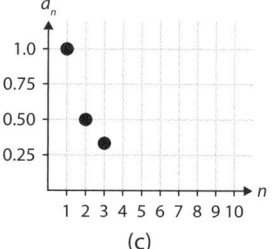

(c)

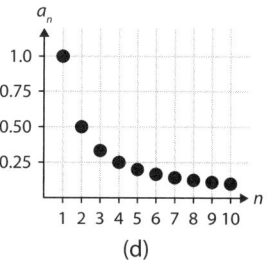

(d)

Figure 4.1: Plots of the sequence $a_n = \frac{1}{n}$, $n \geq 1$, for (a) $n = 1$, (b) $n = 1, 2$, (c) $n = 1, 2, 3$, and (d) $n = 1, 2, \ldots, 10$. Interact with this plot—and plot your own sequence!—at sites.google.com/view/fernandezmath/apps/cs2.

EXAMPLE 4.4   Match each sequence with its graph in figure 4.2:

(i) $\left(3\left(\dfrac{1}{2}\right)^n\right)$ (ii) $\left((-1)^n\right)$ (iii) $\left(1 - \dfrac{1}{2^n}\right)$

** Another approach: The terms of sequence (ii) alternate between 1 and −1, matching graph (c); the terms of sequence (i) decrease as n increases, matching graph (a).

Solution Let's assume that the starting index is $n = 1$ in all three cases. The first terms in the sequences are then (i) $a_1 = 3/2$, (ii) $a_1 = -1$, and (iii) $a_1 = 1 - 1/2 = 1/2$. That's enough to match (i) with (a), (ii) with (c), and (iii) with (b).** ∎

Related Exercises   13–15.

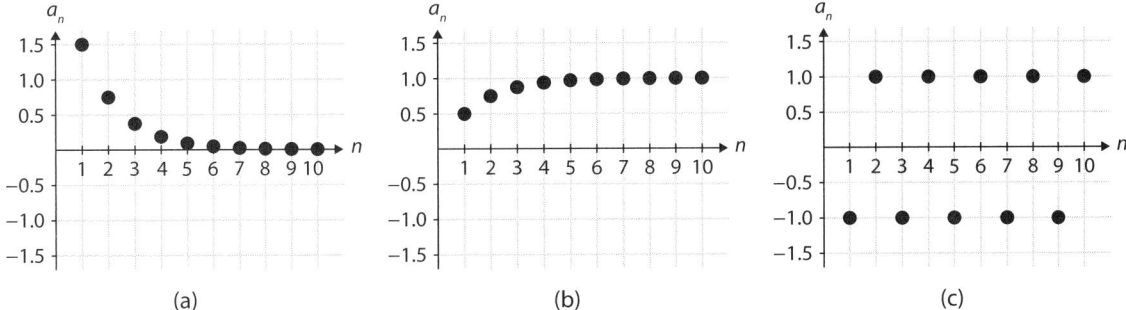

Figure 4.2: Plots of three sequences; see the solutions of example 4.4 for the a_n formulas for each plot.

Limits of Sequences

Figures 4.1(d) and 4.2(a)–(b) hint that perhaps the underlying sequences' values are tending to 0, 0, and 1, respectively, as $n \to \infty$. Expressed in limit language:

$$\lim_{n\to\infty} \frac{1}{n} = 0, \qquad \lim_{n\to\infty} \frac{3}{2^n} = 0, \qquad \lim_{n\to\infty} \left(1 - \frac{1}{2^n}\right) = 1. \qquad (4.3)$$

This turns out to be true (we'll prove it in a minute) and motivates the following definition.

> **DEFINITION 4.3: CONVERGENCE OF A SEQUENCE.** Let $(a_n)_{n=k}^{\infty}$ be a sequence. We say that the sequence **converges** to a number L if
>
> $$\lim_{n\to\infty} a_n = L \qquad (4.4)$$
>
> or, equivalently, if the terms a_n can be made as close to L as desired by making n sufficiently large.* (An alternate notation for this is $a_n \to L$.) If the limit above does not exist, we say that the sequence **diverges**.

* If (a_n) converges to L, then the limit is unique: There's no other number that (a_n) can converge to.

Investigating (4.4) involves evaluating an infinite limit. You did a lot of this in Calculus 1 (appendix C reviews that). You *may* have also learned L'Hôpital's Rule (reviewed in appendix F). A third technique we'll use is *growth arguments*. This approach compares the growth rates of functions in the a_n expression to help determine the limit.** The following theorem summarizes those growth rates.

** Example: e^n grows faster than n^2, so $\frac{n^2}{e^n} \to 0$ as $n \to \infty$.

> **THEOREM 4.1: GROWTH ORDER.** Let p, q, r, s be positive real numbers, and let $b > 1$. Then
>
> $$n^n \gg n! \gg b^n \gg n^{p+s} \gg n^p (\ln n)^r \gg n^p \gg (\ln n)^q.$$
>
> Here "$X \gg Y$" means that $\lim_{n\to\infty} \frac{X}{Y} = \infty$ or, equivalently, $\lim_{n\to\infty} \frac{Y}{X} = 0$.***

*** In other words, "$X \gg Y$" means that X grows much faster than Y as $n \to \infty$.

Now, there's one tiny hurdle to clear before we can apply all those Calculus 1 limit evaluation techniques. You see, in Calculus 1 we learned how to evaluate limits in which the independent variable is a real number, like

$$\lim_{x \to \infty} \frac{x}{x+1}.$$

But the independent variable n in a sequence is *not* a real number—it's a positive integer. The following theorem enables us to use all those Calculus 1 limit techniques to help determine if a sequence converges.

> **THEOREM 4.2:** Let $(a_n)_{n=k}^{\infty}$ be a sequence, and let f be a real-valued function such that $a_n = f(n)$ for $n \geq k$ (i.e., f interpolates the sequence's graph). If $\lim_{x \to \infty} f(x) = L$, then $\lim_{n \to \infty} a_n = L$.

This theorem says that if the graph that the sequence's values are on has an infinite limit L, then the sequence converges to L. Takeaway: *If a sequence's graph approaches the horizontal asymptote $y = L$, then the sequence converges to L.* Figure 4.3 illustrates this—and the theorem—in the context of the sequence $(1/n)$. The dashed curve is the graph of $f(x) = 1/x$. When $x = n$, a positive integer, $f(n) = 1/n$ and thus reproduces the terms in the sequence $(1/n)$. And since we know that $\lim_{x \to \infty} \frac{1}{x} = 0$, theorem 4.2 tells us that the sequence $(1/n)$ converges to 0, verifying the first limit in (4.3).

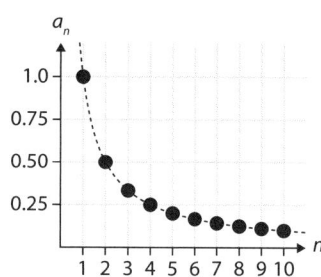

Figure 4.3: Portions of the graphs of $f(x) = 1/x$ (dashed) and the sequence $(1/n)$.

* Often this happens because the infinite limit of f doesn't exist.

Now, what if theorem 4.2 doesn't apply?* In those cases we revert to (4.4) and use the fact that *if a sequence's graph doesn't approach any horizontal asymptote, the sequence diverges.*

EXAMPLE 4.5 (Video)   Determine which of the sequences below converge:

(a) $\left(\dfrac{3}{2^n} \right)$ (b) $((-1)^n)$ (c) $\left(1 - \dfrac{1}{2^n} \right)$

Solution

(a) Since $\lim_{x \to \infty} \frac{3}{2^x} = 0$ and $f(x) = \frac{3}{2^x}$ is a real-valued function, theorem 4.2 applies and we conclude that the sequence converges to 0.

(b) We can't use theorem 4.2 here because the function $f(x) = (-1)^x$ includes "imaginary" numbers. (Example: $f(0.5) = \sqrt{-1}$, which is not a real number.) No problem, though, because as figure 4.2(c) suggests, $(-1)^n$ hops between $+1$ and -1 forever. That graph has no horizontal asymptote, and therefore $a_n = (-1)^n$ diverges.

(c) The sequence converges to 1, by theorem 4.2 and the fact that

$$\lim_{x \to \infty} \left(1 - \frac{1}{2^x} \right) = 1 - \lim_{x \to \infty} \frac{1}{2^x} = 1. \qquad \blacksquare$$

Related Exercises   16–29, A1(d)–A4(d), A5(c), A6(d), D1(e).

The Sequence Laws

In chapter 1, equation (1.3), we determined that the total distance d_n Zeno traveled after walking his nth step was $d_n = 2\left(1 - \frac{1}{2^n}\right)$. In terms of the sequence a_n in part (c) of example 4.5, $d_n = 2a_n$. We showed in that example that $a_n \to 1$. Does it then follow that $d_n \to 2(1) = 2$? Indeed, by the following theorem.

THEOREM 4.3: THE SEQUENCE LAWS. Let $(a_n)_{n=k}^{\infty}$ and $(b_n)_{n=k}^{\infty}$ be convergent sequences, let c be a real number, and let "lim" be a stand-in for $\lim\limits_{n \to \infty}$. Then

1. $\lim [a_n \pm b_n] = \lim a_n \pm \lim b_n$;

2. $\lim [ca_n] = c \lim a_n$;

3. $\lim [a_n b_n] = [\lim a_n][\lim b_n]$;

4. $\lim \dfrac{a_n}{b_n} = \dfrac{\lim a_n}{\lim b_n}$, provided $[\lim b_n] \neq 0$;

5. for $m > 0$ and $a_n > 0$, $\lim a_n^m = [\lim a_n]^m$; and

6. if $\lim a_n = L$, and f is continuous at L, then $\lim f(a_n) = f(L)$.

This theorem is the sequence version of the limit laws (reviewed in appendix C). And like those laws, the ones above are easier to remember when read aloud. The first law, for example, says that the limit of a sum (or difference) of two convergent sequences is the sum (or difference) of the limits of those sequences. These laws reduce convergence/divergence determinations to the convergence/divergence of the "ingredient" sequences a_n and b_n.

One case not covered in the laws above is this: If we determine that the sequence $|a_n|$ converges—that is, the sequence a_n with all its values made nonnegative—does that guarantee that the sequence a_n converges? In general the answer is "no." But the following theorem gives us an instance in which it's "yes!"

THEOREM 4.4: Let (a_n) be a sequence. If $\lim\limits_{n \to \infty} |a_n| = 0$ then $\lim\limits_{n \to \infty} a_n = 0$.

EXAMPLE 4.6 Determine whether the following sequences converge or diverge:

(a) $\left(\dfrac{n}{n+1} + \dfrac{(-1)^n}{n}\right)$ (b) $\left(\dfrac{1 + e^{-n}}{2 + e^{-n}}\right)$ (c) $\left(\sqrt[3]{27 + \dfrac{1}{n^2}}\right)$

Solution

(a) This sequence is the sum of the two sequences $a_n = \dfrac{n}{n+1}$ and $b_n = \dfrac{(-1)^n}{n}$.
Now, since

$$\lim_{x \to \infty} \frac{x}{x+1} = 1, \qquad \text{Since } \frac{x}{x+1} \approx \frac{x}{x} = 1 \text{ for large } x$$

theorem 4.2 tells us that $\lim\limits_{n \to \infty} \dfrac{n}{n+1} = 1$. Thus (a_n) converges to 1. And since

$$\left|\frac{(-1)^n}{n}\right| = \frac{1}{n},$$

which we already showed converges to 0, (b_n) converges to 0 by theorem 4.4. By the first sequence law, the sequence in (a) converges to $1 + 0 = 1$.

(b) Since $e^{-n} \to 0$ as $n \to \infty$, $1 + e^{-n} \to 1$ and $2 + e^{-n} \to 2$. By the fourth sequence law, the sequence in (b) converges to $\frac{1}{2}$.

(c) Denote $a_n = 27 + \frac{1}{n^2}$, so that the sequence is $(f(a_n))$ with $f(x) = \sqrt[3]{x}$. Since $\frac{1}{n^2} \to 0$ as $n \to \infty$, then $a_n \to 27$. And since f is continuous at $x = 27$, we conclude from the sixth sequence law that the sequence in (c) converges to $f(27) = \sqrt[3]{27} = 9$.   ■

The sequences in example 4.2(b) and 4.5(a)–(b) are all geometric sequences. Exercise D2 guides you through a general study of geometric sequences' convergence and helps you prove the following theorem.

THEOREM 4.5: The geometric sequence (ar^n) converges to zero for $-1 < r < 1$, converges to a for $r = 1$, and diverges for all other r-values.

Related Exercises 30–33.

Additional Tips, Tricks, and Takeaways

- Sometimes simplifying the a_n formula before investigating its infinite limit helps. A silly but instructive example is $a_n = (n+1) - n$. If you view

$$\lim_{n \to \infty} [(n+1) - n] \qquad \text{as} \qquad \lim_{n \to \infty} (n+1) - \lim_{n \to \infty} n,$$

then you'd end up with the unhelpful $\infty - \infty$, which *is not* always zero.[1] But $(n+1) - n = 1$, so $a_n = 1$, and thus (a_n) converges to 1.

- Finally, you should know that there are yet more techniques and results for investigating a sequence's convergence. These include the **Monotone Convergence Theorem** and the **Squeeze Theorem for sequences**. These show up infrequently in Calculus 2, so I've skipped them here, but you can find them in any calculus textbook.

   At this point we've learned enough about sequences that we can move on to series. In the chapter preview I foreshadowed the deep connections between these two—sequences and (infinite) series—and it's now time to explore those connections and tackle the Infinite Sum Question from chapter 1: *Does an infinite sum have a sum, and if so, what's the sum?*

4.3   Infinite Series

First things first: Let's define what we'll mean by an "infinite series."

[1] The outcome $\infty - \infty$ is one of many **indeterminate forms**. These are discussed in appendix F.

DEFINITION 4.4: INFINITE SERIES. An **infinite series** is the sum of the values in an infinite sequence $(a_n)_{n=k}^{\infty}$:

$$a_k + a_{k+1} + \cdots = \sum_{n=k}^{\infty} a_n.$$

The number k is called the **starting index** of the series. As a shorthand, we sometimes write $\sum a_n$ for the series.

Let me emphasize the main point once more: *An infinite series is the sum of the values in an infinite sequence.* This intimately connects sequences, and all we've learned about them, to series. We'll leverage this connection soon to extend the notions of convergence/divergence from sequences to series. But first, let's talk about the symbol "$\sum$" in the definition.

Interlude: Sigma Notation

The symbol "$\sum$" is the Greek capital letter sigma. In math it's usually a stand-in for "sum." In particular, the symbol appears in **sigma notation**, a compact way to represent the sum of an expression that depends on one variable, usually called the "index" and denoted by n:

$$\sum_{n=k}^{N} a_n \quad \text{means "sum the outputs generated by } a_n \text{ as } n \text{ ranges from } k \text{ to } N."$$

Here k identifies the lowest index value and N the highest one. Some examples:

$$\sum_{n=1}^{3} n = 1 + 2 + 3, \quad \sum_{n=3}^{10} \frac{1}{n} = \frac{1}{3} + \frac{1}{4} + \cdots + \frac{1}{10},$$

$$\sum_{n=0}^{4} \frac{(-1)^n}{e^{n^2}} = 1 - \frac{1}{e} + \frac{1}{e^4} - \frac{1}{e^9} + \frac{1}{e^{16}}.$$

Sometimes we refer to the values being added in a sigma notation sum by their order in the sum, and other times by the n-value that generated them. For example, in the second sum above (the one with $1/n$), we'd call the value $1/3$ the first term in the sum and also the $n=3$ term, since we get $1/3$ by substituting $n=3$ into $1/n$. We'll also henceforth drop the distinction between "value" and "term" we introduced with sequences.* We'll therefore refer to that same $1/3$ as "the term $1/3$."

Now, since sigma notation indicates a sum, we can manipulate $\sum a_n$ like we would any sum. For example, we can extract a few terms from a sum:

$$\sum_{n=1}^{100} n^2 = 1^2 + 2^2 + \sum_{n=3}^{100} n^2.$$

The terms 1^2 and 2^2 on the right-hand side are the $n=1$ and $n=2$ outputs of n^2. Those two terms been extracted from the sum on the left-hand side, and so the remaining sum must start at $n=3$.

* We made that distinction then to help us understand a sequence's graph. But we won't be graphing series in the same way.

Now Back to Infinite Series

The four examples of sigma notation above are *finite* series. We'll be focusing on *infinite* series in this book, for the same reasons we've focused on infinite sequences. So, henceforth if I say "series" and forget to specify "infinite series," please assume I mean that. And one last comment: As with sequences, the starting index of a series—the lowest n-value—need not be one, but most of the time it is. Thus, we'll often work with infinite series of the form $\displaystyle\sum_{n=1}^{\infty} a_n$.

EXAMPLE 4.7   Write out the first few terms of these infinite series:

(a) $\displaystyle\sum_{n=1}^{\infty} (-1)^n$ (b) $\displaystyle\sum_{n=3}^{\infty} \frac{2^n}{n}$

Solution

(a) $\displaystyle\sum_{n=1}^{\infty} (-1)^n = (-1) + 1 + (-1) + 1 + \cdots .$

(b) $\displaystyle\sum_{n=3}^{\infty} \frac{2^n}{n} = \frac{2^3}{3} + \frac{2^4}{4} + \frac{2^5}{5} + \cdots .$ ■

Convergence of Series

So, does an infinite series have a sum? To explore this, let's first turn on our dynamics mindset. An infinite series $\sum a_n$ is a static object—it's a thing. But if we imagine trying to actually sum this series, we'd have to start with the first term, add the second, etc. *There's the dynamics.*

This thought experiment gets us focused on adding the terms in $\sum a_n$. And that brings us to our application of the Calculus 2 workflow (recall figure 1.3) for the infinite sum Big Question (see figure 1.5, second row). To recap that approach: Approximate the calculus quantity with a finite quantity that depends on n, and then take the limit as $n \to \infty$ to get the calculus result. A very coarse approximation to $\sum a_n$ (the calculus quantity we're after—the infinite sum) is the sum of just its first two terms: $a_1 + a_2$. Let's denote this S_2, for "sum of the first two terms." Replacing "two" by n (i.e., summing the first n terms) gives us a better approximation and brings us to the following definition.

> **DEFINITION 4.5: PARTIAL SUMS.** The Nth **partial sum**, denoted S_N, of the infinite series $\displaystyle\sum_{n=1}^{\infty} a_n$ is the sum of the first N terms in the series:
>
> $$S_N = a_1 + a_2 + \cdots + a_N = \sum_{n=1}^{N} a_n. \tag{4.5}$$

(Note: We can generalize this definition to a starting index $n \neq 1$.)

We've already studied one partial sum—d_n, the total distance Zeno traveled after having taken his nth step, in equation (1.1) in chapter 1:

$$d_1 = 1, \quad d_2 = 1.5, \quad d_3 = 1.75, \quad \ldots, \quad d_n = 2 - \frac{1}{2^{n-1}}, \qquad (4.6)$$

where the last equation comes from (1.3). I hope this *list of numbers* reminds you of a concept we studied recently: *sequences*! That's right: *The partial sums of an infinite series form a sequence*! If we're talking about *all* the partial sums, then we're talking about the *infinite* sequence (S_N). Cue the last step in the Calculus 2 workflow: take the infinite limit. Taking the infinite limit of (4.5) yields

$$\lim_{N \to \infty} S_N = \lim_{N \to \infty} \sum_{n=1}^{N} a_n. \qquad (4.7)$$

If this limit exists, we assign its value to the infinite series:

$$\sum_{n=1}^{\infty} a_n = \lim_{N \to \infty} \sum_{n=1}^{N} a_n. \qquad (4.8)$$

Takeaway: *An infinite series is equal to the infinite limit of its sequence of partial sums when that limit exists.* In other words, to sum the infinite series on the left, we sum its first N terms—obtaining S_N—and then take the infinite limit of that. When that limit exists (i.e., yields a number), we say the series *converges* and call the value of the limit "the" sum of the series.*

* We did this in chapter 1 with the partial sums shown in (4.6) to get the result (1.2), which was our first summation of an infinite series.

> **DEFINITION 4.6: CONVERGENCE OF SERIES.** Let $\sum a_n$ be an infinite series, and let (S_N) be its sequence of partial sums. We say that the series
>
> - **converges** if (S_N) is a convergent sequence, that is, if $\lim_{N \to \infty} S_N$ exists;
>
> - **diverges** if (S_N) is a divergent sequence, that is, if $\lim_{N \to \infty} S_N$ does not exist.
>
> If the series converges and $\lim_{N \to \infty} S_N = S$, we call S the **sum** of the series.

EXAMPLE 4.8 (Video) Calculate the first three partial sums of the series below, and then find a formula for S_N. Finally, use that formula to determine whether each series converges or diverges:

(a) $\displaystyle\sum_{n=1}^{\infty} 1$ (b) $\displaystyle\sum_{n=1}^{\infty} (-1)^n$ (c) $\displaystyle\sum_{n=1}^{\infty} \left(\frac{1}{n} - \frac{1}{n+1} \right)$

Solution

(a) We have

$$S_1 = 1, \quad S_2 = 1 + 1 = 2, \quad S_3 = 1 + 1 + 1 = 3,$$

etc. Observing the pattern here, $S_N = N$. Since $\lim_{N \to \infty} N$ does not exist (it approaches infinity), the series diverges.

(b) We have

$$S_1 = -1, \quad S_2 = -1 + 1 = 0, \quad S_3 = -1 + 1 - 1 = -1,$$

$$S_4 = -1 + 1 - 1 + 1 = 0.$$

The pattern here: $S_N = -1$ if N is odd, and $S_N = 0$ if N is even. Once again, $\lim\limits_{N \to \infty} S_N$ does not exist (because the value of S_N oscillates between 0 and -1 forever). Thus, the series diverges.

(c) We have

$$S_1 = 1 - \frac{1}{2}, \quad S_2 = S_1 + \left(\frac{1}{2} - \frac{1}{3} \right) = 1 - \frac{1}{3},$$

$$S_3 = S_2 + \left(\frac{1}{3} - \frac{1}{4} \right) = 1 - \frac{1}{4}.$$

Continuing with these calculations shows that $S_N = 1 - \frac{1}{N+1}$. And since $\lim\limits_{N \to \infty} \left(1 - \frac{1}{N+1} \right) = 1$, the series converges and sums to 1:

$$\sum_{n=1}^{\infty} \left(\frac{1}{n} - \frac{1}{n+1} \right) = 1. \qquad \blacksquare$$

Related Exercises   34–37.

The factoid I used in part (c) of this example is worth noting: The Nth partial sum S_N is a_N plus the $(N-1)$th partial sum S_{N-1}:

$$S_N = a_1 + a_2 + \cdots + a_{N-1} + a_N$$

$$= S_{N-1} + a_N. \tag{4.9}$$

This fact speeds up partial sum calculations and will also come in handy in a minute.

The Divergence Test

We can't find S_N explicitly for many series (we'll study some in the next section). So using the definition of convergence—definition 4.6—is infeasible for most series. Luckily, there's a simple test for determining whether a series *diverges*: if its terms don't tend to zero as $n \to \infty$.

> **THEOREM 4.6: THE DIVERGENCE TEST.** Let $\sum a_n$ be an infinite series. If $\lim\limits_{n \to \infty} a_n \neq 0$, then the series diverges.

Exercise D3 guides you through the proof of this theorem, and exercise D4 through the proof of its contrapositive (which is also true): *If $\sum a_n$ converges, then $a_n \to 0$.*

To use the divergence test in practice we calculate $\lim\limits_{n \to \infty} a_n$ and see if we get a nonzero answer. *If $a_n \to 0$, however, this **does not** tell you the series converges.*[2] In the next section we'll meet a *convergent* series whose terms tend to zero, as well as a *divergent* series whose terms tend to zero. Takeaway: Use the divergence test only when $a_n \not\to 0$.

---

[2] The divergence test *assumes* $a_n \not\to 0$ (the symbol "$\not\to$" is short for "does not tend to"), not $a_n \to 0$.

EXAMPLE 4.9 (Video)   For each series below, determine whether the divergence test applies. If it does, use it to show that the series diverges.

(a) $\displaystyle\sum_{n=1}^{\infty} \frac{2^n}{n}$ (b) $\displaystyle\sum_{n=1}^{\infty} \frac{n^2+7}{3n^3+2n+1}$ (c) $\displaystyle\sum_{n=1}^{\infty} \cos(n)$

(d) $\displaystyle\sum_{n=1}^{\infty} \frac{n^2}{\sqrt{n^2+n+1}}$

Solution

(a) Since 2^n grows much faster than n (recall the growth order theorem 4.1), $a_n = \frac{2^n}{n} \to \infty$. Thus, $a_n \not\to 0$ so the series diverges.

(b) Let's replace the ns with xs and apply theorem 4.2:

$$\lim_{x\to\infty} \frac{x^2+7}{3x^3+2x+1} = 0, \qquad \text{Since for large } x,\ \frac{x^2+7}{3x^3+2x+1} \approx \frac{x^2}{3x^3} = \frac{1}{3x} \to 0$$

so by the theorem, $a_n \to 0$ too. The $a_n \not\to 0$ hypothesis of the divergence test is not met, and so we cannot use the divergence test.

(c) Since $a_n = \cos n$ oscillates between -1 and $+1$ forever, $a_n \not\to 0$, so the series diverges.

(d) Applying theorem 4.2 again,

$$\lim_{x\to\infty} \frac{x^2}{\sqrt{x^2+x+1}} = \infty. \qquad \text{Since for large } x,\ \frac{x^2}{\sqrt{x^2+x+1}} \approx \frac{x^2}{\sqrt{x^2}} = x \to \infty$$

Therefore $a_n \not\to 0$, so the series diverges.   ■

Related Exercises   38–43.

4.4   Special Series and the Series Laws

We now know what infinite series are, what it means for them to converge (or diverge), and how we might show this in some cases. In this section we'll tour four series families that we can easily determine convergence or divergence for. Then we'll generalize the sequence laws to series. The result will help us determine convergence/divergence for combinations of series.

p-Series

For some series we can tell whether they converge or diverge just by looking at them. The best example of that is the p-series family.

> **DEFINITION 4.7: P-SERIES.** Suppose that $p \geq 0$. A **p-series** is a series of the form
> $$\sum_{n=1}^{\infty} \frac{1}{n^p} = 1 + \frac{1}{2^p} + \frac{1}{3^p} + \cdots .$$

We note here that p is a nonnegative *real number* and so could be rational or even irrational. Some examples of p-series (from left to right, $p = 3$, $p = 1$, and $p = 1/2$):

$$\sum_{n=1}^{\infty} \frac{1}{n^3} = 1 + \frac{1}{2^3} + \frac{1}{3^3} + \cdots, \quad \sum_{n=1}^{\infty} \frac{1}{n} = 1 + \frac{1}{2} + \frac{1}{3} + \cdots,$$

$$\sum_{n=1}^{\infty} \frac{1}{\sqrt{n}} = 1 + \frac{1}{\sqrt{2}} + \frac{1}{\sqrt{3}} + \cdots. \tag{4.10}$$

The next theorem says that we can determine a p-series's convergence or divergence right from its p-value. To understand intuitively why this might be true, recall from the previous section that if the series' Nth partial sum (S_N) converges then the series converges. In a p-series with, say, $p = 100$, the third term is $1/3^{100}$, already a minuscule number. This term—and subsequent terms in the series—add little to nothing to the partial sums of the series. We'd therefore expect the partial sums of a $p = 100$ p-series to change very little as we add more of its terms, and to thus converge. But what if $p = 50$? Or $p = 10$? How low can we go and still get convergence? The following theorem gives the answer.

THEOREM 4.7: CONVERGENCE OF *P*-SERIES. The p-series

$$\sum_{n=1}^{\infty} \frac{1}{n^p} = 1 + \frac{1}{2^p} + \frac{1}{3^p} + \cdots$$

* In chapter 5 we'll prove this theorem using a convergence test called the integral test.

converges for $p > 1$ and diverges for $0 \le p \le 1$.*

EXAMPLE 4.10 Determine whether the following series converge or diverge:

(a) $\displaystyle\sum_{n=1}^{\infty} \frac{1}{n^{1.2}}$ (b) $\displaystyle\sum_{n=1}^{\infty} \frac{1}{\sqrt{n}}$ (c) $\displaystyle\sum_{n=1}^{\infty} \frac{1}{n^3}$

Solution Each of these is a p-series, so we apply theorem 4.7: (a) converges since $p = 1.2 > 1$; (b) diverges since $p = 0.5 < 1$; and (c) converges since $p = 3 > 1$.   ∎

** $p > 1$ implies convergence; $p \le 1$ implies divergence.

Notice that $p = 1$ is the "convergence boundary" for p-series.** The corresponding $p = 1$ p-series—the second series in (4.10), known as the **harmonic series**—therefore plays a special role in series land; we'll see it arise in future convergence discussions.

Telescoping Series

Next up on our tour of special series is a series family characterized by *lots* of helpful term cancellations.

DEFINITION 4.8: TELESCOPING SERIES. A **telescoping series** is an infinite series of the form

$$(b_1 - b_2) + (b_2 - b_3) + (b_3 - b_4) + \cdots = \sum_{n=1}^{\infty} (b_n - b_{n+1}).$$

The cancellations in successive terms in a telescoping series simplify its partial sums (I've highlighted in gray the terms that cancel):

$$S_1 = b_1 - b_2,$$

$$S_2 = (b_1 - b_2) + (b_2 - b_3) = b_1 - b_3,$$

$$S_3 = (b_1 - b_2) + (b_2 - b_3) + (b_3 - b_4) = b_1 - b_4,$$

etc. The pattern here: $S_N = b_1 - b_{N+1}$. Using this in our definition of convergence (definition 4.6) yields the following result.

THEOREM 4.8: CONVERGENCE OF TELESCOPING SERIES. Let $\sum a_n$ be a telescoping series. Then its Nth partial sum is

$$S_N = b_1 - b_{N+1}.$$

Furthermore, the series diverges if $\lim_{N \to \infty} b_{N+1}$ does not exist and converges if it does exist. When the series converges, its sum is $b_1 - \lim_{N \to \infty} b_{N+1}$.

EXAMPLE 4.11 (Video)   Determine which of the following telescoping series converge:

(a) $\displaystyle\sum_{n=1}^{\infty} [\cos(n) - \cos(n+1)]$ (b) $\displaystyle\sum_{n=1}^{\infty} \left(\frac{2}{3^n} - \frac{2}{3^{n+1}} \right)$

(c) $\displaystyle\sum_{n=1}^{\infty} [n - (n+1)]$ (d) $\displaystyle\sum_{n=1}^{\infty} \ln \left(\frac{n}{n+1} \right)$

Solution

(a) Here $b_1 = \cos 1$, so $S_N = \cos 1 - \cos(N+1)$. But since $\lim_{N \to \infty} \cos(N+1)$ doesn't exist (the values oscillate forever), the series diverges.

(b) Here $b_1 = 2/3$, and since

$$S_N = \frac{2}{3} - \frac{2}{3^{N+1}} \quad \text{and} \quad \lim_{N \to \infty} \frac{2}{3^{N+1}} = 0,$$

we conclude that the series converges, and it's sum is $2/3$.

(c) Here $b_1 = 1$, but since $S_N = 1 - (N+1) = -N \to -\infty$, the series diverges.

(d) The series isn't given in telescoping form. But we can put it into that form by using $\ln \left(\frac{n}{n+1} \right) = \ln n - \ln(n+1)$. The resulting series,

$$\sum_{n=1}^{\infty} [\ln n - \ln(n+1)],$$

is a telescoping series with $b_1 = \ln 1 = 0$. Hence $S_N = 0 - \ln(N+1) = -\ln(N+1)$. This tends to $-\infty$, so the series diverges. ■

Related Exercises   44–47.

Geometric Series

Let's end this tour of series families with perhaps the best-known family of series: *geometric series*.

> **DEFINITION 4.9: GEOMETRIC SERIES.** Let $a \neq 0$ and r be real numbers. The series
>
> $$\sum_{n=0}^{\infty} ar^n = a + ar + ar^2 + \cdots \tag{4.11}$$
>
> is called a **geometric series**.

Let me give you one quick insight regarding this definition before we start working with geometric series. Recall that every infinite series is the sum of the terms in an infinite sequence (definition 4.4). A *geometric* series is the sum of the terms in a *geometric* sequence.* Therefore, each term in a geometric series is a multiple r of the previous term. This makes it easy to spot—and rule out—geometric series: if *each* term is (or is not) a multiple r of the previous term, then the series is (respectively, is not) a geometric series.

* We discussed that family of sequences in section 4.1.

We actually already ran into two geometric series in chapter 1:

$$1 + \frac{1}{2} + \frac{1}{4} + \cdots = 2, \qquad \frac{1}{2} + \frac{1}{4} + \frac{1}{8} + \cdots = 1. \tag{4.12}$$

The left one is the infinite sum of the distances Zeno walked, and the right one is the infinite sum illustrated geometrically in figure 1.4(b). In the first geometric series, $a = 1$ and $r = 1/2$. In the second, $a = 1/2$ and $r = 1/2$. We showed that the first series converged by taking the infinite limit of its nth partial sum—equation (1.3), which we denoted d_n—to get 2. I promised then that we'd later learn how to calculate that partial sum. Here's how to do that for a general geometric series.

First, let's write out the Nth partial sum of a geometric series:

$$S_N = \sum_{n=0}^{N-1} ar^n = a + ar + ar^2 + \cdots + ar^{N-1}.$$

Multiplying this equation by r yields

$$rS_N = ar + ar^2 + ar^3 + \cdots + ar^{N-1} + ar^N,$$

and subtracting this from S_N yields

$$S_N - rS_N = a + ar + ar^2 + \cdots + ar^{N-1} - \left(ar + ar^2 + ar^3 + \cdots + ar^{N-1} + ar^N \right)$$
$$= a - ar^N,$$

since all the highlighted terms cancel out. Now, factoring $S_N - rS_N = a - ar^N$,

** Applying this formula to the first geometric series in (4.12) yields the partial sum d_n of (1.3) in chapter 1.

$$(1 - r)S_N = a(1 - r^N) \quad \Rightarrow \quad S_N = \frac{a}{1 - r}\left(1 - r^N\right).** \tag{4.13}$$

This is the Nth partial sum of a geometric series in the standard form (4.11). Taking the infinite limit and using theorem 4.5 yields the following theorem.

THEOREM 4.9: CONVERGENCE OF GEOMETRIC SERIES. The geometric series

$$\sum_{n=0}^{\infty} ar^n = a + ar + ar^2 + \cdots$$

converges to $\dfrac{a}{1-r}$ if $|r| < 1$. The series diverges if $|r| \geq 1$.

EXAMPLE 4.12 (Video)   Determine which of the geometric series below converge. (Below, x is a fixed number.) For those that do, find their sum.

(a) $\displaystyle\sum_{n=0}^{\infty} \frac{5}{4^n}$ (b) $\displaystyle\sum_{n=0}^{\infty} \left(\frac{5}{4}\right)^n$ (c) $\displaystyle\sum_{n=0}^{\infty} \left(-\frac{4}{5}\right)^n$ (d) $\displaystyle\sum_{n=0}^{\infty} x^n$

Solution   All of these geometric series are in the standard form (4.11). The a-value is therefore the $n = 0$ term, and the r-value is the $n = 1$ term divided by a. In (a), $a = 5$ and $r = 1/4$. Since $|r| < 1$, the series converges, and its sum is $\frac{5}{1-(1/4)} = 20/3$. In (b), $a = 1$ and $r = 5/4$. Since $|r| > 1$, the series diverges. In (c), $a = 1$ and $r = -4/5$. Since $|r| = 4/5 < 1$, the series converges, and its sum is $\frac{1}{1-(-4/5)} = 5/9$. In (d), $a = 1$ and $r = x$. By theorem 4.9 the series will diverge if $|x| \geq 1$ and converge if $|x| < 1$. When it converges, its sum is $\frac{1}{1-x}$.   ∎

Related Exercises   48–53, A8–A9, D5.

APPLICATIONS   Applied exercise A9 explores the *multiplier effect* in economics, which describes how $\$T$ spent into a community can generate *more than* $\$T$ worth of spending by the community's members.

Let me record the results of that last example here for future reference:

$$\sum_{n=0}^{\infty} x^n = \frac{1}{1-x}, \quad |x| < 1. \tag{4.14}$$

I mentioned earlier that geometric series are the best-known series family. I bet you've seen them (in disguise) as early as in middle school. That's because they show up in repeating decimals, as the following example illustrates.

EXAMPLE 4.13   Express these repeating decimals as geometric series: (a) $0.\overline{3}$, (b) $0.\overline{12}$. Then, express the series as ratios of two integers.

Solution

(a) We have

$$0.\overline{3} = 0.333\ldots = \frac{3}{10} + \frac{3}{10^2} + \frac{3}{10^3} + \cdots$$

$$= \frac{3}{10}\left(1 + \frac{1}{10} + \frac{1}{10^2} + \cdots\right)$$

$$= \sum_{n=0}^{\infty} \frac{3}{10}\left(\frac{1}{10}\right)^n.$$

In this geometric series $a = \frac{3}{10}$, $r = \frac{1}{10}$. Since $|r| < 1$ it converges, and its sum is $\frac{3/10}{1-(1/10)} = \frac{3/10}{9/10} = \frac{1}{3}$.

(b) We have

$$0.\overline{12} = 0.1212\ldots = \frac{12}{100} + \frac{12}{100^2} + \cdots$$

$$= \frac{12}{100}\left(1 + \frac{1}{100} + \cdots\right)$$

$$= \sum_{n=0}^{\infty} \frac{12}{100}\left(\frac{1}{100}\right)^n.$$

In this geometric series $a = \frac{12}{100}$, $r = \frac{1}{100}$. Since $|r| < 1$ it converges, and its sum is $\frac{12/100}{1-(1/100)} = \frac{12/100}{99/100} = 12/99$.   ■

Related Exercises   54–55.

Series Properties

In section 4.2 we saw how the sequence laws (theorem 4.3) help us to determine whether certain combinations of convergent sequences converge. A similar result holds for series, as the theorem below shows.

> **THEOREM 4.10:   THE SERIES LAWS.** Let $\sum a_n$ and $\sum b_n$ be convergent series, and let c be a real number and k a nonnegative integer. Then
>
> $$\sum_{n=k}^{\infty} c a_n = c\sum_{n=k}^{\infty} a_n, \quad \sum_{n=k}^{\infty}(a_n + b_n) = \sum_{n=k}^{\infty} a_n + \sum_{n=k}^{\infty} b_n,$$
>
> $$\sum_{n=k}^{\infty}(a_n - b_n) = \sum_{n=k}^{\infty} a_n - \sum_{n=k}^{\infty} b_n.^\star$$

* These equations imply that $\sum c a_n$ and $\sum(a_n \pm b_n)$ converge, and converge to the stated combinations of the sums of $\sum a_n$ and $\sum b_n$.

EXAMPLE 4.14   For each of the series below, determine whether theorem 4.10 applies. If it does, use it to determine which of them converge, and if possible, find their sum.

(a) $\displaystyle\sum_{n=0}^{\infty}[(0.1)^n + (0.6)^n]$ (b) $\displaystyle\sum_{n=1}^{\infty}\left(\frac{1}{n} + \frac{1}{n^2}\right)$ (c) $\displaystyle\sum_{n=1}^{\infty}\left(\frac{1}{n^2} - \frac{1}{2^n}\right)$

Solution Theorem 4.10 requires that each "ingredient" series converges. The ingredient series $\sum\frac{1}{n}$ in (b) is the divergent harmonic series, so we can't apply the theorem to the series in (b). For the other two we *can* apply the theorem:

(a) The series $\sum(0.1)^n$ and $\sum(0.6)^n$ are both convergent geometric series ($r = 0.1$ and $r = 0.6$, respectively). So by theorem 4.10 the series in (a) converges, and its sum is $\frac{1}{1-0.1} + \frac{1}{1-0.6} = \frac{10}{9} + \frac{10}{4} = \frac{65}{18}$.

(c) The series $\sum\frac{1}{n^2}$ is a p-series with $p = 2 > 1$ and so converges by the p-series theorem 4.7. The series $\sum\frac{1}{2^n} = \sum\left(\frac{1}{2}\right)^n$, a geometric series with $a = \frac{1}{2}$ and

$r = 0.5$, converges since $|r| < 1$. So by theorem 4.10 the series in (c) converges. We don't know how to sum $\sum \frac{1}{n^2}$, so we can't calculate the sum of the series in (c).[3]
∎

Related Exercises 56–59, D7.

Additional Tips, Tricks, and Takeaways

I've given you the simplest definitions for both telescoping series and geometric series. But sometimes you'll encounter variants of those definitions.

- For telescoping series, we defined them (definition 4.8) to have cancellations occurring in each pair of successive terms: $(b_1 - b_2) + (b_2 - b_3) = b_1 - b_3$. But in some variants the cancellations take longer to occur. For example, in the series

$$\sum_{n=1}^{\infty} \left(\frac{1}{n} - \frac{1}{n+2} \right) = \left(1 - \frac{1}{3} \right) + \left(\frac{1}{2} - \frac{1}{4} \right) + \left(\frac{1}{3} - \frac{1}{5} \right) + \cdots, \quad (4.15)$$

the cancellations occur every other term (e.g., the first and third terms have a quantity that cancels: the 1/3). Exercise D6 guides you through the generalization of the telescoping series convergence theorem 4.8 to handle these telescoping series variants.

- For geometric series, we defined them with an $n = 0$ starting index (definition 4.9). But sometimes they show up with a starting index of $n = k > 0$. We can convert these into the $n = 0$ form by factoring out the first term from the series. Example:

$$\sum_{n=3}^{\infty} 2 \left(\frac{4}{7} \right)^n = 2 \left(\frac{4}{7} \right)^3 + 2 \left(\frac{4}{7} \right)^4 + 2 \left(\frac{4}{7} \right)^5 + \cdots$$

$$= 2 \left(\frac{4}{7} \right)^3 \left[1 + \left(\frac{4}{7} \right) + \left(\frac{4}{7} \right)^2 + \cdots \right] = 2 \left(\frac{4}{7} \right)^3 \sum_{n=0}^{\infty} \left(\frac{4}{7} \right)^n.$$

In the original infinite sum, $a = 2(4/7)^3$ and $r = 4/7$. The final infinite sum is $2(4/7)^3$ times a geometric series in the standard $n = 0$ form, with $a = 1$ and $r = 4/7$.

One final important takeaway I'll leave you with pertains to p-series, how fast they converge, and what that teaches us about convergence in general. Recall that $\sum \frac{1}{n^p}$ converges for $p > 1$ and diverges for $p \le 1$. In all the $p > 0$ p-series, the $a_n = 1/n^p$ terms tend to zero as $n \to \infty$. But evidently these terms *don't* tend to zero

---

[3] A bit of math history here to illustrate the far-reaching implications of fully understanding p-series. Calculating $\sum \frac{1}{n^2}$ was a problem posed in 1650 and solved by Euler in 1734 when he was 28 years old. (The last exercise in chapter 5 shows that $\sum \frac{1}{n^2} = \frac{\pi^2}{6}$.) In 1859 Riemann (from "Riemann sums" fame) expanded on Euler's solution and connected the sums of "complexified" p-series—p-series where $p = a + b\sqrt{-1}$, with a and b real numbers—to the distribution of prime numbers. Riemann then conjectured what's now the most famous unsolved problem in mathematics, the **Riemann hypothesis**: If $p = a + b\sqrt{-1}$, then $\sum \frac{1}{n^p} = 0$ only when $p = -2, -4, -6, \ldots$, or when $a = \frac{1}{2}$. (By the way, if you can prove this you'll be paid \$1 million by the Clay Mathematics Institute.)

fast enough to guarantee convergence if $0 < p \leq 1$, and *do* if $p > 1$. Put another way: If a p-series's terms tend to zero faster than the harmonic series's terms (i.e., faster than $1/n$), then it converges. Otherwise, it diverges. Over the next few sections we'll discover that this insight generalizes—a general series indeed *converges only when its terms tend to zero fast enough*—and quantify how fast is "fast enough."

We've made a lot of progress on series. But the series results we've developed thus far are limited in applicability. If we can't calculate the Nth partial sum S_N for the series, and if the series isn't one of the special ones studied in this section, we have no way of investigating its convergence. In the next few sections we'll address this issue by developing several "convergence tests" to help us determine when a series converges or diverges.

4.5   The Limit and Direct Comparison Tests

Consider the two series

$$\sum_{n=1}^{\infty} \frac{1}{n^2}, \qquad \sum_{n=1}^{\infty} \frac{1}{n^2+1}. \tag{4.16}$$

The first is a convergent p-series series ($p = 2 > 1$). The second is not a p-series. But its terms go to zero *just a little bit faster* than the first one's terms do. Based on the insight at the end of the previous section, we therefore expect the second series to converge. How can we establish this? Well, what we just did was compare a series we'd *like to* establish convergence for to a series *we already know* converges. The new convergence tests we'll discuss in this section formalize these types of comparisons in two different ways.

The Direct Comparison Test

Let's start by comparing the denominators in (4.16). Notice that $n^2 + 1 > n^2$ for every n-value. Taking the reciprocal of this inequality,

$$\frac{1}{n^2+1} < \frac{1}{n^2} \quad \Longrightarrow \quad \sum_{n=1}^{N} \frac{1}{n^2+1} < \sum_{n=1}^{N} \frac{1}{n^2}.$$

Translation: Each partial sum of the series $\sum \frac{1}{n^2+1}$ is less than the corresponding partial sum of the series $\sum \frac{1}{n^2}$. Since the latter series converges, and so its partial sums tend to a finite limit, we suspect that the partial sums of the former series also tend to a finite limit, and therefore that $\sum \frac{1}{n^2+1}$ converges as well. The following theorem confirms this and adds the necessary hypotheses to make such arguments work in general.

THEOREM 4.11: THE DIRECT COMPARISON TEST. Let $\sum a_n$ and $\sum b_n$ be two series. Then

1. If $\sum b_n$ converges, and $0 < a_n \leq b_n$, then $\sum a_n$ converges.

2. If $\sum b_n$ diverges, and $0 < b_n \leq a_n$, then $\sum a_n$ diverges.

A few notes on this theorem:

- Both parts of the theorem require working with inequalities. In practice, verifying the particular inequality we need (e.g., $0 < a_n \le b_n$) begins by finding relationships in the series's terms—like the $n^2 + 1 > n^2$ inequality we started with above—and then manipulating those to get the inequality we need.

- Note the directionality in the two parts of the theorem. To show $\sum a_n$ *converges* we need to find a series $\sum b_n$ we already know converges and show that its terms are "larger" (i.e., $0 < a_n \le b_n$). To show $\sum a_n$ *diverges* we need to find a series $\sum b_n$ we already know diverges and show that its terms are "smaller" (i.e., $0 < b_n \le a_n$). The rough intuition: *If $\sum a_n$ is bounded above by a convergent series, then $\sum a_n$ converges; if $\sum a_n$ is bounded below by a divergent series, then $\sum a_n$ diverges.*

- Finally, note that in some contexts we'll end up showing that $0 < a_n < b_n$ or $0 < b_n < a_n$, stronger conditions than the "$\le$" conditions in the theorem.

EXAMPLE 4.15 (Video) Use the direct comparison test to investigate the convergence of each series:

(a) $\displaystyle\sum_{n=1}^{\infty} \frac{\cos^2 n}{n^2}$
(b) $\displaystyle\sum_{n=0}^{\infty} \frac{1}{2^n + 1}$
(c) $\displaystyle\sum_{n=1}^{\infty} \frac{1}{3n - 2}$

(d) $\displaystyle\sum_{n=2}^{\infty} \frac{5^n}{4^n + 7}$

Solution

(a) The largest $\cos^2 x$ ever gets is 1. So the series here is bounded above by the p-series $\sum \frac{1}{n^2}$, which converges since $p = 2 > 1$. Let's therefore compare to $\sum \frac{1}{n^2}$, setting

$$a_n = \frac{\cos^2 n}{n^2}, \quad b_n = \frac{1}{n^2}.$$

Both of these terms are always positive.[4] Furthermore, since $\cos^2 n < 1$, dividing through by n^2 yields

$$\frac{\cos^2 n}{n^2} < \frac{1}{n^2}.$$

Thus, $a_n < b_n$, and so the series in (a) converges.

(b) This series resembles the geometric series $\sum \frac{1}{2^n} = \sum \left(\frac{1}{2}\right)^n$, which converges ($r = 1/2$, so $|r| < 1$). So let's set

$$a_n = \frac{1}{2^n + 1}, \quad b_n = \frac{1}{2^n}.$$

These are both always positive. And since $2^n + 1 > 2^n$, taking reciprocals yields

$$\frac{1}{2^n + 1} < \frac{1}{2^n},$$

establishing $a_n < b_n$. Thus, the series in (b) converges.

---

[4]Solving $\cos^2 n = 0$ yields $n = \pm\frac{\pi}{2}, \pm\frac{3\pi}{2}, \ldots$. But $n = 1, 2, \ldots$ in a series.

(c) If the 2 in the denominator weren't there, this series would be $\sum \frac{1}{3n}$, which is a multiple of the divergent (harmonic) series $\sum \frac{1}{n}$ and therefore diverges. So let's compare with this series, setting

$$a_n = \frac{1}{3n-2}, \quad b_n = \frac{1}{3n}.$$

Since $n \geq 1$ (the starting index of the series), these terms are always positive. From $3n > 3n - 2$, taking reciprocals yields

$$\frac{1}{3n} < \frac{1}{3n-2},$$

verifying $b_n < a_n$. Therefore, the series in (c) diverges.

(d) If the 7 in the denominator weren't there, the series would be the divergent geometric series $\sum \frac{5^n}{4^n} = \sum \left(\frac{5}{4}\right)^n$ ($r = 5/4 > 1$). To set up the comparison, note that since $n \geq 2$ (the range of n-values in the series in (d)), then

$$4^n + 7 < 4^n + 4^n = 2 \cdot 4^n.$$

Taking reciprocals yields

$$\frac{1}{4^n + 7} > \frac{1}{2 \cdot 4^n}.$$

Multiplying by 5^n yields

$$\frac{5^n}{4^n + 7} > \frac{5^n}{2 \cdot 4^n}.$$

Setting $a_n = \frac{5^n}{4^n+7}$ and $b_n = \frac{5^n}{2 \cdot 4^n}$, we've shown that $b_n < a_n$. And since these two sequences' values are all positive, the direct comparison test tells us that the series in (d) diverges. ■

Each of the examples above was intended to expose you to a different strategy for generating the inequality you'll need to apply the direct comparison test. In (a) the strategy was to use information about one of the functions' values (e.g., $|\cos x| \leq 1$); in (b) and (c) it was to drop appropriate numbers (the 1 and -2, respectively) from the a_n expressions; in (d) it was to replace an appropriate number with one of the functions already in the a_n expression (the 7 with 4^n). Keep these strategies in mind as you work through the exercises suggested below.

Related Exercises   60–63.

The Limit Comparison Test

We developed the direct comparison test by comparing the *values* of the terms in the series in (4.16). Let's now instead compare those terms' *growth rates*. Notice that, for large n-values,

$$\frac{1}{n^2 + 1} \approx \frac{1}{n^2}.$$

Takeaway: The terms in both the series in (4.16) are tending to zero at the same rate. We know that $\frac{1}{n^2} \to 0$ fast enough to trigger convergence ($p = 2 > 1$). Does this then imply that $\sum \frac{1}{n^2+1}$ converges? The following theorem formalizes this new type of comparison test.

> **THEOREM 4.12: THE LIMIT COMPARISON TEST.** Let $\sum a_n$ and $\sum b_n$ be two series, with $a_n > 0$ and $b_n > 0$. If
>
> $$\lim_{n\to\infty} \frac{a_n}{b_n} = L, \qquad \text{where } L > 0 \text{ is finite,}$$
>
> then $\sum a_n$ and $\sum b_n$ either both converge or both diverge.

To use this theorem we again need to identify a series we know converges or diverges. A tip for that: *Start with the series you're investigating, and keep only the terms that grow the fastest.** Let's now work through examples of doing this and of using theorem 4.12.

* Use the Growth Order Theorem, theorem 4.1, for help with this.

EXAMPLE 4.16 (Video)   Use the limit comparison test to investigate the convergence of each series:

(a) $\displaystyle\sum_{n=0}^{\infty} \frac{1}{2^n + 1}$ (b) $\displaystyle\sum_{n=1}^{\infty} \frac{\sqrt{n}}{n^2 + 1}$ (c) $\displaystyle\sum_{n=1}^{\infty} \frac{n^2 + n + 1}{n^3 + 7n}$

Solution

(a) For large n,

$$\frac{1}{2^n + 1} \approx \frac{1}{2^n},$$

so let's use the convergent geometric series $\sum \frac{1}{2^n}$ ($r = 1/2$) for our comparison. With $a_n = \frac{1}{2^n+1} > 0$ and $b_n = \frac{1}{2^n} > 0$,

$$\lim_{n\to\infty} \frac{a_n}{b_n} = \lim_{n\to\infty} \left[\frac{1}{2^n + 1} \div \frac{1}{2^n} \right]$$

$$= \lim_{n\to\infty} \left[\frac{1}{2^n + 1} \cdot \frac{2^n}{1} \right] \qquad \text{Since } x \div \frac{1}{y} = x \cdot y$$

$$= \lim_{n\to\infty} \frac{2^n}{2^n + 1}$$

$$= \lim_{n\to\infty} \frac{1}{1 + \frac{1}{2^n}} = 1,$$

where I've divided the numerator and denominator by 2^n to help evaluate the limit. Since the resulting limit value is a positive number, we conclude that the series in (a) converges.

(b) For large n,

$$\frac{\sqrt{n}}{n^2 + 1} \approx \frac{\sqrt{n}}{n^2} = \frac{1}{n^{3/2}}.$$

We know that $\sum \frac{1}{n^{3/2}}$ is a convergent p-series ($p = 3/2 > 1$). This hints that the series in (b) converges. With $a_n = \frac{\sqrt{n}}{n^2+1} > 0$ and $b_n = \frac{1}{n^{3/2}} > 0$,

$$\lim_{n\to\infty} \frac{a_n}{b_n} = \lim_{n\to\infty} \frac{\sqrt{n}}{n^2 + 1} \cdot n^{3/2} \qquad \text{Since } \frac{1}{b_n} = n^{3/2}$$

$$= \lim_{n \to \infty} \frac{n^2}{n^2 + 1}$$

$$= \lim_{n \to \infty} \frac{1}{1 + \frac{1}{n^2}} = 1,$$

where I've divided the numerator and denominator by n^2 to help evaluate the limit. The limit is a positive number again, so the series in (b) indeed converges.

(c) For large n,

$$\frac{n^2 + n + 1}{n^3 + 7n} \approx \frac{n^2}{n^3} = \frac{1}{n}.$$

We know that $\sum \frac{1}{n}$ is the divergent harmonic series. This hints that the series in (c) diverges. With $a_n = \frac{n^2 + n + 1}{n^3 + 7n} > 0$ and $b_n = \frac{1}{n} > 0$,

$$\lim_{n \to \infty} \frac{a_n}{b_n} = \lim_{n \to \infty} \frac{n^2 + n + 1}{n^3 + 7n} \cdot n \qquad \text{Since } \frac{1}{b_n} = n$$

$$= \lim_{n \to \infty} \frac{n^3 + n^2 + n}{n^3 + 7n}$$

$$= \lim_{n \to \infty} \frac{1 + \frac{1}{n} + \frac{1}{n^2}}{1 + \frac{7}{n^2}} = 1,$$

where I've divided the numerator and denominator by n^3 to help evaluate the limit. The limit value is a positive number again, so the series in (c) indeed diverges. ∎

Related Exercises   64–67.

Additional Tips, Tricks, and Takeaways

- The series in example 4.16(a) is the same as the one in example 4.15(b). Takeaway: Sometimes we can use more than one convergence test to determine the convergence/divergence of a series. Which test works best depends in part on your comfort with verifying the tests' hypotheses (e.g., evaluating limits versus manipulating inequalities).

- Other times one test won't apply but the other will. Example: Using the limit comparison test for the series in example 4.15(a), with $a_n = \frac{\cos^2 n}{n^2}$ and $b_n = \frac{1}{n^2}$, involves

$$\lim_{n \to \infty} \frac{a_n}{b_n} = \lim_{n \to \infty} \frac{\cos^2 n}{n^2} \cdot n^2 = \lim_{n \to \infty} (\cos^2 n),$$

which doesn't exist. This isn't finite and positive, so the limit comparison test can't be used.

- Finally, sometimes one test is easier to use than the other. For example, it'd be difficult to set up the requisite inequalities to apply the direct comparison test to the series in example 4.16(c). Using the limit comparison test, by contrast, involved evaluating a reasonably straightforward limit.

4.6   Alternating Series

Section 4.1 defined an *alternating sequence* as one whose values have the form $a_n = (-1)^n b_n$ or $a_n = (-1)^{n+1} b_n$, where $b_n > 0$. When you add up the values in an alternating *sequence* you get an alternating *series*.

> **DEFINITION 4.10: ALTERNATING SERIES.** An **alternating series** is a series of the form
>
> $$\sum_{n=1}^{\infty} (-1)^n b_n \quad \text{or} \quad \sum_{n=1}^{\infty} (-1)^{n+1} b_n, \qquad \text{where } b_n > 0.$$

Two simple examples of alternating series:

$$\sum_{n=1}^{\infty} (-1)^n, \qquad \sum_{n=1}^{\infty} \frac{(-1)^{n+1}}{n}.$$

In the first series $b_n = 1$. We showed that this series diverges in example 4.8(b). The second series is the **alternating harmonic series** (the alternating version of the divergent harmonic series $\sum \frac{1}{n}$). In this one, $b_n = \frac{1}{n}$. We know that the harmonic series *diverges*. Are the cancellations in the alternating harmonic series—arising from the $(-1)^n$ factor—enough to trigger *convergence*? The following theorem helps us answer this question.

> **THEOREM 4.13:   THE ALTERNATING SERIES TEST.** Let $b_n > 0$. The alternating series
>
> $$\sum_{n=1}^{\infty} (-1)^n b_n \quad \text{and} \quad \sum_{n=1}^{\infty} (-1)^{n+1} b_n$$
>
> converge when both (1) $\lim_{n \to \infty} b_n = 0$ and (2) $b_{n+1} \leq b_n$ for all n.*

* These two conditions can be condensed into one: (b_n) must be a non-increasing sequence that tends to zero. "Nonincreasing" means that the values of successive terms in the sequence decrease or remain constant.

Two quick comments on this theorem. First things first: The alternating series test only applies to alternating series. Thus, for example, it doesn't apply to the series

$$\sum_{n=0}^{\infty} \cos(n) = \cos 0 + \cos 1 + \cos 2 + \cdots \approx 1 + 0.54 + \cdots,$$

because this series's terms don't *alternate* sign. Finally, as we'll shortly see in the examples, verifying the $b_{n+1} \leq b_n$ hypothesis is typically done in one of two ways: using either the **difference method** (verify that $b_n - b_{n+1} \geq 0$) or the **ratio method**: verify that $\frac{b_{n+1}}{b_n} \leq 1$).

EXAMPLE 4.17 (Video)   First, verify that the series below are alternating series. Then, use the alternating series test to show that they converge.

(a) $\displaystyle\sum_{n=1}^{\infty} \frac{(-1)^{n+1}}{n}$ (b) $\displaystyle\sum_{n=1}^{\infty} \frac{n}{(-2)^{n-1}}$

Solution

(a) Here $a_n = (-1)^{n+1} b_n$ with $b_n = \frac{1}{n} > 0$, so it's indeed an alternating series. And since $\frac{1}{n} \to 0$ as $n \to \infty$, the first requirement of the alternating series test is verified. To verify the second we'll use the difference method:

$$b_n - b_{n+1} = \frac{1}{n} - \frac{1}{n+1}$$

$$= \frac{n+1}{n(n+1)} - \frac{n}{n(n+1)} \qquad \text{Common denominator}$$

$$= \frac{(n+1) - n}{n(n+1)} = \frac{1}{n(n+1)}, \qquad \text{Combining into a single fraction}$$

which is positive for all the series's n-values ($n \geq 1$). Thus, the series in (a) converges.

(b) Since

$$\frac{n}{(-2)^{n-1}} = \frac{n}{(-1)^{n-1} 2^{n-1}}$$

$$= \frac{n}{(-1)^{n-1} 2^{n-1}} \cdot \frac{(-1)^{n+1}}{(-1)^{n+1}}$$

$$= \frac{(-1)^{n+1} n}{(-1)^{2n-2} 2^{n-1}} = \frac{(-1)^{n+1} n}{2^{n-1}},$$

the series is the same as $\sum (-1)^{n+1} \frac{n}{2^{n-1}}$. Here $b_n = \frac{n}{2^{n-1}} > 0$, and since

$$\lim_{n \to \infty} \frac{n}{2^{n-1}} = 0, \qquad \text{Since } 2^{n-1} \text{ grows much faster than } n$$

the first requirement of the alternating series test is met. Let's use the ratio method to verify the second:

$$\frac{b_{n+1}}{b_n} = \frac{n+1}{2^n} \cdot \frac{2^{n-1}}{n}$$

$$= \frac{n+1}{2n} \qquad \text{Since } \frac{2^{n-1}}{2^n} = \frac{1}{2}$$

$$= \frac{1}{2} + \frac{1}{2n} \leq 1.^{\star} \qquad (4.17)$$

Thus, $\frac{b_{n+1}}{b_n} \leq 1$, and so the series in (b) converges.   ∎

* When $n=1$, $\frac{1}{2} + \frac{1}{2} = 1$. For $n > 1$, $\frac{1}{2n} < \frac{1}{2}$, so $\frac{1}{2} + \frac{1}{2n} < 1$.

Related Exercises   68(a)–73(a) ((a) only).

Absolute and Conditional Convergence

Part (a) of this last example answers the question I posed right before the alternating series test: The cancellations in the alternating harmonic series *are* enough to trigger convergence. Let's now look at this in the opposite direction: If we *start* with the convergent series $\sum \frac{(-1)^{n+1}}{n}$ and make all its negative terms positive, we get the divergent series $\sum \frac{1}{n}$. Might there be a general relationship between the convergence/divergence of the two series $\sum a_n$ and $\sum |a_n|$? Let's find out. We'll begin with some new terminology.

> **DEFINITION 4.11: ABSOLUTE AND CONDITIONAL CONVERGENCE.** We say that a series $\sum a_n$ is
>
> 1. **Absolutely convergent** if $\sum |a_n|$ converges.
>
> 2. **Conditionally convergent** if $\sum a_n$ converges but $\sum |a_n|$ diverges.

In this new terminology, the alternating harmonic series, $\sum \frac{(-1)^{n+1}}{n}$ is conditionally convergent. The following theorem adds to our understanding of how these new concepts interrelate.

> **THEOREM 4.14: THE ABSOLUTE CONVERGENCE TEST.** If the series $\sum |a_n|$ converges, then the series $\sum a_n$ also converges.*

* Takeaway: Absolute convergence implies convergence.

Exercise D8 guides you through this theorem's proof. To use it in practice for a series $\sum a_n$, we first examine $\sum |a_n|$. If that converges, then $\sum a_n$ converges. (If $\sum |a_n|$ diverges, however, that tells us nothing about the convergence/divergence of $\sum a_n$.) All this works for an alternating series, too: $\sum (-1)^n b_n$ converges if $\sum |(-1)^n b_n| = \sum |b_n|$ converges. This therefore gives us a second convergence test for alternating series, in addition to the alternating series test.

EXAMPLE 4.18 Determine whether these series are convergent, absolutely convergent, or conditionally convergent:

(a) $\displaystyle\sum_{n=1}^{\infty} \frac{\sin n}{n^2}$ (b) $\displaystyle\sum_{n=1}^{\infty} \frac{(-1)^n}{\sqrt{n}}$

Solution

(a) Here $\sum |a_n| = \sum \frac{|\sin n|}{n^2}$. Since $0 < |\sin n| < 1$, dividing this inequality by n^2 yields

$$\frac{|\sin n|}{n^2} < \frac{1}{n^2}.$$

And because $\sum \frac{1}{n^2}$ converges (p-series with $p = 2 > 1$), $\sum \frac{|\sin n|}{n^2}$ converges by the direct comparison test (theorem 4.11). Thus, the series in (a) is absolutely convergent and therefore converges.

(b) Here $\sum |a_n| = \sum \frac{1}{\sqrt{n}}$ (since $|(-1)^n| = 1$), which diverges (p-series with $p = 0.5 < 1$). Thus the series in (b) is *not* absolutely convergent. To see if it's conditionally convergent, let's try the alternating series test (theorem 4.13). With $b_n = \frac{1}{\sqrt{n}} > 0$, $\lim_{n \to \infty} \frac{1}{\sqrt{n}} = 0$. To show that $b_{n+1} \le b_n$ let's use the ratio method:

$$\frac{b_{n+1}}{b_n} = \frac{1}{\sqrt{n+1}} \cdot \sqrt{n} = \sqrt{\frac{n}{n+1}} < 1. \qquad \text{Since } n + 1 > n$$

All conditions of the alternating series test are satisfied, so the series in (b) converges. The series is therefore conditionally convergent. ∎

Related Exercises   68(b)–73(b) ((b) only), D9.

EXPLORATIONS Exercise D9 introduces you to the **Riemann Rearrange-ment Theorem**, which says that if $\sum a_n$ is conditionally convergent, then one can find a rearrangement of the series that sums to *whatever real number you'd like!*

4.7   The Ratio Test

The intuition we've developed thus far is that for $\sum a_n$ to converge we need $a_n \to 0$ as $n \to \infty$, *and it needs to do so fast enough.* We've quantified "fast enough" for *p*-series (those require $p > 1$ for convergence) and, via the direct and limit comparison tests, have used this to help determine convergence/divergence for other series. The alternating series test added a twist: The series may still converge even though its terms don't tend to zero fast enough, provided there are sufficient cancellations in the series. (The alternating harmonic series is a great example.) Recall that one way to verify the nonincreasing-terms requirement of the alternating series test is to use the ratio method. The following theorem develops that method into its own convergence test.

> **THEOREM 4.15: THE RATIO TEST.** Let $\sum a_n$ be a series with nonzero terms. Then:
>
> 1. The series $\sum a_n$ converges absolutely when $\lim\limits_{n \to \infty} \left| \dfrac{a_{n+1}}{a_n} \right| < 1$.
>
> 2. The series $\sum a_n$ diverges when
>
> $$\lim_{n \to \infty} \left| \frac{a_{n+1}}{a_n} \right| > 1 \quad \text{or} \quad \lim_{n \to \infty} \left| \frac{a_{n+1}}{a_n} \right| = \infty.$$

* This can be relaxed—all that matters is that above a certain *n*-value the series's terms are nonzero.

The ratio test is one of the most useful (and general) convergence tests—its only hypothesis is that $\sum a_n$ has nonzero terms.* Therefore, we can apply the test to series containing negative terms, including alternating series. To apply the test we evaluate the infinite limit of $|a_{n+1}/a_n|$. If we get a number that's not 1, or if we get infinity, the test gives us a conclusion. If we get a limit of 1, however, the test is inconclusive. Note that in some cases the limits you end up with will require using L'Hôpital's Rule (reviewed in appendix F).

EXAMPLE 4.19 (Video)   Determine convergence/divergence using the ratio test.

(a) $\displaystyle\sum_{n=1}^{\infty} \frac{n}{2^n}$ (b) $\displaystyle\sum_{n=1}^{\infty} (-1)^n \frac{3^n}{n^2}$ (c) $\displaystyle\sum_{n=1}^{\infty} \frac{1}{n^2}$

Solution

(a) Here $a_n = n/2^n$, so

$$\left| \frac{a_{n+1}}{a_n} \right| = \frac{n+1}{2^{n+1}} \cdot \frac{2^n}{n} = \frac{n+1}{2n}.$$

And since

$$\lim_{n\to\infty} \frac{n+1}{2n} = \frac{1}{2} < 1,$$

the series converges absolutely.

(b) This is an alternating series: $a_n = (-1)^n b_n$, where $b_n = 3^n/n^2$. The ratio test's limit takes absolute values, so the $(-1)^n$ here will go away. Let's therefore look only at the ratio

$$\frac{b_{n+1}}{b_n} = \frac{3^{n+1}}{(n+1)^2} \cdot \frac{n^2}{3^n} = 3\left(\frac{n}{n+1}\right)^2.$$

Then

$$\lim_{n\to\infty} \left|\frac{a_{n+1}}{a_n}\right| = 3 \lim_{n\to\infty} \left(\frac{n}{n+1}\right)^2$$

$$= 3 \left(\lim_{n\to\infty} \frac{n}{n+1}\right)^2 = 3(1)^2 > 1,$$

so the series diverges.

(c) This is a convergent p-series ($p = 2 > 1$). With $a_n = 1/n^2$, the ratio test, however, gives

$$\lim_{n\to\infty} \left|\frac{a_{n+1}}{a_n}\right| = \lim_{n\to\infty} \frac{n^2}{(n+1)^2} = \left(\lim_{n\to\infty} \frac{n}{n+1}\right)^2 = 1,$$

which means that the test is inconclusive. ∎

Related Exercises 74–76.

One of the contexts in which the ratio test is most useful is series involving factorials (definition 4.2).

EXAMPLE 4.20 (Video) Determine convergence/divergence using the ratio test.

(a) $\displaystyle\sum_{n=0}^{\infty} \frac{2^n}{n!}$ (b) $\displaystyle\sum_{n=0}^{\infty} (-1)^n \frac{4}{(2n)!}$ (c) $\displaystyle\sum_{n=1}^{\infty} \frac{n^n}{n!}$

Solution

(a) Here $a_n = 2^n/n!$, so

$$\frac{a_{n+1}}{a_n} = \frac{2^{n+1}}{(n+1)!} \cdot \frac{n!}{2^n} = \frac{2n!}{(n+1)!}.$$

Recalling now that $(n+1)! = (n+1)n!$, from (4.2),

$$\frac{2n!}{(n+1)!} = \frac{2n!}{(n+1)n!} = \frac{2}{n+1}.$$

Then

$$\lim_{n\to\infty} \left|\frac{a_{n+1}}{a_n}\right| = \lim_{n\to\infty} \frac{2}{n+1} = 0 < 1,$$

so the series converges absolutely.

(b) This is another alternating series: $a_n = (-1)^n b_n$, where $b_n = 4/(2n)!$. So we need only look at the ratio

$$\frac{b_{n+1}}{b_n} = \frac{4}{[2(n+1)]!} \cdot \frac{(2n)!}{4} = \frac{(2n)!}{(2n+2)!}$$

$$= \frac{(2n)!}{(2n+2)(2n+1)(2n)!}$$

$$= \frac{1}{(2n+2)(2n+1)}.$$

Then

$$\lim_{n\to\infty} \left| \frac{a_{n+1}}{a_n} \right| = \lim_{n\to\infty} \frac{1}{(2n+2)(2n+1)} = 0 < 1,$$

so the series converges absolutely.

(c) Here $a_n = n^n/n!$, so

$$\frac{a_{n+1}}{a_n} = \frac{(n+1)^{n+1}}{(n+1)!} \cdot \frac{n!}{n^n} = \frac{(n+1)^n}{n^n} \cdot \frac{(n+1)n!}{(n+1)!}$$

$$= \left(\frac{n+1}{n}\right)^n \cdot \frac{(n+1)n!}{(n+1)n!} = \left(1 + \frac{1}{n}\right)^n.$$

Then

$$\lim_{n\to\infty} \left| \frac{a_{n+1}}{a_n} \right| = \lim_{n\to\infty} \left(1 + \frac{1}{n}\right)^n = e > 1,$$

so the series diverges. ∎

Related Exercises   77–79.

Additional Tips, Tricks, and Takeaways

We've learned a lot of convergence/divergence tests in this chapter. Figure 4.4 summarizes those tests in the order we've learned them. (The root test is discussed in exercise D10.)

I wouldn't try using these tests in the order given in figure 4.4, though. Box 4.2 provides a procedure that balances the difficulty of the calculations involved in using each test with the number of steps needed to verify the test's hypotheses.

> ### Box 4.2: A Procedure for Testing Series for Convergence/Divergence
>
> To help determine if $\sum a_n$ converges/diverges, try the following procedure.
>
> 1. Is it a p-series? If so, it converges if $p > 1$ and diverges if $p \leq 1$.
>
> 2. Is it a geometric series? If so, it converges if $|r| < 1$ and sums to $\frac{a}{1-r}$ (if the series's starting index $k = 0$), and diverges if $|r| \geq 1$.
>
> 3. Is it a telescoping series? If so, try the Telescoping Series Theorem.
>
> 4. Does $a_n \to 0$? If not, the series diverges by the divergence test.

5. Does the ratio test's limit yield a finite (and not 1) value, or infinity? If so, the ratio test will tell you whether the series converges or diverges.

6. Is it an alternating series? If so, try the alternating series test.

7. Does the series contain negative terms but is not an alternating series? If so, try the absolute convergence test.

8. Is there a series you know that converges/diverges that you can compare the series to? If so, try the limit comparison test or the direct comparison test.

Test	Series	Comments	Conclusions
Divergence test	$\sum a_n$		Diverges if $\lim\limits_{n \to \infty} a_n \neq 0$
p-series	$\sum \dfrac{1}{n^p}$ $(p \geq 0)$		Converges if $p > 1$
			Diverges if $0 \leq p \leq 1$
Telescoping series	$\sum (b_n - b_{n+1})$	$S_N = b_1 - b_{N+1}$	If $\lim\limits_{n \to \infty} b_{N+1}$ exists, converges to $b_1 - \lim\limits_{n \to \infty} b_{N+1}$; else diverges
Geometric series	$\sum\limits_{n=0}^{\infty} ar^n$ $(a \neq 0)$		If $|r| < 1$, converges to $\dfrac{a}{1-r}$
			If $|r| \geq 1$, diverges
Direct comparison test	$\sum a_n$ $(a_n > 0)$	Requires choosing $\sum b_n$ $(b_n > 0)$	Converges if $\sum b_n$ converges and $0 < a_n \leq b_n$
			Diverges if $\sum b_n$ diverges and $0 < b_n \leq a_n$
Limit comparison test	$\sum a_n$ $(a_n > 0)$	Requires choosing $\sum b_n$ $(b_n > 0)$	If $\lim\limits_{n \to \infty} \dfrac{a_n}{b_n} = L > 0$ is finite, then both series converge or both series diverge
Alternating series test	$\sum (-1)^n b_n$ $(b_n > 0)$		Converges if $b_{n+1} \leq b_n$ and $\lim\limits_{n \to \infty} b_n = 0$
Absolute convergence test	$\sum a_n$		Converges if $\sum |a_n|$ converges
Ratio test	$\sum a_n$ $(a_n > 0)$	Let $\lim\limits_{n \to \infty} \left| \dfrac{a_{n+1}}{a_n} \right| = L$	Converges if $L < 1$
			Diverges if $L > 1$ or $L = \infty$
Root test	$\sum a_n$ $(a_n \geq 0)$	Let $\lim\limits_{n \to \infty} |a_n|^{1/n} = L$	Converges if $L < 1$
			Diverges if $L > 1$ or $L = \infty$

Gray highlighted text indicates that the convergence is absolute convergence.

Figure 4.4: A summary of this chapter's convergence/divergence tests.

Chapter 1's Infinite Sum Question asked: *Does an infinite sum have a sum, and if so, what's the sum?* Seven sections later, we've now developed a plethora of convergence tests that help us answer the first part of that question. But we've made less progress on the "what's the sum?" portion of the question. We can only calculate the sums of (convergent) geometric series and telescoping series. What if—here comes the pivotal insight—we dropped the desire to calculate the *exact* sum of a convergent series and relaxed that to merely *approximating* the sum of a convergent series? The Approximation Question in chapter 1 is all about accurately approximating quantities, so could the problem of accurately approximating the sum of a convergent infinite series be related to the problem of accurately approximating a function? Indeed it is.[5] In the next section we'll learn how to approximate the sum of an alternating series to any desired accuracy. That will get us into the Approximation Question mindset and lay the foundation for working out how to accurately approximate a function by polynomials (via Taylor polynomials) in section 4.9. We'll then push that accuracy to *exactness* (via Taylor series) in section 4.13, yielding explicit formulas for summing convergent infinite series (section 4.15) and therefore circling back to answer that second part of the Infinite Sum Question.

4.8   Approximating the Sum of an Alternating Series

Suppose that $\sum(-1)^{n+1}b_n$ is a convergent alternating series with sum S. One approximation of S is S_N, the Nth partial sum of the alternating series. This approximation of S will generate some error R_N, which we define as the difference $S - S_N$:

$$R_N = S - S_N. \tag{4.18}$$

Since S is the sum of the entire infinite series and S_N is the sum of the first N terms, R_N is the sum of the terms from $N + 1$ onward:

$$R_N = \sum_{n=1}^{\infty}(-1)^{n+1}b_n - \sum_{n=1}^{N}(-1)^{n+1}b_n = \sum_{n=N+1}^{\infty}(-1)^{n+1}b_n.$$

To get a sense of the maximum this error can be, let's look at $|R_N|$:

$$|R_N| = |(-1)^{N+2}b_{N+1} + (-1)^{N+3}b_{N+2} + \cdots|$$
$$= |(-1)^{N+2}[b_{N+1} - b_{N+2} + b_{N+3} - \cdots]| \quad \text{Factoring out } (-1)^{N+2}$$
$$= |(-1)^{N}[b_{N+1} - b_{N+2} + b_{N+3} - \cdots]| \quad \text{Since } (-1)^{N+2} = (-1)^{N}(-1)^{2}$$
$$\qquad\qquad\qquad\qquad\qquad\qquad\qquad\qquad = (-1)^{N}$$
$$= |b_{N+1} - b_{N+2} + b_{N+3} - \cdots| \quad \text{Since } |(-1)^{N}| = 1$$
$$= |b_{N+1} - (b_{N+2} - b_{N+3}) - \cdots|. \quad \text{Factoring out } -1$$

Now, suppose the alternating series satisfies $b_{n+1} \leq b_n$, the second requirement of the alternating series test (theorem 4.13). This means that the values of the

---

[5]This is how progress is made sometimes in mathematics: We notice that one problem may have something to do with another and hope that in solving one problem we end up with insights into the other.

successive terms in the sequence (b_n) either stay the same or decrease, and so the largest value in the sequence $(b_{N+1}, b_{N+2}, b_{N+3}, \ldots)$ is b_{N+1}. Therefore the largest the quantity $b_{N+1} - (b_{N+2} - b_{N+3}) - (b_{N+4} - b_{N+5}) - \cdots$ can get is b_{N+1}. Using this in the $|R_N|$ equation above yields

$$|R_N| \leq b_{N+1}. \tag{4.19}$$

Translation: Under the hypotheses we've made, the *maximum* error incurred in approximating the sum of an alternating series by its Nth partial sum is *at most* the magnitude of the next term: b_{N+1}. We've just proven the following theorem.

> **THEOREM 4.16: ALTERNATING SERIES REMAINDER.** Let $\sum (-1)^{n+1} b_n$ be an alternating series that converges to S, and let S_N denote the Nth partial sum of the series. If $b_{n+1} \leq b_n$, then
>
> $$|S - S_N| \leq b_{N+1}. \tag{4.20}$$

If we unravel the inequality (4.20), we get*

$$-b_{N+1} \leq S - S_N \leq b_{N+1} \quad \Longrightarrow \quad S_N - b_{N+1} \leq S \leq S_N + b_{N+1}. \tag{4.21}$$

Translation: In a convergent alternating series with $b_{n+1} \leq b_n$, its sum S is *at least* $S_N - b_{N+1}$ and *at most* $S_N + b_{N+1}$. Therefore, if we'd like to approximate S with an error of at most, say, 0.01, then by solving $b_{N+1} \leq 0.01$ for N we'll know how many terms of the series to add up to guarantee a sum within 0.01 of the actual sum S. This is an Approximation Question type of insight: without knowing S, (4.21) helps us approximate it to whatever desired accuracy.

EXAMPLE 4.21 (Video)   Consider the alternating series $\displaystyle\sum_{n=1}^{\infty} \frac{(-1)^{n+1}}{n^4}$.

(a) Verify that the series converges.

(b) Find lower and upper bounds for the sum of the series if only the third partial sum is used to approximate the sum.

(c) Determine the number of terms required to approximate the sum of the series with an error of less than 0.001.

Solution

(a) Here $b_n = \frac{1}{n^4}$. This tends to zero as $n \to \infty$. And since

$$\frac{b_{n+1}}{b_n} = \frac{1}{(n+1)^4} \cdot n^4 = \left(\frac{n}{n+1}\right)^4 < 1,$$

because $n < n + 1$, then the hypotheses of the alternating series test (theorem 4.13) are satisfied, and thus the series converges.

(b) You can verify that

$$S_3 = 1 - \frac{1}{2^4} + \frac{1}{3^4} = \frac{1{,}231}{1{,}296}.$$

Noting that $b_4 = \frac{1}{4^4}$, from the rightmost inequalities in (4.21) we then get

$$\frac{1{,}231}{1{,}296} - \frac{1}{4^4} \le S \le \frac{1{,}231}{1{,}296} + \frac{1}{4^4}.$$

Rounding the numbers gives $0.9459 \le S \le 0.9538$.

(c) Since $|R_N| \le b_{N+1}$ (from (4.19)), if we find N such that $b_{N+1} \le 0.001$, then we'll have assured that $|R_N| \le 0.001$. Since $b_n = \frac{1}{n^4}$, we need to solve

$$\frac{1}{(N+1)^4} \le 0.001.$$

Taking reciprocals yields $(N+1)^4 \ge 1{,}000$, from which we get $N \ge \sqrt[4]{1000} - 1 \approx 4.6$. So, summing the first five terms in the series yields a value that's within 0.001 of the actual sum of the series.   ∎

Related Exercises   80–81.

4.9 Taylor Polynomials

We can now approximate the sum of a convergent alternating series (satisfying $b_{n+1} \le b_n$) to any desired accuracy. That puts us in Approximation Question land: *Without knowing the exact value of a function, can we accurately approximate it?* In this section we'll tackle this third Big Question; in the next one we'll discuss the accuracy of the approximation we'll develop. We'll start, as usual, by shifting to a dynamics mindset.

Figure 4.5(a) shows the graph of a function f. To approximate it near $x = c$, imagine tracing the graph near the point $A = (c, f(c))$. To do so you could put your finger at point A and then trace the portion of the curve near that point. To do that accurately, your finger would need to curve as the graph does. We use derivatives to measure the curviness of graphs, so this dynamics-mindset take is foreshadowing the presence of derivatives somewhere in our approximation.

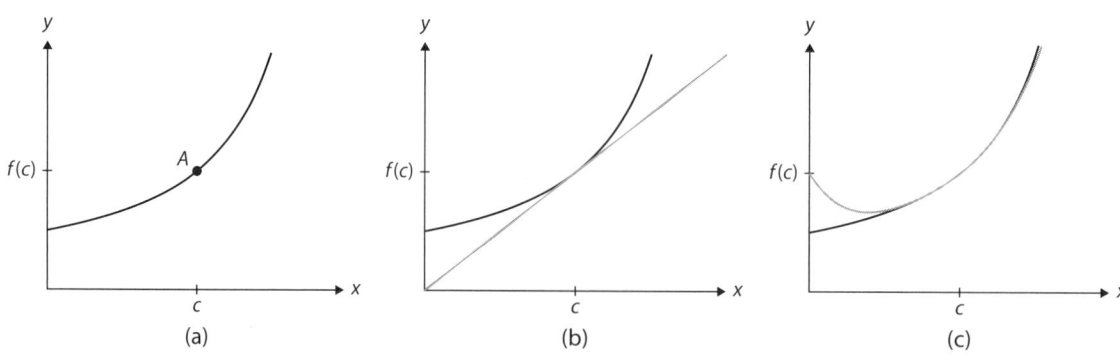

Figure 4.5: Plots of part of the graphs of a function $f(x)$ (black curves) and its approximations near $x = c$ by a linear polynomial (b) and a fifth-degree polynomial (c) (gray curves).

To mathematize this insight, let's return to our tracing thought experiment. The coarsest approximation to $f(x)$ at $x = c$ would be to start at point A and then trace out the line whose slope matches the "slope of the graph" at $x = c$, which is $f'(c)$. Figure 4.5(b) shows the resulting line. The equation $T_1(x)$ of this line is our coarse approximation of $f(x)$ near $x = c$:[6]

$$f(x) \approx T_1(x) = a_0 + a_1(x - c). \tag{4.22}$$

(The "T" here foreshadows our endpoint—Taylor polynomials—and the "1" keeps track of the degree n of the polynomial: $n = 1$.) I chose the particular form "$a_0 + a_1(x - c)$" because it's the point-slope form of the equation of a line, and because we know the x-value of the point: $x = c$. To find the coefficients a_0 and a_1 we go back to how we constructed the line in figure 4.5(b): We insisted that $T_1(x)$ have the same y-value as $f(x)$ at $x = c$ and the same derivative as $f(x)$ at $x = c$. In other words, we insisted that

$$f(c) = T_1(c), \quad f'(c) = T_1'(c). \tag{4.23}$$

From (4.22), $T_1(c) = a_0 + a_1(c - c) = a_0$, and since $T_1'(x) = a_1$, $T_1'(c) = a_1$ too. Using this in (4.23) yields $a_0 = f(c)$ and $a_1 = f'(c)$. Substituting these into (4.22) then yields

$$f(x) \approx T_1(x) = f(c) + f'(c)(x - c).^\star$$

> $\star$ This is the linear approximation of f at $x = c$. For a review of this, see appendix C; see also *Calculus Simplified* [2], section 4.2.

A better approximation to $f(x)$ near $x = c$ would replace the linear function (a first-degree polynomial) with an nth degree polynomial having the same form as $T_1(x)$:

$$T_n(x) = a_0 + a_1(x - c) + a_2(x - c)^2 + a_3(x - c)^3 + \cdots + a_n(x - c)^n. \tag{4.24}$$

To determine the coefficients $a_0, a_1, \ldots, a_n$, we generalize the requirements (4.23) and insist that

$$f(c) = T_n(c), \quad f'(c) = T_n'(c), \quad f''(c) = T_n''(c), \quad \ldots, \quad f^{(n)}(c) = T_n^{(n)}(c). \tag{4.25}$$

This ensures that $T_n(x)$ has the same y-value as f at $x = c$, the same tangent line slope as f at $x = c$, the same concavity as f at $x = c$, etc., up to the same nth derivative as f at $x = c$.

To impose the requirements in (4.25), we begin with (4.24). Substituting $x = c$ into that equation yields $T_n(c) = a_0$. Using this in the first equation in (4.25) yields $a_0 = f(c)$. To impose the other requirements we differentiate (4.24):

$$T_n'(x) = a_1 + 2a_2(x - c) + 3a_3(x - c)^2 + 4a_4(x - c)^3 + \cdots + na_n(x - c)^{n-1}$$

$$T_n''(x) = 2a_2 + 3!a_3(x - c) + 4 \cdot 3a_4(x - c)^2 + \cdots + n(n - 1)a_n(x - c)^{n-2}$$

$$T_n'''(x) = 3!a_3 + 4!a_4(x - c) + \cdots + n(n - 1)(n - 2)a_n(x - c)^{n-3}$$

$$\vdots \quad \vdots \vdots$$

$$T_n^{(n)}(x) = [n(n - 1) \cdots 1]a_n = n!a_n$$

---

[6] In equation (4.22), the symbol "$\approx$" means "is approximately equal to."

Substituting $x = c$ into these equations yields

$$T_n'(c) = a_1, \quad T_n''(c) = 2a_2, \quad T_n'''(c) = 3!a_3, \quad \ldots, \quad T_n^{(n)}(c) = n!a_n.$$

Substituting these into (4.25) and solving for the a_i yields

$$a_1 = f'(c), \quad a_2 = \frac{f''(c)}{2!}, \quad a_3 = \frac{f'''(c)}{3!}, \quad \ldots, \quad a_n = \frac{f^{(n)}(c)}{n!}.$$

Finally, substituting these coefficients into (4.24) yields polynomial (4.26) in the theorem below.

> **DEFINITION 4.12: TAYLOR AND MACLAURIN POLYNOMIALS.** Suppose $f(x)$ has n derivatives at $x = c$. Then the nth degree polynomial
>
> $$T_n(x) = f(c) + f'(c)(x - c) + \frac{f''(c)}{2!}(x - c)^2 + \cdots + \frac{f^{(n)}(c)}{n!}(x - c)^n \quad (4.26)$$
>
> is called the nth **Taylor polynomial of f at c**. The $c = 0$ case is called the nth **Maclaurin polynomial of f**, denoted $M_n(x)$:
>
> $$M_n(x) = f(0) + f'(0)x + \frac{f''(0)}{2!}x^2 + \cdots + \frac{f^{(n)}(0)}{n!}x^n. \quad (4.27)$$
>
> In summation form, these become
>
> $$T_n(x) = \sum_{k=0}^{n} \frac{f^{(k)}(c)}{k!}(x - c)^k, \quad M_n(x) = \sum_{k=0}^{n} \frac{f^{(k)}(0)}{k!}x^k. \quad (4.28)$$

EXAMPLE 4.22 Calculate the $n = 1, 2, 3$ Maclaurin polynomials for $f(x) = \dfrac{1}{1 - x}$.

Solution We're asked for Maclaurin polynomials, so $c = 0$ here. And since n goes up to 3, we'll need $f(0), f'(0), f''(0)$, and $f'''(0)$. With $f(x) = (1 - x)^{-1}$, the Chain Rule gives

$$f'(x) = -(1 - x)^{-2}(-1) = (1 - x)^{-2},$$

$$f''(x) = -2(1 - x)^{-3}(-1) = 2(1 - x)^{-3},$$

$$f'''(x) = -3 \cdot 2(1 - x)^{-4}(-1) = 3!(1 - x)^{-4}.$$

Thus,

$$f'(0) = 1, \quad f''(0) = 2, \quad f'''(0) = 3!.$$

Using these and $f(0) = \frac{1}{1-0} = 1$ in (4.27) yields the Maclaurin polynomials

$$M_1(x) = f(0) + f'(0)x = 1 + x,$$

$$M_2(x) = f(0) + f'(0)x + \frac{f''(0)}{2}x^2 = 1 + x + \frac{2}{2!}x^2 = 1 + x + x^2,$$

and

$$M_3(x) = f(0) + f'(0)x + \frac{f''(0)}{2!}x^2 + \frac{f'''(0)}{3!}x^3 = 1 + x + \frac{2}{2!}x^2 + \frac{3!}{3!}x^3$$

$$= 1 + x + x^2 + x^3. \qquad \blacksquare$$

Related Exercises 82(a)–87(a) ((a) only).

Notice from this example that *calculating Taylor and Maclaurin polynomials boils down to calculating the derivatives $f^{(k)}(c)$ for $k = 0, 1, \ldots, n$ and substituting those into (4.27) or (4.28).** *

* The "zero-th derivative" $f^{(0)}(c)$ is just $f(c)$.

EXAMPLE 4.23 (Video)   Calculate the nth Maclaurin polynomial of each function:

(a) $f(x) = e^x$ (b) $f(x) = \sin x$

Solution

(a) We'll need the values of $f^{(k)}(0)$ for $k = 0, \ldots, n$. We start with $f(0) = e^0 = 1$. Then, since $f'(x) = e^x$, $f'(0) = 1$ as well. In fact, since $f^{(k)}(x) = e^x$, then $f^{(k)}(0) = 1$ for each $k = 1, \ldots, n$. Using these results in (4.27) then yields

$$M_n(x) = 1 + x + \frac{x^2}{2!} + \cdots + \frac{x^n}{n!} = \sum_{k=0}^{n} \frac{x^k}{k!}.$$

(b) Here $f(0) = \sin 0 = 0$, and since $f'(x) = \cos x, f''(x) = -\sin x, f'''(x) = -\cos x$, and $f^{(4)}(x) = \sin x$, we see that after every four differentiations we're back to $f(x)$. These results yield $f'(0) = \cos 0 = 1$, $f''(0) = -\sin 0 = 0$, $f'''(0) = -\cos 0 = -1$, and $f^{(4)}(0) = \sin 0 = 0$. These data yield the Maclaurin polynomials

$$M_1(x) = x, \quad M_2(x) = M_1(x), \quad M_3(x) = x - \frac{x^3}{3!}, \quad M_4(x) = M_3(x),$$

since $f''(0) = f^{(4)}(0) = 0$. Continuing the calculations above yields

$$M_{2n+1}(x) = x - \frac{x^3}{3!} + \frac{x^5}{5!} - \cdots + (-1)^n \frac{x^{2n+1}}{(2n+1)!}, \quad M_{2n+2}(x) = M_{2n+1}(x),$$

for each $n \geq 0$. $\blacksquare$

Related Exercises 82(b)–87(b) ((b) only), A10, A11.

APPLICATIONS Applied exercise A10 employs Maclaurin polynomials to derive the **Rule of 70**, a simple formula that approximates how long it takes an interest-bearing account to double in value. Applied exercise A11 reveals how the acceleration of gravity depends on where on Earth you're standing, and uses Maclaurin polynomials to explore that acceleration near the equator.

EXAMPLE 4.24 (Video)   Calculate the nth Taylor polynomial of $f(x) = \ln x$ at $c = 1$.

Solution The first few derivatives of f are

$$f'(x) = \frac{1}{x}, \quad f''(x) = (-1)\frac{1}{x^2}, \quad f'''(x) = (-1)^2\frac{2}{x^3}, \quad f^{(4)}(x) = (-1)^3\frac{3!}{x^4}.$$

Following the pattern, $f^{(k)}(x) = (-1)^{k-1}\frac{(k-1)!}{x^k}$. Therefore, $f^{(k)}(1) = (-1)^{k-1}(k-1)!$. Using this, along with $f(1) = \ln 1 = 0$, in (4.26) yields

$$T_n(x) = (x-1) - \frac{1}{2}(x-1)^2 + \frac{2}{3!}(x-1)^3 - \cdots + (-1)^{n-1}\frac{(n-1)!}{n!}(x-1)^n$$

$$= \sum_{k=1}^{n}\frac{(-1)^{k-1}(k-1)!}{k!}(x-1)^k = \sum_{k=1}^{n}\frac{(-1)^{k-1}}{k}(x-1)^k,$$

using $k! = k(k-1)!$. ∎

Related Exercises 82(c)–87(c) ((c) only).

Figure 4.6 plots the first three Taylor polynomials for the example above in graph (a) and the first three Maclaurin polynomials we calculated for $f(x) = \sin x$ in example 4.23 in graph (b).

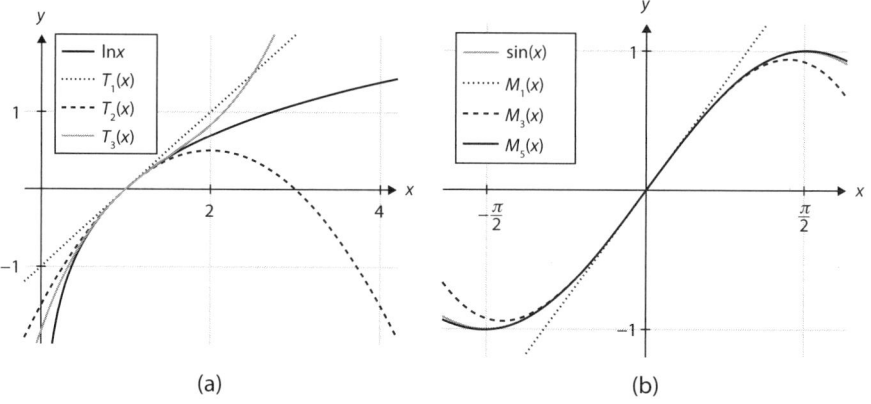

(a)   (b)

⟲**Figure 4.6:** Portions of the graphs of (a) $f(x) = \ln x$ (black) and its first three $c = 1$ Taylor polynomials, and (b) $f(x) = \sin x$ (gray) and its first three distinct Maclaurin polynomials. Interact with this plot—and plot your own Taylor polynomials—at sites.google.com/view/fernandezmath/apps/cs2.

We've also already seen the Maclaurin polynomial for $f(x) = e^x$ calculated in example 4.23(a)—I graphed $M_n(x)$ for $n = 1, 2, 3$ in the third row of figure 1.5 (the gray curves) in chapter 1.* All these plots suggest that the accuracy of the Taylor/Maclaurin polynomial approximations increases as n increases. How do we know for sure? Cue the next section.

* Those Maclaurin polynomials are the ones I referenced in my explanation of the second row in figure 1.5.

4.10 Taylor's Theorem

We can now approximate a sufficiently differentiable function f by polynomials (the Taylor polynomials): $f(x) \approx T_n(x)$. But how accurate is this approximation? Well, the error in that approximation is the difference between the exact value $f(x)$ and the approximated value $T_n(x)$, what we'll call the **error function** $R_n(x)$:

$$R_n(x) = f(x) - T_n(x), \qquad \text{so that} \qquad f(x) = T_n(x) + R_n(x).$$

This is reminiscent of the error R_n incurred in approximating the sum S of an alternating series by its Nth partial sum.* The following theorem might also feel familiar.

* Recall that $R_N = S - S_N$ (4.18).

> **THEOREM 4.17: TAYLOR'S THEOREM.** Let I be an interval containing c, and let f be differentiable up to order $n + 1$ on I. Then, for each x in I, there exists a z between x and c such that
>
> $$f(x) = T_n(x) + \frac{f^{(n+1)}(z)}{(n+1)!}(x - c)^{n+1}. \qquad (4.29)$$

This theorem tells us that

$$R_n(x) = \frac{f^{(n+1)}(z)}{(n+1)!}(x - c)^{n+1}. \qquad (4.30)$$

This is *almost* the next term in the Taylor polynomial $T_n(x)$. (It would be were $z = c$.) Equation (4.29) and the $R_n(x)$ formula above, therefore, are conceptually similar to the alternating series remainder inequality (4.20) and the error bound (4.19), respectively. So, like we did then, we'll be able to use (4.30) to help us determine what n-value—equivalently, what degree Taylor polynomial—we'll need to approximate a function to a desired accuracy. We'll soon see examples of that. But first, let's get comfortable with Taylor's Theorem.

EXAMPLE 4.25   Write out Taylor's Theorem for $f(x) = \sin x$ and calculate the error function in approximating $f(x) = \sin x$ with its third Maclaurin polynomial.

Solution   Let I be an interval containing 0 (the c-value for a Maclaurin polynomial). We're using the third Maclaurin polynomial to approximate f, so $n = 3$. The fourth derivative of $\sin x$ is $f^{(4)}(x) = \sin x$, which is defined for all x, and so also on the interval I. We've therefore met all the hypotheses of Taylor's Theorem. Equation (4.29) then becomes

$$\sin x = M_3(x) + \frac{f^{(4)}(z)}{4!}(x - 0)^4 = x - \frac{x^3}{6} + \frac{\sin z}{4!}x^4, \qquad (4.31)$$

for some z between x and 0, since $M_3(x) = x - \frac{x^3}{6}$ (from example 4.23(b)). The highlighted term above is the error in approximating $\sin x$ with $M_3(x)$.   ∎

Estimating the Error in a Taylor Approximation

Let's now rearrange the Taylor's Theorem equation (4.29) and take absolute values:

$$|f(x) - T_n(x)| = \frac{\left| f^{(n+1)}(z) \right|}{(n+1)!}|x - c|^{n+1}. \qquad (4.32)$$

* In practice we
choose $I = [c, x]$ if
$x \geq c$, and $I = [x, c]$
if $x < c$.

Now, since f is assumed differentiable up to order $n+1$ on I (the hypothesis in Taylor's Theorem), and since differentiability implies continuity (from Calculus 1), if we choose I to be a closed interval containing c* then the Extreme Value Theorem kicks in[7] and tells us that $f^{(n+1)}(x)$ has a maximum and a minimum value on I. We can combine these extrema into the new quantity

$$U_{n+1} = \max \left| f^{(n+1)}(z) \right| \text{ for } z \text{ in } I.$$

Thus the right-hand side in (4.32) is bounded above as follows:

$$|f(x) - T_n(x)| \leq \frac{U_{n+1}|x - c|^{n+1}}{(n+1)!}. \qquad (4.33)$$

(This is the Taylor polynomial analogue of the alternating series remainder inequality (4.20).) Conclusion: Given x and c, if we can find the maximum of $|f^{(n+1)}(z)|$ for z between x and c, then the right-hand side of (4.33) is the maximum error in the approximation $f(x) \approx T_n(x)$.

EXAMPLE 4.26 Calculate the maximum error in approximating $\sin(0.3)$ by $M_3(0.3)$.

Solution We verified the hypotheses of Taylor's Theorem in example 4.25. In our present example, the interval $I = [c, x] = [0, 0.3]$, and once again $n = 3$. Thus,

$$U_4 = \max \left| f^{(4)}(z) \right| = \max |\sin z|,$$

where $0 \leq z \leq 0.3$ (recall that "z is between c and x"). Since $-1 \leq \sin z \leq 1$ for all z, we can safely argue that $|\sin z| \leq 1$ and so choose $U_4 = 1$.[8] Therefore, (4.33) becomes

$$\left| \sin x - M_3(x) \right| \leq \frac{x^4}{4!}.$$

When $x = 0.3$ the right-hand side evaluates to $(0.3)^4/4! = 0.0003375$. Thus, $M_3(0.3) = 0.3 - \frac{(0.3)^3}{3!}$ approximates $\sin(0.3)$ with an error of at most 0.0003375.**

** The actual
magnitude of this
error is about
0.00002.

Related Exercises 88(a), 89(a).

EXAMPLE 4.27 (Video) Use Taylor's Theorem to determine the degree of the Maclaurin polynomial guaranteeing that the error in approximating $\sin 1$ by $M_n(1)$ is at most 0.1.

Solution From (4.33), the error $\left| \sin 1 - M_n(1) \right|$ is at most

$$\frac{U_{n+1}|1 - 0|^{n+1}}{(n+1)!} = \frac{U_{n+1}}{(n+1)!},$$

---

[7]The Extreme Value Theorem (from Calculus 1) says that a continuous function defined on a closed interval attains both a maximum value and a minimum value on that interval.
[8]Since $0 \leq z \leq 0.3$ and $\sin z$ is increasing on that interval, we could also have chosen $U_4 = \sin 0.3$. But that would yield an error function value that involves the quantity we're trying to approximate.

since $x = 1$ here. And because the derivatives of $\sin x$ are $\pm \sin x$ or $\pm \cos x$, and $|\pm \sin x| \leq 1$ and $|\pm \cos x| \leq 1$, we can choose $U_{n+1} = 1$. Thus,

$$|\sin 1 - M_n(1)| \leq \frac{1}{(n+1)!}.$$

To ensure that this error is at most 0.1, we can insist that $\frac{1}{(n+1)!} \leq 0.1$, or $(n+1)! \geq$ 10. Trial and error yields $n + 1 \geq 4$ as the first solution. Thus, $n = 3$ is the lowest degree Maclaurin polynomial that'll do the job.* ■

* The actual error, $|\sin 1 - M_3(1)|$, is about 0.008.

Related Exercises 88(b), 89(b)

We've succeeded in addressing the Approximation Question. (Woohoo!) Let's now apply the Calculus 2 workflow (figure 1.3) to extract the new calculus result— Taylor series—from Taylor polynomials by taking the infinite limit of $T_n(x)$. Using the summation form of $T_n(x)$ (equation 4.28),

$$\lim_{n \to \infty} T_n(x) = \lim_{n \to \infty} \sum_{k=0}^{n} \frac{f^{(k)}(c)}{k!}(x-c)^k = \sum_{k=0}^{\infty} \frac{f^{(k)}(c)}{k!}(x-c)^k. \qquad (4.34)$$

The series we've obtained is a series of *functions*, not *numbers*. More precisely, it's a series of the form $\sum a_n (x-c)^n$, an infinite sum of multiples of *powers* of $x - c$. Such "power series" are new to us. At the same time, (4.34) might remind you of when we expressed an infinite series as the limit of its Nth partial sum (recall (4.8)). After doing that, we studied infinite series in general before discussing special cases (e.g., geometric series). We'll follow this same approach to understanding (4.34) and so focus first on learning about general "power series" and their convergence. Then in section 4.13 we will view (4.34) as a special power series. Our results up to then will help us finish working out the third row of chapter 1's figure 1.5. This will culminate in the incredible result that some functions are *equal* to their "infinite-degree" Taylor polynomials.

4.11 Power Series and Their Convergence

Let's start by formally defining power series.

DEFINITION 4.13: POWER SERIES. Let c be a real number and (a_n) an infinite sequence. A **power series centered at** c is an infinite series of the form

$$\sum_{n=0}^{\infty} a_n(x-c)^n = a_0 + a_1(x-c) + a_2(x-c)^2 + \cdots. \qquad (4.35)$$

Examples:

$$\sum_{n=0}^{\infty} \frac{x^n}{n+1} = 1 + \frac{x}{2} + \frac{x^2}{3} + \cdots \qquad \text{Center: } c = 0; \; a_n = \frac{1}{n+1}$$

$$\sum_{n=0}^{\infty} 3^n(x-2)^n = 1 + 3(x-2) + 3^2(x-2)^2 + \cdots \quad \text{Center: } c = 2; a_n = 3^n$$

We defined an infinite sequence to be an infinite *list* of numbers (definition 4.1). We defined an infinite series to be the *sum* of the infinite sequence's values (definition 4.4). Now (4.35) defines a power series to be an infinite series whose terms are the *products* $a_n(x-c)^n$.

Going the other way, we can reduce a power series to an infinite series by substituting an x-value into the power series. Example:

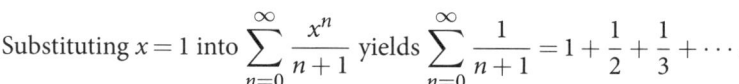

$$\text{Substituting } x = 1 \text{ into } \sum_{n=0}^{\infty} \frac{x^n}{n+1} \text{ yields } \sum_{n=0}^{\infty} \frac{1}{n+1} = 1 + \frac{1}{2} + \frac{1}{3} + \cdots.$$

This is the divergent harmonic series. Therefore, the power series $\sum \frac{x^n}{n+1}$ diverges when $x = 1$. This example hints that we can use what we know about infinite series' convergence to formulate a notion of convergence for power series. Cue the following definition.

DEFINITION 4.14: CONVERGENCE OF POWER SERIES. We say that a power series $\sum_{n=0}^{\infty} a_n(x-c)^n$ **converges at** $x = p$ if the infinite series $\sum_{n=0}^{\infty} a_n(p-c)^n$ converges, and **diverges at** $x = p$ if the same infinite series diverges. The **convergence set** of the power series is the set of x-values for which a power series converges.

Under this definition we'd say that the power series $\sum \frac{x^n}{n+1}$ diverges at $x = 1$. In a minute we'll learn how to find the convergence set for a general power series. But let me point out now that we've already met one power series (though we didn't call it that at the time), complete with its convergence set: the series (4.14):

$$\sum_{n=0}^{\infty} x^n = \frac{1}{1-x}, \quad |x| < 1 \tag{4.36}$$

The convergence set here is $|x| < 1$, or $-1 < x < 1$. Thus, for example, the power series converges at $x = 0.5$ and diverges at $x = 2$.

Now, notice that the convergence set $|x| < 1$ of the power series in (4.36) can be expressed as $|x - \mathbf{0}| < 1$, where I've bolded the 0 because it's the center of the power series. This last inequality has the form $|x - c| < R$, where $c = 0$ and $R = 1$. The next theorem relates the convergence set of a power series to sets of this "$|x - c| < R$" form.

THEOREM 4.18: CONVERGENCE OF A POWER SERIES. For a power series centered at c, only one of the following is true:

1. The series converges only at c.

2. There exists a real number $R > 0$ such that the series converges absolutely for $|x - c| < R$ and diverges for $|x - c| > R$.

3. The series converges absolutely for all x.

In a moment I'll help you visualize the results of this theorem. But given the important role the R-value plays in the theorem, let's introduce a few additional concepts first.

DEFINITION 4.15: RADIUS OF CONVERGENCE, INTERVAL OF CONVERGENCE. The R in theorem 4.18 is called the **radius of convergence** of the power series. When a power series converges only at c, we'll say that $R = 0$. When the power series converges for all x, we'll say that $R = \infty$. When $R > 0$ or $R = \infty$, the convergence set is called the **interval of convergence** of the power series.

Figure 4.7 visualizes this definition and theorem 4.18. Note that there are six possible convergence sets for a power series. Not all of these are intervals—the $R = 0$ convergence set contains only one x-value—but most of them are. Notice also that there are four possibilities when $R > 0$. Which of those four turns out to be the convergence set depends on the power series's convergence (or not) at the endpoints $x = c \pm R$. Per definition 4.14, we determine this by substituting each of those endpoint x-values into the power series and then using a convergence/divergence test on the resulting infinite series. More generally, we use those same convergence/divergence tests to determine whether (and for what x-values) a power series converges, as the following example shows.

Radius of convergence	Convergence set	Number line representation
$R = 0$	$\{c\}$	$\bullet$ at c
$R = \infty$	$(-\infty, \infty)$	line at c
$R > 0$	$|x - c| < R$	$(\ )$ at $c - R$, c, $c + R$
$R > 0$	$|x - c| \le R$	$[\ ]$ at $c - R$, c, $c + R$
$R > 0$	$-R < x - c \le R$	$(\ ]$ at $c - R$, c, $c + R$
$R > 0$	$-R \le x - c < R$	$[\ )$ at $c - R$, c, $c + R$

Figure 4.7: The six possible convergence sets of a power series.

EXAMPLE 4.28 (Video)   Find the radius of convergence of these power series:

(a) $\displaystyle\sum_{n=0}^{\infty} 3(x - 2)^n$ (b) $\displaystyle\sum_{n=0}^{\infty} \frac{x^{2n+1}}{(2n+1)!}$

Solution

(a) Let's apply the ratio test (theorem 4.15), with $a_n = 3(x - 2)^n$:

$$\lim_{n\to\infty} \left| \frac{a_{n+1}}{a_n} \right| = \lim_{n\to\infty} \left| \frac{3(x - 2)^{n+1}}{3(x - 2)^n} \right| = \lim_{n\to\infty} |x - 2| = |x - 2|,$$

since $|x - 2|$ doesn't depend on n. By the ratio test, the power series converges absolutely for $|x - 2| < 1$. Comparing this to $|x - c| < R$ tells us that $R = 1$.

(b) With $a_n = \frac{x^{2n+1}}{(2n+1)!}$, we have

$$a_{n+1} = \frac{x^{2(n+1)+1}}{[2(n+1)+1]!}$$

$$= \frac{x^{2n+3}}{(2n+3)!} = \frac{x^{2n+3}}{(2n+3)(2n+2)(2n+1)!}.$$

Then

$$\lim_{n \to \infty} \left| \frac{a_{n+1}}{a_n} \right| = \lim_{n \to \infty} \left| \frac{x^{2n+3}}{(2n+3)(2n+2)(2n+1)!} \cdot \frac{(2n+1)!}{x^{2n+1}} \right|$$

$$= \lim_{n \to \infty} \left| \frac{x^2}{(2n+2)(2n+3)} \right| = 0.$$

Because this result is less than 1 for all values of x, the ratio test says that the series converges absolutely for all x-values. Therefore, $R = \infty$.   ∎

Related Exercises   90(a)–93(a) ((a) only).

To find the interval of convergence, we first find the radius of convergence and then examine the power series at the endpoints using a convergence test.

EXAMPLE 4.29 (Video)   Find the interval of convergence of $\displaystyle\sum_{n=1}^{\infty} \frac{x^n}{n}$.

Solution Let's first find the radius of convergence. With $a_n = \frac{x^n}{n}$ and $a_{n+1} = \frac{x^{n+1}}{n+1}$,

$$\lim_{n \to \infty} \left| \frac{a_{n+1}}{a_n} \right| = \lim_{n \to \infty} \left| \frac{x^{n+1}}{n+1} \cdot \frac{n}{x^n} \right| = \lim_{n \to \infty} \left| \frac{nx}{n+1} \right| = |x| \lim_{n \to \infty} \frac{n}{n+1} = |x|.$$

By the ratio test, the power series converges absolutely for $|x| < 1$. Thus, $R = 1$ here. The endpoints we need to investigate now are $x = \pm 1$. At $x = -1$, the power series becomes $\sum \frac{(-1)^n}{n}$, the convergent alternating harmonic series. At $x = 1$, the power series becomes $\sum \frac{1}{n}$, the divergent harmonic series. Thus, the interval of convergence is $[-1, 1)$.   ∎

Related Exercises   90(b)–93(b) ((b) only).

4.12 Power Series as Functions

We know that, for each x-value in the convergence set of a power series, substituting that x-value into the power series yields a convergent infinite series. Recall now that an infinite series converges when its *limit* of partial sums exists (definition 4.6). We know from Calculus 1 that when a limit exists it's unique. Putting these two facts together tells us that, *for each input* of an x-value in the convergence set of a power series, *we get a unique output*—the sum of the resulting infinite series. This meets the textbook definition of a function. So, *a convergent power series is a function, with domain equal to its convergence set!* Hereafter I'll denote these special functions with the function notation $S(x)$ and put their domain right next to them, like this:

$$S(x) = \sum_{n=0}^{\infty} a_n (x-c)^n, \quad |x-c| < R.^*$$

* Here I'm illustrating just one of the six possible convergence sets from figure 4.7.

Now, I've chosen the $S(x)$ notation to remind you that if we substitute an x-value in the convergence set of the power series (say, $x = p$) into the power series, we get an infinite series whose sum is $S(p)$. In other words, *the sum of a power series is literally a function of x.* For example, recalling (4.36) (which shows that $\sum x^n = \frac{1}{1-x}$ for $|x| < 1$),

$$S(0.5) = \sum_{n=0}^{\infty} (0.5)^n = \frac{1}{1-0.5} = 2, \quad S(0.1) = \sum_{n=0}^{\infty} (0.1)^n = \frac{1}{1-0.1} = \frac{10}{9}.$$

Put another way, the power series $\sum_{n=0}^{\infty} x^n$ is the function $S(x) = \frac{1}{1-x}$ with domain $|x| < 1$. One *super-cool* ramification: We can *graph the power series $\sum x^n$!* Figure 4.8 shows part of that graph. The takeaway from this discussion: A power series $\sum a_n(x-c)^n$ with convergence set I defines a function $S(x)$ with domain I.

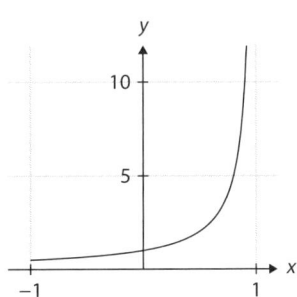

Figure 4.8: $S(x) = \sum_{n=0}^{\infty} x^n$ for $|x| < 1$.

Representing Functions by Power Series

We've just shown that a power series defines a function within its convergence set. It turns out we can also go backward: given a function, we can sometimes represent it by a power series. We call the resulting power series the **power series representation** of the function f. The next example illustrates this.

EXAMPLE 4.30 (Video)   Find a power series representation for each function, and determine its interval of convergence:

(a) $f(x) = \dfrac{1}{1+x^3}$ (b) $g(x) = \dfrac{2x^2}{1+x^3}$

Solution

(a) Start from $\frac{1}{1+x^3} = \frac{1}{1-(-x^3)}$, and then substitute "$-x^3$" for "$x$" everywhere in (4.36):

$$\frac{1}{1-(-x^3)} = \sum_{n=0}^{\infty} (-x^3)^n = \sum_{n=0}^{\infty} (-1)^n x^{3n}, \quad |-x^3| < 1.$$

* The series is
$\sum(-1)^n$ when
$x=1$, and $\sum 1$
when $x=-1$. Both
diverge by the
divergence test.

Now $|-x^3| < 1 \Rightarrow |x|^3 < 1 \Rightarrow |x| < 1$. When $x = \pm 1$, the resulting series diverge.* Thus, the interval of convergence remains $|x| < 1$.

(b) Note that $g(x) = 2x^2 f(x)$, where f is the function from (a). Therefore,

$$g(x) = 2x^2 \sum_{n=0}^{\infty} (-1)^n x^{3n} = \sum_{n=0}^{\infty} 2(-1)^n x^2 \cdot x^{3n} = \sum_{n=0}^{\infty} 2(-1)^n x^{3n+2}.$$

Since $f(x)$ is defined on $|x| < 1$ and $2x^2$ is defined for all x, $g(x)$ is defined on the overlap of the two sets. That overlap is $|x| < 1$, which is then the interval of convergence of g. ∎

Related Exercises 94–95.

Differentiating Power Series

The next theorem lays the foundation for a more general approach to finding a function's power series representation.

> **THEOREM 4.19: DIFFERENTIATION OF POWER SERIES.** Let
>
> $$S(x) = \sum_{n=0}^{\infty} a_n(x-c)^n = a_0 + a_1(x-c) + a_2(x-c)^2 + a_3(x-c)^3 + \cdots$$
>
> be a power series with radius of convergence $R > 0$. Then, inside the corresponding interval of convergence $(c-R, c+R)$, f is differentiable, and
>
> $$S'(x) = \sum_{n=0}^{\infty} na_n(x-c)^{n-1} = a_1 + 2a_2(x-c) + 3a_3(x-c)^2 + \cdots.$$
>
> Moreover, the radius of convergence of this series is also R. Convergence at the endpoints $x = c \pm R$ may not be the same as that of S's.

In a nutshell, this theorem says that for a power series with radius of convergence R, (1) we can differentiate it, and its derivative series, $S'(x)$, also has radius of convergence R; and (2) the differentiation is done term by term.

Sometimes you'll see the $S'(x)$ series written with a starting index of 1, not 0. The two approaches yield the same series because the $n = 0$ term in $S'(x)$ is $0a_0(x-c)^{0-1} = 0$. Similarly, sometimes you'll also see the series differentiation expressed in Leibniz notation:

$$S'(x) = \frac{d}{dx}S(x) = \frac{d}{dx}\left\{\sum_{n=0}^{\infty} a_n(x-c)^n\right\} = \sum_{n=0}^{\infty}\left\{\frac{d}{dx}[a_n(x-c)^n]\right\}$$

$$= \sum_{n=1}^{\infty} na_n(x-c)^{n-1}.$$

The move from before the highlighted equals sign to after it can be described this way: The derivative of the infinite sum of $a_n(x-c)^n$ is the infinite sum of the derivatives of $a_n(x-c)^n$. This is a useful new take on theorem 4.19: *For a convergent*

power series, the derivative operator d/dx can be interchanged with the infinite sum operator $\sum$.

Let's now see how to use theorem 4.19 to find power series representations of functions.

EXAMPLE 4.31   The ratio test shows that $S(x) = \sum_{n=0}^{\infty} \frac{x^n}{n!}$ converges absolutely for all x. Calculate the derivative of this series and determine its convergence set.

Solution   According to theorem 4.19,

$$S'(x) = \sum_{n=0}^{\infty} \left[\frac{d}{dx} \left(\frac{x^n}{n!} \right) \right]$$

$$= \sum_{n=0}^{\infty} \frac{nx^{n-1}}{n!}$$

$$= \frac{0 \cdot x^{0-1}}{0!} + \sum_{n=1}^{\infty} \frac{nx^{n-1}}{n!}$$

$$= \sum_{n=1}^{\infty} \frac{x^{n-1}}{(n-1)!}. \qquad \text{Using } n! = n(n-1)!$$

Since $R = \infty$ in the original series, this new series also has $R = \infty$. Therefore, the new series converges absolutely for all x.   ∎

Related Exercises   96–97.

The series $S'(x)$ we just calculated is hiding a secret. To discover it, let's write out the series and compare it to the original series in the example's prompt:

$$S'(x) = \sum_{n=1}^{\infty} \frac{x^{n-1}}{(n-1)!} = 1 + x + \frac{x^2}{2!} + \frac{x^3}{3!} + \cdots$$

$$S(x) = \sum_{n=0}^{\infty} \frac{x^n}{n!} = 1 + x + \frac{x^2}{2!} + \frac{x^3}{3!} + \cdots$$

They're the same! Conclusion: $S'(x) = S(x)$; that is, the derivative of the function is itself.[9] Note also that $S(0) = 1$. Dig back into your Calculus 1 knowledge and answer the following question: Can you think of (the only) function f whose derivative is itself (i.e., $f'(x) = f(x)$) and $f(0) = 1$? Yup: e^x. The surprising conclusion:

$$e^x = \sum_{n=0}^{\infty} \frac{x^n}{n!}, \quad \text{valid for all } x: (-\infty, \infty). \tag{4.37}$$

---

[9] We can also transform the sum in $S'(x)$ into the one in $S(x)$ to see this directly. To do so, let $m = n - 1$. When $n = 1, m = 0$, and when $n = \infty, m = \infty$. Thus, $S'(x) = \sum_{n=1}^{\infty} \frac{x^{n-1}}{(n-1)!} = \sum_{m=0}^{\infty} \frac{x^m}{m!} = S(x)$, since the particular letter we use for the index doesn't matter.

Additional Tips, Tricks, and Takeaways

Equation (4.37)—and the route we took to derive it—illustrates how theorem 4.19 can be used to generate power series representations of functions. Over the next two sections we'll study an even more general approach.

Before then, let me use (4.37) to illustrate the interplay between the Infinite Sum Question and the Approximation Question we discussed at the end of section 4.7. For example, say we'd like to sum the infinite series

$$1 + 1 + \frac{1}{2!} + \frac{1}{3!} + \frac{1}{4!} + \cdots,$$

an Infinite Sum Question question. We can now recognize this as the power series in (4.37) with $x = 1$ substituted in. And since $x = 1$ is in the convergence set of that power series, and at $x = 1$ the power series equals e^1, we conclude that

$$1 + 1 + \frac{1}{2!} + \frac{1}{3!} + \frac{1}{4!} + \cdots = e.$$

We've just summed an infinite series. And it wasn't one we knew how to sum before now. But I'll be honest, we got lucky. The methods we're using aren't general enough to connect a power series to a function we know. For that we need to return to Taylor polynomials. Cue the next section.

4.13   Taylor Series

In (4.34) in section 4.10 we took the limit as $n \to \infty$ of a Taylor polynomial $T_n(x)$. Flip back there now and you'll be able to see the result in a new light—you'll see a power series with specific coefficients a_n. That special power series is aptly called a *Taylor series*.

> **DEFINITION 4.16: TAYLOR SERIES.** Let f be a function with derivatives of all orders at $x = c$. Then
>
> $$\sum_{n=0}^{\infty} \frac{f^{(n)}(c)}{n!}(x-c)^n = f(c) + f'(c)(x-c) + \cdots + \frac{f^{(k)}(c)}{k!}(x-c)^k + \cdots$$
>
> $$\text{(4.38)}$$
>
> is called the **Taylor series of** $f(x)$ **at** $x = c$, denoted $T.S.(f)$. When $c = 0$ the series is called the **Maclaurin series of** f, denoted $M.S.(f)$.

The connection with the Taylor polynomials $T_N(x)$ here is worth emphasizing again:

$$T.S.(f) = \sum_{n=0}^{\infty} \frac{f^{(n)}(c)}{n!}(x-c)^n = \lim_{N \to \infty} \sum_{n=0}^{N} \frac{f^{(n)}(c)}{n!}(x-c)^n = \lim_{N \to \infty} T_N(x).$$

This shows that the Taylor *series* of f is the limit of the Taylor *polynomial* of f as that polynomial's degree tends to infinity. (Loosely speaking, the Taylor series is the "infinite degree" Taylor polynomial.) It also shows that the Nth Taylor polynomial $T_N(x)$ of f is the Nth partial sum of the Taylor series f.

EXAMPLE 4.32 Verify the Maclaurin series and interval of convergence.

$$M.S.(e^x) = \sum_{n=0}^{\infty} \frac{x^n}{n!}, \qquad -\infty < x < \infty.$$

Solution We calculated the Nth Maclaurin polynomial of e^x in example 4.23(a):

$$M_N(x) = \sum_{n=0}^{N} \frac{x^n}{n!}.$$

From (4.38), the associated Maclaurin series is just the $N = \infty$ version of this, yielding the Maclaurin series in the prompt. To calculate the interval of convergence, we use the ratio test. The Maclaurin series terms are $a_n = x^n/n!$, so

$$\lim_{n \to \infty} \left| \frac{a_{n+1}}{a_n} \right| = \lim_{n \to \infty} \left| \frac{x^{n+1}}{(n+1)!} \cdot \frac{n!}{x^n} \right|$$

$$= \lim_{n \to \infty} \left| \frac{x^{n+1}}{x^n} \cdot \frac{n!}{(n+1)n!} \right| \qquad \text{Using } (n+1)! = (n+1)n!$$

$$= \lim_{n \to \infty} \left| \frac{x}{n+1} \right| = 0, \tag{4.39}$$

regardless of the x-value. By the ratio test, the Maclaurin series converges absolutely for every x-value (interval of convergence $(-\infty, \infty)$). ∎

EXAMPLE 4.33 (Video) Verify the Maclaurin series and interval of convergence below:

$$M.S.(\sin x) = \sum_{n=0}^{\infty} \frac{(-1)^n x^{2n+1}}{(2n+1)!}, \qquad -\infty < x < \infty.$$

Solution We calculated the Nth Maclaurin polynomial of $\sin x$ in example 4.23(b):

$$M_{2N+1}(x) = \sum_{n=0}^{N} \frac{(-1)^n x^{2n+1}}{(2n+1)!}.$$

From (4.38), the associated Maclaurin series is just the $N = \infty$ version of this, yielding the Maclaurin series in the prompt. To calculate the interval of convergence, we use the ratio test again. The Maclaurin series terms are $a_n = \frac{(-1)^n x^{2n+1}}{(2n+1)!}$, so

$$\left| \frac{a_{n+1}}{a_n} \right| = \left| \frac{(-1)^{n+1} x^{2(n+1)+1}}{(2(n+1)+1)!} \cdot \frac{(2n+1)!}{(-1)^n x^{2n+1}} \right|$$

$$= \left| \frac{(2n+1)!}{(2n+3)!} \cdot \frac{x^{2n+3}}{x^{2n+1}} \right| \qquad \text{Since } \left| \frac{(-1)^{n+1}}{(-1)^n} \right| = 1$$

$$= \frac{(2n+1)!|x^2|}{(2n+3)(2n+2)(2n+1)!}$$

$$= \frac{x^2}{(2n+3)(2n+2)},$$

since $|x^2| = x^2$ (because $x^2 \geq 0$). It follows that

$$\lim_{n\to\infty} \left| \frac{a_{n+1}}{a_n} \right| = \lim_{n\to\infty} \frac{x^2}{(2n+3)(2n+2)} = x^2 \left[\lim_{n\to\infty} \frac{1}{(2n+3)(2n+2)} \right] = 0,$$

regardless of the x-value. By the ratio test, the Maclaurin series converges absolutely for every x-value, making its interval of convergence $(-\infty, \infty)$. ∎

The table in figure 4.9 lists the most common Maclaurin series along with their intervals of convergence. In the next section we'll verify the last column of the table—that on their convergence sets all of the listed Maclaurin series converge to the functions they came from.

Related Exercises 98–101.

Function $f(x)$	Maclaurin series	Interval of convergence I	Converges to $f(x)$ on I?
e^x	$\displaystyle\sum_{n=0}^{\infty} \frac{x^n}{n!} = 1 + x + \frac{x^2}{2!} + \frac{x^3}{3!} + \frac{x^4}{4!} + \cdots$	$(-\infty, \infty)$	Yes
$\sin x$	$\displaystyle\sum_{n=0}^{\infty} \frac{(-1)^n x^{2n+1}}{(2n+1)!} = x - \frac{x^3}{3!} + \frac{x^5}{5!} - \frac{x^7}{7!} + \cdots$	$(-\infty, \infty)$	Yes
$\cos x$	$\displaystyle\sum_{n=0}^{\infty} \frac{(-1)^n x^{2n}}{(2n)!} = 1 - \frac{x^2}{2!} + \frac{x^4}{4!} - \frac{x^6}{6!} + \cdots$	$(-\infty, \infty)$	Yes
$\dfrac{1}{1-x}$	$\displaystyle\sum_{n=0}^{\infty} x^n = 1 + x + x^2 + x^3 + x^4 + \cdots$	$-1 < x < 1$	Yes
$\dfrac{1}{1+x}$	$\displaystyle\sum_{n=0}^{\infty} (-1)^n x^n = 1 - x + x^2 - x^3 + x^4 - \cdots$	$-1 < x < 1$	Yes
$\ln(1+x)$	$\displaystyle\sum_{n=1}^{\infty} \frac{(-1)^{n+1} x^n}{n} = x - \frac{x^2}{2} + \frac{x^3}{3} - \frac{x^4}{4} + \cdots$	$-1 < x \leq 1$	Yes
$\arctan(x)$	$\displaystyle\sum_{n=0}^{\infty} \frac{(-1)^n x^{2n+1}}{2n+1} = x - \frac{x^3}{3} + \frac{x^5}{5} - \frac{x^7}{7} + \cdots$	$-1 \leq x \leq 1$	Yes

Figure 4.9: A table of common Maclaurin series and their intervals of convergence. On that interval, each of the listed series converges to the function used to calculate the series. (We'll discuss this in the next section.)

One final thing to know is that we don't always have to use the definition of a Taylor series to derive the Taylor series. Often we can use known Maclaurin/Taylor series and manipulate them instead.[10] Example: To find $M.S.(e^{2x})$, we could return to the result of example 4.32 and replace x everywhere in the Maclaurin series of e^x by $2x$:

$$M.S.(e^{2x}) = \sum_{n=0}^{\infty} \frac{(2x)^n}{n!}, \qquad -\infty < 2x < \infty$$

Dividing $-\infty < 2x < \infty$ by 2 yields the new interval of convergence: $-\infty < x < \infty$. Simplifying the series yields

$$M.S.(e^{2x}) = \sum_{n=0}^{\infty} \frac{2^n x^n}{n!}, \qquad -\infty < x < \infty.$$

The next theorem generalizes these types of manipulations to generate new Maclaurin series from known ones.

THEOREM 4.20: Let $R > 0$, and suppose that $f(x) = \sum_{n=0}^{\infty} a_n x^n$ and $g(x) = \sum_{n=0}^{\infty} b_n x^n$ converge absolutely for $|x| < R$. Then:

1. $f(x) \pm g(x) = \sum_{n=0}^{\infty} (a_n \pm b_n) x^n$, and this series converges absolutely for $|x| < R$.

2. $f(x)g(x) = \left(\sum_{n=0}^{\infty} a_n x^n \right) \left(\sum_{n=0}^{\infty} b_n x^n \right)$, and this series converges absolutely for $|x| < R$.

3. If $h(x)$ is a polynomial, then $f(h(x)) = \sum_{n=0}^{\infty} a_n [h(x)]^n$, and this series converges absolutely for $|h(x)| < R$.

The $M.S.(e^{2x})$ example we just did illustrates part 3 of this theorem. Let's now see an illustration of part 2.

EXAMPLE 4.34 Find the first two terms in the Maclaurin series for $e^x \sin x$.

Solution Let's first substitute in the Maclaurin series for e^x from figure 4.9:

$$e^x \sin x = \left(1 + x + \frac{x^2}{2!} + \cdots \right) \sin x.$$

Distributing the $\sin x$ now yields

$$1 \cdot \sin x + x \cdot \sin x + \frac{x^2}{2!} \cdot \sin x + \cdots.$$

---

[10]This process parallels what we did in example 4.30 to find new power series representations from known ones.

Substituting in now the Maclaurin series for $\sin x$ from figure 4.9 yields

$$1 \cdot \left(x - \frac{x^3}{3!} + \frac{x^5}{5!} - \cdots \right) + x \cdot \left(x - \frac{x^3}{3!} + \frac{x^5}{5!} - \cdots \right) + \cdots .$$

Distributing the "x" in front of the second set of parentheses yields

$$\left(x - \frac{x^3}{3!} + \frac{x^5}{5!} - \cdots \right) + \left(x^2 - \frac{x^4}{3!} + \frac{x^6}{5!} - \cdots \right) + \cdots ,$$

whose first two terms are $x + x^2$.

Related Exercises   102–104.

4.14 Convergence of Taylor Series

As promised, let's now work out the last column in figure 4.9. That is, let's figure out how to determine whether a Taylor/Maclaurin series equals (converges to) the function used to calculate it. We've already done that in two instances: equation (4.37) (which shows that e^x is equal to its Maclaurin series for all x) and equation (4.36) (which shows that $\frac{1}{1-x}$ is equal to its Maclaurin series for $|x| < 1$). To begin, recall from section 4.10 that

$$f(x) = T_n(x) + R_n(x),$$

where $T_n(x)$ is the nth Taylor polynomial of $f(x)$ at $x = c$ and $R_n(x)$ is the error function. Taking the infinite limit of this equation yields

$$f(x) = \lim_{n \to \infty} T_n(x) + \lim_{n \to \infty} R_n(x),$$

provided each limit exists. The first limit yields the Taylor series of f (from the previous section), so

$$f(x) = T.S.(f) + \lim_{n \to \infty} R_n(x).$$

Were the second limit zero, we'd be left with $f(x) = T.S.(f)$, which would tell us that the Taylor series of f converges to f. This is exactly what the following theorem says.

THEOREM 4.21: If $\lim_{n \to \infty} R_n(x) = 0$ for all x in an interval I, then the Taylor series of $f(x)$ at $x = c$ converges to $f(x)$, and

$$f(x) = \sum_{n=0}^{\infty} \frac{f^{(n)}(c)}{n!} (x - c)^n \quad \text{for all } x \text{ in } I.$$

Great! But how do we show $R_n(x) \to 0$ in practice? To do that we return to Taylor's Theorem. Recall from (4.33) that

$$|R_n(x)| = |f(x) - T_n(x)| \le \frac{U_{n+1} |x - c|^{n+1}}{(n+1)!}, \tag{4.40}$$

where we recall that here $U_{n+1} = \max |f^{(n+1)}(z)|$ for z in I, a closed interval containing c. If we can show that the right-hand side of the inequality above tends to zero as $n \to \infty$, then it'll follow that $|R_n(x)| \leq 0$, which implies that $R_n(x) = 0$, and theorem 4.21 will kick in. Let me summarize this for easy reference later.

Box 4.3: How to Show That a Taylor Series Converges to Its Function

To show the Taylor series of a function $f(x)$ converges to $f(x)$ on an interval I:

1. Calculate U_{n+1} (the maximum of $|f^{(n+1)}(z)|$ on the interval I).

2. Show that $\displaystyle\lim_{n \to \infty} \frac{U_{n+1}|x - c|^{n+1}}{(n+1)!} = 0$ for all x-values in I.

EXAMPLE 4.35 (Video)   Show that the Maclaurin series for e^x and $\sin x$ converge to e^x and $\sin x$, respectively, for all x.

Solution   Let's follow the procedure in box 4.3. To start, let $I = [-a, a]$, where $a \neq 0$, and let's first consider $f(x) = e^x$.

- *Step 1: Calculate U_{n+1}.* Since $f'(x) = e^x$, $f''(x) = e^x$, ..., and $f^{(n+1)}(x) = e^x$, then $|f^{(n+1)}(z)| = |e^z| = e^z$, since $e^x \geq 0$. And since $f(x) = e^x$ is an increasing function, $e^z \leq e^a$ on the interval I we've chosen. Therefore, $U_{n+1} = e^a$.

- *Step 2: Evaluate the limit in box 4.3.* In this case, that limit is

$$\lim_{n \to \infty} \frac{e^a |x - 0|^{n+1}}{(n+1)!} = e^a \lim_{n \to \infty} \frac{|x|^{n+1}}{(n+1)!}.$$

For a given x-value the numerator is an exponential function of n. But since factorials grow faster than exponentials,* the limit is zero. The a-value didn't matter, so the conclusion is that the Maclaurin series of $f(x) = e^x$ converges to e^x for all x-values.

> * Recall Growth Order Theorem 4.1.

Next let's consider $f(x) = \sin x$.

- *Step 1: Calculate U_{n+1}.* Since the derivatives of $\sin x$ are $\pm \sin x$ or $\pm \cos x$, and because $|\pm \sin x| \leq 1$ and $|\pm \cos x| \leq 1$, then $|f^{(n+1)}(z)| \leq 1$. Therefore, $U_{n+1} = 1$.

- *Step 2: Evaluate the limit in box 4.3.* In this case, that limit is

$$\lim_{n \to \infty} \frac{|x - 0|^{n+1}}{(n+1)!} = \lim_{n \to \infty} \frac{|x|^{n+1}}{(n+1)!}.$$

Because factorials grow faster than exponentials (theorem 4.1), we once again conclude that the limit yields zero, regardless of the x-value.   ∎

Related Exercises   105–106.

Let me pause the teaching for a minute to celebrate how far we've come. In 14 sections, we've studied sequences, series, and now how to represent functions as infinite-degree polynomials. And we've also succeeded in answering most of the last two Big Questions from chapter 1 in precise and deep ways. (I'm proud of you for working through all that!) In section 4.7 I promised that a new way to sum infinite series would emerge from our Approximation Question work. The next section delivers on that promise.

4.15   Applications of Taylor Series

First up, let's work out a new Maclaurin series that helps approximate quantities of the form $(1 + x)^m$ (this helps, as we'll see, approximate things like $\sqrt{1.05}$).

The Binomial Series

The **binomial series** is the power series representation of the function $f(x) = (1 + x)^m$, where $m \neq 0$. To derive it, let's make use of the first few derivatives of f:

$$f'(x) = m(1 + x)^{m-1}$$
$$f''(x) = m(m - 1)(1 + x)^{m-2}$$
$$f'''(x) = m(m - 1)(m - 2)(1 + x)^{m-3}$$

Thus, $f'(0) = m$, $f''(0) = m(m - 1)$, and $f'''(0) = m(m - 1)(m - 2)$, so that

$$M.S.[(1 + x)^m] = 1 + mx + \frac{m(m - 1)x^2}{2!} + \frac{m(m - 1)(m - 2)x^3}{3!} + \cdots .$$

In more detail:

$$M.S.[(1 + x)^m] = 1 + mx + \frac{m(m - 1)x^2}{2} + \cdots +$$
$$\frac{m(m - 1) \cdots (m - (n - 1))x^n}{n!} + \cdots$$
$$= \sum_{n=0}^{\infty} \binom{m}{n} x^n, \qquad (4.41)$$

where $\binom{m}{n}$ is the **binomial coefficient**, defined as

$$\binom{m}{n} = \frac{m(m - 1) \cdots (m - (n - 1))}{n!} .$$

Let's now find the radius of convergence of this series.

EXAMPLE 4.36 (Video)   Show that the binomial series (4.41) converges absolutely for $|x| < 1$ (i.e., it has radius of convergence $R = 1$).

Solution   Let's first note that if m is a positive integer k (1, 2, 3, etc.), then $f^{(k+1)}(x) = 0$, and so the binomial series is a finite-degree polynomial.[11] There's no

---

[11] Example: For $f(x) = (1 + x)^2$, every higher-order derivative above $f''(x)$ is zero.

infinite series there, so there's no convergence question to answer. So let's therefore assume that $m \neq 0$, and also not a positive integer, and employ the ratio test. With $a_n = \frac{m(m-1)\cdots(m-(n-1))x^n}{n!}$, we have

$$\left|\frac{a_{n+1}}{a_n}\right| = \left|\frac{m(m-1)\cdots(m-([n+1]-1))x^{n+1}}{(n+1)!} \cdot \frac{n!}{m(m-1)\cdots(m-(n-1))x^n}\right|$$

$$= \left|\frac{x^{n+1}}{x^n}\right| \cdot \frac{n!}{(n+1)!} \cdot \left|\frac{m(m-1)\cdots(m-n)}{m(m-1)\cdots(m-(n-1))}\right|$$

$$= |x| \cdot \frac{n!}{(n+1)n!} \cdot |m-n|,$$

since $m(m-1)\cdots(m-n) = [m(m-1)\cdots(m-(n-1))](m-n)$.[12] Thus,

$$\lim_{n\to\infty}\left|\frac{a_{n+1}}{a_n}\right| = \lim_{n\to\infty}\frac{n!|m-n||x|}{(n+1)n!} = |x|\lim_{n\to\infty}\frac{|m-n|}{n+1} = |x|\lim_{n\to\infty}\frac{n-m}{n+1} = |x|.^{\star}$$

$\star$ When $n \to \infty$, n eventually becomes larger than m, so $|m-n| = n - m$.

By the ratio test, the series converges absolutely when $|x| < 1$. ∎

By using the methods from the previous section, we can show that the binomial series converges to $(1+x)^m$ for $|x| < 1$, so that

$$(1+x)^m = \sum_{n=0}^{\infty}\binom{m}{n}x^n, \quad |x| < 1. \tag{4.42}$$

(For some m-values the binomial series also converges at $x = \pm 1$.)

One particularly useful implication of this is

$$(1+x)^m \approx 1 + mx, \quad \text{for } |x| \ll 1 \text{ (``much less than 1'').} \tag{4.43}$$

This approximation is used throughout the sciences and engineering; the applied exercises suggested below showcase some of those instances.

EXAMPLE 4.37 Use the binomial series to express $\sqrt{1.05}$ as an infinite series. How close of an approximation do you obtain if only the first three terms in the series are used? (For comparison, $\sqrt{1.05} = 1.02470\ldots$.)

Solution Since $\sqrt{1.05} = (1+x)^m$ for $x = 0.05$ and $m = 1/2$, substituting these into (4.42) yields

$$(1.05)^{1/2} = 1 + \frac{1}{2}(0.05) + \frac{\frac{1}{2}\left(\frac{1}{2}-1\right)}{2}(0.05)^2 + \cdots.$$

Keeping only these first three terms yields the approximation $\sqrt{1.05} \approx 1.0246875$, which is accurate to three decimal places. ∎

Related Exercises   107–108, A12–A13.

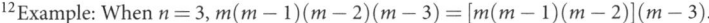

---

[12]Example: When $n = 3$, $m(m-1)(m-2)(m-3) = [m(m-1)(m-2)](m-3)$.

APPLICATIONS Applied exercise A12 explores Einstein's theory of time dilation, which shows that the passage of time is relative. Applied exercise A13 studies Earth's elliptical orbit around the Sun and how closely circular it is.

Summing Infinite Series via Taylor Series

Let's finally return to the second half of chapter 1's Infinite Sum Question (calculating the sum of an infinite series). To use Maclaurin/Taylor series to sum infinite series, we need to recognize the infinite series as the result of substituting a particular x-value into a known Maclaurin/Taylor series. (We did something similar to this at the end of section 4.12.) Here's an example of that.

EXAMPLE 4.38 (Video)   Use an appropriate Taylor series to find the sum of these series:

(a) $1 - \dfrac{\pi^2}{2!} + \dfrac{\pi^4}{4!} - \dfrac{\pi^6}{6!} + \cdots$ (b) $1 - \dfrac{1}{2} + \dfrac{1}{3} - \dfrac{1}{4} + \dfrac{1}{5} - \cdots$

Solution

(a) From the third row in figure 4.9:

$$\cos \pi = 1 - \frac{\pi^2}{2!} + \frac{\pi^4}{4!} - \frac{\pi^6}{6!} + \cdots .$$

This is valid since the Maclaurin series of $\cos x$ converges to $\cos x$ for all x. Therefore, the series in the prompt converges to $\cos \pi = -1$.

(b) From the sixth row in figure 4.9:

$$\ln(1+1) = 1 - \frac{1}{2} + \frac{1}{3} - \frac{1}{4} + \frac{1}{5} - \cdots .$$

This is valid since the Maclaurin series of $\ln(1+x)$ converges to $\ln(1+x)$ at $x = 1$. Therefore, the series in the prompt converges to $\ln(1+1) = \ln 2$. By the way, the series in (b) is the alternating harmonic series. We've just shown that its sum is $\ln 2$. Cool!   ■

Related Exercises   109–111.

Successfully using Maclaurin/Taylor series to sum an infinite series requires (a) identifying the infinite series as a known Taylor series evaluated at a certain x-value and (b) confirming that the x-value is in the Taylor series's interval of convergence. The second task is fairly straightforward. One tip for the first task is to look at the features of the infinite series: Does it alternate? Does it have factorials (or not)? This can help rule out—or help suggest—particular Maclaurin/Taylor series to consider.

Using Taylor Series to Evaluate Limits

As a bonus, let's illustrate how Taylor series can help us evaluate limits. We'll do so by evaluating two limits familiar from Calculus 1 that took a lot of work to compute back then. As we'll see now, these limits can be evaluated much faster with the help of Taylor series.

EXAMPLE 4.39 In Calculus 1 you learned that

$$\lim_{x \to 0} \frac{\sin x}{x} = 1, \qquad \lim_{x \to 0} \frac{1 - \cos x}{x} = 0.$$

Rederive these using appropriate Maclaurin series.

Solution For the first limit, we take the Maclaurin series for $\sin x$ and divide it by x:

$$\frac{\sin x}{x} = \frac{x - \frac{x^3}{3!} + \frac{x^5}{5!} - \cdots}{x} = 1 - \frac{x^2}{3!} + \frac{x^4}{5!} - \cdots.$$

(This expansion is valid for $x \neq 0$.) It then follows that

$$\lim_{x \to 0} \frac{\sin x}{x} = \lim_{x \to 0} \left[1 - \frac{x^2}{3!} + \frac{x^4}{5!} - \cdots \right] = 1.$$

For the second limit, we substitute in the Maclaurin series for $\cos x$:

$$\frac{1 - \cos x}{x} = \frac{1 - \left(1 - \frac{x^2}{2!} + \frac{x^4}{4!} - \cdots \right)}{x} = \frac{1 - 1 + \frac{x^2}{2!} - \frac{x^4}{4!} + \cdots}{x}$$

$$= \frac{x}{2!} - \frac{x^3}{4!} + \frac{x^5}{6!} - \cdots.$$

(This expansion is valid for $x \neq 0$.) It follows that

$$\lim_{x \to 0} \frac{1 - \cos x}{x} = \lim_{x \to 0} \left[\frac{x}{2!} - \frac{x^3}{4!} + \frac{x^5}{6!} - \cdots \right] = 0. \qquad \blacksquare$$

Related Exercises 112–113.

The uses and applications of Taylor series don't stop at the three studied in this lesson. Section 5.2 will revisit Taylor series and use them to help approximate definite integrals.

4.16 Parting Thoughts

We've now answered all of the Infinite Sum Question as well as the Approximation Question. Midway through this chapter we discovered a nice connection between these two seemingly unconnected questions: We suspected that by learning how to approximate functions we'd be able to calculate the sums of infinite series. We've now shown how to do that via Taylor series, and in addition we've learned how they can help evaluate limits and approximate numbers difficult to compute by hand (e.g., $\sqrt{1.05}$). The success we've had in connecting those two Big Questions by thinking in terms of approximations shouldn't be surprising, given the role that approximations of calculus results by finite quantities played in our earliest attempts to make sense of Calculus 2's Big Questions (recall figure 1.5 and the Calculus 2 workflow, figure 1.3).

If you're heading to chapter 2 next, you'll find therein a detailed discussion of how to approximate and evaluate definite integrals, all of which will come in handy in chapter 3, which tackles the Geometry Question from chapter 1. If you're moving on to chapter 5 next, then you'll experience a continuation of this chapter's theme of interconnection. Chapter 5 focuses on the interrelationships between integration, volumes of revolution, and this chapter's content. That chapter is intended to be both an opportunity to review what you've learned and a capstone chapter that'll help you make new connections between those concepts. See you in the next chapter (whichever one that is)!

Chapter 4   Exercises

1–6: Find a formula for the nth term in the sequence, and give the starting index k.

1. $(\cos(\pi), \cos(2\pi), \cos(3\pi), \ldots)$

2. $(1, 5, 9, \ldots)$

3. $\left(\left(\frac{2}{3}\right)^2, \left(\frac{3}{4}\right)^3, \left(\frac{4}{5}\right)^4, \ldots \right)$

4. $\left(-\frac{1}{2}, \frac{1}{4}, -\frac{1}{8}, \ldots \right)$

5. $(2e^2, 2e^3, 2e^4, \ldots)$

6. $\left(-\frac{4}{3}, \frac{9}{4}, -\frac{16}{5}, \ldots \right)$

7–8: Write out the second and third terms of the recursively defined sequence.

7. $a_1 = 2, \; a_{n+1} = (a_n)^2$

8. $a_1 = 10, \; a_{n+1} = na_n - 1$

9–12: Simplify the factorial expression.

9. $\dfrac{(3!)(4!)}{6!}$

10. $\dfrac{n!}{(n+1)!}$

11. $\dfrac{(n-1)!}{(n+1)!}$

12. $\dfrac{(2n+2)!}{(2n)!}$

13–15: Match each graph below with the correct sequence.

(a)

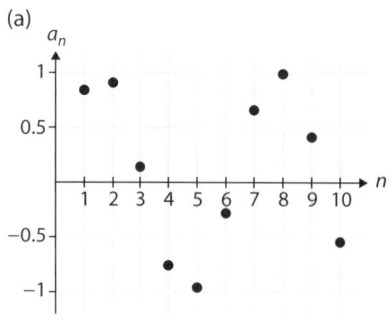

(b)

(c)

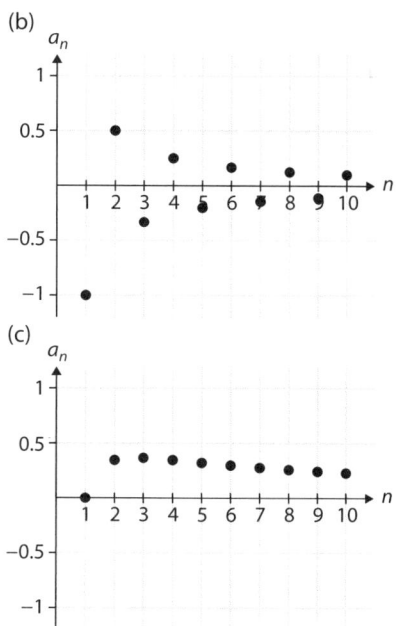

13. $a_n = \dfrac{\ln n}{n}$ **14.** $a_n = \dfrac{(-1)^n}{n}$

15. $a_n = \sin n$

16–33: Determine whether the sequence with the given nth term converges (and if so, find its limit) or diverges.

16. $a_n = 3^{-n}$ **17.** $a_n = \dfrac{2n}{2n+3}$

18. $a_n = \dfrac{2^n}{3^{n+1}}$ **19.** $a_n = \dfrac{n^{100}}{e^n}$

20. $a_n = (-1)^n \left(\dfrac{n+2}{n+1} \right)$

21. $a_n = \dfrac{\pi n^2}{(n+7)^2}$

22. $a_n = \ln(2n+4) - \ln n$

23. $a_n = \dfrac{n^2}{\sqrt{4n^4 - 3n}}$

24. $a_n = \dfrac{2n}{\ln(n^2)}$

25. $a_n = \dfrac{\sqrt[4]{n+1}}{\sqrt[3]{n+2}}$

26. $a_n = \dfrac{n!}{(n+1)!}$

27. $a_n = \dfrac{(-1)^n}{10\sqrt{n+2}}$

28. $a_n = \dfrac{(-2)^n}{n!}$

29. $a_n = \dfrac{e^n}{e^n + 2}$

30. $a_n = \sin n + \dfrac{1}{n}$

31. $a_n = \dfrac{n}{e^n} + e^{-n}$

32. $a_n = \dfrac{(-1)^n}{\ln n} + 10$

33. $a_n = e^{1 + \frac{2}{n}}$

34–35: Calculate the first three partial sums of the series.

34. $\displaystyle\sum_{n=1}^{\infty} 2n^2$ **35.** $\displaystyle\sum_{n=2}^{\infty} \dfrac{n}{n^2 + 1}$

36–37: The Nth partial sum of a series is given. Does the series converge (if so, calculate its sum) or diverge?

36. $S_N = \dfrac{N^2 + N}{2N + 3}$

37. $S_N = \ln(3N) - \ln(N + 5)$

38–43: Determine if the divergence test applies. If so, use it to show the series diverges.

38. $\displaystyle\sum_{n=0}^{\infty} \dfrac{1}{(n+7)^2}$ **39.** $\displaystyle\sum_{n=1}^{\infty} \dfrac{n!}{2^n}$

40. $\displaystyle\sum_{n=1}^{\infty} \dfrac{n}{3n+2}$ **41.** $\displaystyle\sum_{n=1}^{\infty} \dfrac{(-1)^n n^2}{n^3 + 1}$

42. $\displaystyle\sum_{n=0}^{\infty} \left(\dfrac{4}{5}\right)^n$ **43.** $\displaystyle\sum_{n=1}^{\infty} \dfrac{(-1)^n n}{n + 8}$

44–53: Determine whether the series converges (and if so, find its sum) or diverges.

44. $\displaystyle\sum_{n=0}^{\infty} (e^n - e^{n+1})$

45. $\displaystyle\sum_{n=1}^{\infty} \left(\dfrac{1}{2n - 1} - \dfrac{1}{2n}\right)$

46. $\displaystyle\sum_{n=2}^{\infty} \left(\dfrac{1}{\sqrt{n}} - \dfrac{1}{\sqrt{n+1}}\right)$

47. $\displaystyle\sum_{n=1}^{\infty} \left(e^{-n} - e^{-(n+1)}\right)$

48. $\displaystyle\sum_{n=0}^{\infty} \left(\dfrac{4}{5}\right)^n$ **49.** $\displaystyle\sum_{n=0}^{\infty} 2\left(-\dfrac{1}{4}\right)^n$

50. $\displaystyle\sum_{n=0}^{\infty} 5\,(1.1)^n$ **51.** $\displaystyle\sum_{n=0}^{\infty} 3\pi^n$

52. $\displaystyle\sum_{n=0}^{\infty} 2\left(-\dfrac{11}{13}\right)^n$ **53.** $\displaystyle\sum_{n=0}^{\infty} \dfrac{2^{n+1}}{3^{2n}}$

54–55: Express the repeating decimal in terms of a geometric series. Then, express the series as a ratio of two integers.

54. $0.\overline{5}$ **55.** $3.\overline{14}$

56–59: Determine whether the series laws apply. If so, use them to determine whether the series converges.

56. $\displaystyle\sum_{n=0}^{\infty} \left[\left(\dfrac{4}{5}\right)^n + 2\left(-\dfrac{1}{4}\right)^n\right]$

57. $\displaystyle\sum_{n=0}^{\infty} \left(3\pi^n + \dfrac{1}{n^3}\right)$

58. $\displaystyle\sum_{n=0}^{\infty} \left[2\left(-\dfrac{11}{13}\right)^n + \dfrac{2^{n+1}}{3^{2n}}\right]$

59. $\displaystyle\sum_{n=0}^{\infty} (e^{-n} + n^{-e})$

60–63: Use the direct comparison test to determine whether the series converges or diverges.

60. $\displaystyle\sum_{n=0}^{\infty} \frac{1}{3^n + 2}$ **61.** $\displaystyle\sum_{n=0}^{\infty} \frac{4^n}{3^n + 2}$

62. $\displaystyle\sum_{n=1}^{\infty} \frac{1}{\sqrt{n^5 + 10}}$ **63.** $\displaystyle\sum_{n=3}^{\infty} \frac{1}{\sqrt{n} - 1}$

64–67: Use the limit comparison test to determine whether the series converges or diverges.

64. $\displaystyle\sum_{n=1}^{\infty} \frac{5^n + 3}{3^n + 3}$ **65.** $\displaystyle\sum_{n=1}^{\infty} \frac{1}{\sqrt{n^2 + 6}}$

66. $\displaystyle\sum_{n=1}^{\infty} \frac{n}{(3n + 5)7^n}$ **67.** $\displaystyle\sum_{n=3}^{\infty} \frac{3n^2 + 7n - 1}{4n^4 + 2n^2 + n}$

68–73: For each series, (a) verify that the alternating series test applies and then use it to show the series converges; and (b) determine whether the series converges absolutely or conditionally, or whether $\sum |a_n|$ diverges. Identify at least one convergence test used to reach this conclusion.

68. $\displaystyle\sum_{n=0}^{\infty} (-1)^n \frac{n}{n^2 + 5}$

69. $\displaystyle\sum_{n=1}^{\infty} (-1)^n \frac{1}{\sqrt{n}}$

70. $\displaystyle\sum_{n=1}^{\infty} (-1)^{n+1} \frac{n}{n^4 + 2}$

71. $\displaystyle\sum_{n=1}^{\infty} (-1)^n \frac{1}{2n + 1}$

72. $\displaystyle\sum_{n=1}^{\infty} (-1)^n \frac{1}{\sqrt{n(n + 1)}}$

73. $\displaystyle\sum_{n=0}^{\infty} (-1)^n \frac{1}{n^2 + 2}$

74–79: Use the ratio test to determine whether the series converges absolutely or diverges.

74. $\displaystyle\sum_{n=1}^{\infty} (-1)^n \frac{n}{5^{n-1}}$ **75.** $\displaystyle\sum_{n=0}^{\infty} \frac{n^2}{(\ln 2)^n}$

76. $\displaystyle\sum_{n=1}^{\infty} (-1)^{n+1} \frac{1}{n7^n}$ **77.** $\displaystyle\sum_{n=1}^{\infty} \frac{n!}{5^n}$

78. $\displaystyle\sum_{n=1}^{\infty} (-1)^n \frac{1}{(n!)^2}$ **79.** $\displaystyle\sum_{n=1}^{\infty} \frac{3^{2n}}{(n + 1)!}$

80–81: (a) Verify that the series converges. (b) Find lower and upper bounds for the sum of the series if only the third partial sum is used to approximate the sum. (c) Use the Alternating Series Remainder Theorem to determine the smallest number of terms required to approximate the sum of the series with an error of less than 0.001.

80. $\displaystyle\sum_{n=1}^{\infty} (-1)^n \frac{1}{n^2}$

81. $\displaystyle\sum_{n=1}^{\infty} (-1)^n \frac{1}{n^3}$

82–87: (a) Calculate the function's first three Maclaurin polynomials. (b) Calculate the function's nth Maclaurin polynomial. (c) Calculate the function's first three Taylor polynomials at $c = 1$.

82. $f(x) = \cos x$

83. $f(x) = xe^x$

84. $f(x) = \dfrac{1}{x + 1}$

85. $f(x) = \arctan x$

86. $f(x) = \sin\left(\dfrac{\pi}{2}x\right)$

87. $f(x) = \ln(1 + x)$

88–89: Taylor's Theorem applies to each function f. Use it to (a) calculate the maximum error in approximating $f(a)$ for the given a-value by $M_n(a)$ for the given n-value, and (b) determine the degree of the Maclaurin polynomial guaranteeing that the error in approximating $f(a)$ by $M_n(a)$ is at most 0.001.

88. $f(x) = \cos x$, $a = 0.3$, $n = 4$

89. $f(x) = e^x$, $a = 1$, $n = 5$

90–93: Find each series's (a) radius of convergence and (b) interval of convergence. (Hint: Use the root test for exercise 93(a).)

90. $\displaystyle\sum_{n=1}^{\infty} \frac{(2x)^n}{n^2}$ **91.** $\displaystyle\sum_{n=1}^{\infty} \frac{nx^n}{5^n}$

92. $\displaystyle\sum_{n=1}^{\infty} \frac{(x-3)^n}{n}$ **93.** $\displaystyle\sum_{n=1}^{\infty}(-1)^n \frac{(x-2)^n}{3^n}$

94–95: Find the power series representation and interval of convergence for the function.

94. $\dfrac{1}{1-3x}$ **95.** $\dfrac{x^4}{1+x^2}$

96–97: Differentiate the power series representation of the function f and use it to obtain the power series representation of the function g. Give the interval of convergence of the series for g.

96. $f(x) = \dfrac{1}{1-x}, \quad g(x) = \dfrac{1}{(1-x)^2}$

97. $f(x) = \dfrac{1}{1+x^2}, \quad g(x) = \dfrac{x}{(1+x^2)^2}$

98–101: (a) Calculate the Maclaurin series of f. (b) Determine the series's interval of convergence.

98. $f(x) = e^{-x}$

99. $f(x) = \dfrac{e^x + e^{-x}}{2}$

100. $f(x) = \dfrac{e^x - e^{-x}}{2}$

101. $f(x) = \arctan(x)$

102–104: Manipulate known Maclaurin series to generate the Maclaurin series for the function.

102. $\ln(1+x^3)$ **103.** $\cos x^{3/2}$

104. $\ln\left(\dfrac{1+x}{1-x}\right)$

105. Prove that the Maclaurin series calculated in exercise 98 converges to e^{-x} for all x.

106. Prove that the Maclaurin series calculated in exercise 100 converges to the function in the exercise for all x.

107–108: (a) Write out the first four nonzero terms in the binomial expansion of the function. (b) Use the result in (a) to approximate the number.

107. $f(x) = \sqrt[4]{1+x}, \ \sqrt[4]{1.1}$

108. $f(x) = (1+x)^{-2/3}, \ (1.1)^{-2/3}$

109–111: Find the sum of the infinite series by using an appropriate Maclaurin series.

109. $1 - \dfrac{1}{3} + \dfrac{1}{5} - \cdots$

110. $1 + 2 + \dfrac{2^2}{2!} + \dfrac{2^3}{3!} + \cdots$

111. $1 - \dfrac{1}{2!} + \dfrac{1}{3!} - \dfrac{1}{4!} + \cdots$

112–113: Use an appropriate Maclaurin series to evaluate the limit.

112. $\displaystyle\lim_{x\to 0} \frac{e^x - 1}{x}$

113. $\displaystyle\lim_{x\to 0} \frac{\arctan(2x)}{x}$

A1–A4: For each exercise: (a) calculate the first three terms of the sequence; (b) find a formula for the nth term in the sequence and classify the type of sequence obtained (e.g., arithmetic); (c) find a recurrence relation involving the sequence's terms; and (d) calculate the infinite limit of the sequence's values—if finite, interpret the result in the context of the problem.

A1. The bouncing basketball problem A basketball is thrown 10 feet into the air. With each bounce it rebounds to 0.5 times its previous height. Denote by y_n the maximum height the ball reaches after the nth bounce ($n \geq 1$).

A2. Monthly bills You've signed up for a new monthly subscription. The service has an activation fee of \$10 and costs \$30 per month after that. Denote by T_n the total you've spent on the subscription service after $n \geq 0$ months.

A3. Population growth Since 2010, the U.S. population has been growing by about 0.75% each year. Its population in 2010 was 309.3 million.

Denote by P_n its population (in millions) $n \geq 0$ years later.

A4. Inflation Inflation is defined as the general increase in the prices of goods and services in an economy. In the United States, the Consumer Price Index (CPI) is one broad measure of inflation. The base value of the CPI was set at 100 in 1984. Assuming that the CPI has increased by 3% every year since, $100 earned in 1984 buys approximately $G_n = 100 \cdot \frac{100}{C_n}$ dollars of goods and services $n \geq 0$ years after 1984, where C_n is the CPI value. (For this problem, focus only on the sequence (G_n).)

A5. Compounding interest in a savings account Suppose M_n is the balance (in dollars) of a savings account n years after opening it with an initial deposit of M_0 dollars (and that no subsequent deposits are made). Let r denote the yearly interest rate earned—where r is expressed as a decimal—and suppose that the savings account compounds interest t times per year.

(a) Show that $M_1 = M_0 \left(1 + \frac{r}{t}\right)$.

(b) Find a formula for M_n.

(c) Evaluate $\lim_{t \to \infty} M_n$ and interpret your result in the context of the problem.

A6. Archimedes's method of exhaustion This exercise works through how the Greek mathematician Archimedes (ca. 287–ca. 212 BCE) used sequences and early calculus ideas to approximate the area of a circle. Suppose we inscribe $n \geq 4$ isosceles triangles of equal base length r in a circle of radius r, in the manner illustrated below for one such triangle.

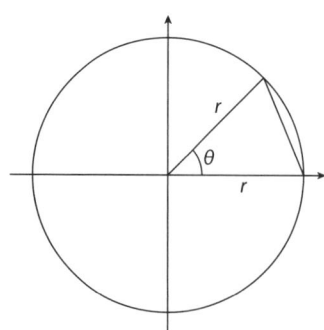

(a) Explain why $\theta = \frac{2\pi}{n}$.

(b) Let A_n be the sum of the areas of the n inscribed triangles. Show that

$$A_n = \frac{1}{2}nr^2 \sin\left(\frac{2\pi}{n}\right).$$

(c) Show that the substitution $m = 2\pi/n$ converts A_n into

$$A_m = \pi r^2 \left(\frac{\sin m}{m}\right).$$

(d) From Calculus 1 we know that $\lim_{x \to \infty} \frac{\sin x}{x} = 1$. Use this, along with theorem 4.2, to show that $A_m \to \pi r^2$, and interpret your result in the context of the problem.

A7. The geometric sequences in music The musical note A above middle C has a frequency of 440 Hertz (Hz). The **chromatic scale**, which forms the foundation for Western music, then determines the frequencies of the subsequent notes on a musical scale according to the following 12th root of 2 progression:

$$f_0 = 440 \cdot 2^{\frac{0}{12}}, \quad f_1 = 400 \cdot 2^{\frac{1}{12}}, \quad \ldots$$

Here f_n is the frequency of the note produced n semitones (a twelfth of an octave) above the note A above middle C. The corresponding musical notes are A (f_0), A♯ ("A sharp," f_1), etc., up to A2 (f_{12}), the note with frequency 880 Hz that is one octave above A. (Each of these notes corresponds to a key on a piano between A and A2.)

(a) Find a formula for the sequence (f_n), and show that the sequence is a geometric sequence. (Takeaway: *We hear geometric sequences!*)

(b) Show that $f_{12n} = 2^n f_0$, and interpret your result.

A8. The bouncing basketball problem, part 2 Returning to exercise A1, let d_n be the round-trip vertical distance covered by the bouncing basketball between the nth and $(n+1)$th bounce, $n \geq 1$. Calculate $\sum_{n=1}^{\infty} d_n$, and interpret your result in the context of this problem.

A9. The multiplier effect Suppose that each citizen of a community is given an equal amount of money, totaling $T for the whole community. Suppose further that no money enters or leaves the community, and that all spending by community members is redistributed throughout the community.

(a) If each month every citizen saves a fraction $0 < x < 1$ of their wealth and spends the remaining $1 - x$ into the community, find a formula for how much of the $T is spent after month n ($n \geq 1$).

(b) Show that if this experiment continues forever, the result is a total spending of $T \left(\frac{1-x}{x} \right)$.

(c) Show that the total spending is larger than T if $x < \frac{1}{2}$ (economists call this the **multiplier effect**), and interpret your result in the context of this problem.

A10. Doubling time and the Rule of 70 After t years, the balance $M(t)$ of a savings account wherein interest is compounded yearly and no subsequent deposits or withdrawals are made is

$$M(t) = M_0(1 + r)^t,$$

where M_0 is the initial deposit, r is the yearly interest rate expressed as a decimal.

(a) One can show that the time T required for the balance to equal double the initial deposit (i.e., $M(T) = 2M_0$), the **doubling time**, is

$$T = \frac{\ln 2}{\ln(1 + r)}.$$

Real-world interest rates for savings accounts are typically in the range $0 \leq r \leq 0.1$. Use this information and an appropriate Maclaurin polynomial to show that $T \approx \frac{\ln 2}{r}$.

(b) Let $R = 100r$. (R is then the interest rate expressed as a number between 0 and 100.) Use part (a) to show that

$$T \approx \frac{70}{R}.$$

This is known as the **Rule of 70**. Use this rule to estimate the doubling time of a savings account earning 3% interest per year.

A11. The acceleration due to gravity as a function of latitude The acceleration of gravity, g, is often approximated by $g \approx 9.8$ m/s^2 (in SI units). A more accurate formula for g is the *geodetic reference formula* of 1967:

$$g(x) = a(1 + b \sin^2 x - c \sin^4 x) \text{ m/s}^2,$$

where x is the latitude (in degrees) north or south of the equator, and $a = 9.7803185$, $b = 0.005278895$, and $c = 0.000023462$. Using an appropriate Maclaurin polynomial, show that $g(1) \approx a(1 + b - c)$, and interpret this approximation in the context of the problem.

A12. Time dilation In 1905 Albert Einstein discovered that measurements of some physical quantities, such as time and length, depend on the frame of reference used and how fast the frame is moving relative to the speed of light. For example, suppose you are sitting on a train moving at speed v. Einstein's special theory of relativity says that the passage of t seconds *relative to you* is equivalent to the passage of T seconds *relative to a stationary observer* outside the train, where

$$T(v) = \frac{t}{\sqrt{1 - v^2/c^2}}.$$

(Here c is the speed of light.) Notice, for example, that when $v = 0.5c$ (your train is traveling at half the speed of light), 1 second of time passage to you would be perceived as $T(0.5c) = \frac{2}{\sqrt{3}} \approx 1.15$ seconds to the external observer. This is the famous *time dilation* effect: Moving clocks run slow relative to stationary clocks.

(a) Use an appropriate Maclaurin series to show that
$$T(v) \approx t \left(1 + \frac{v^2}{2c^2} \right)$$
for $v \ll c$ ("much less than c").

(b) Discuss how the approximation in (a) implies the "moving clocks run slow" interpretation of time dilation (for speeds $v \ll c$).

(c) Discuss how the approximation in (a) explains why we don't perceive this time dilation effect in our everyday lives.

A13. The shape of planetary orbits One of the triumphs of Newton's universal law of gravity was to explain why planets in our solar system orbit the Sun in elliptical orbits. Newton's law shows that a planet's distance r to the Sun as it orbits the Sun in the ecliptic plane can be modeled by the angular version of the ellipse equation:

$$r(\theta) = \frac{a(1 - e^2)}{1 + e \cos \theta},$$

where $0 \leq \theta \leq 2\pi$ is the angle of the planet from the x-axis, $0 \leq e \leq 1$ is the **eccentricity** of the orbit, and $a > 0$ is the ellipse's semimajor axis length.

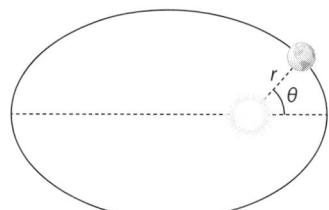

Use an appropriate Maclaurin series to show that $r(\theta) \approx a(1 - e^2)(1 - e \cos \theta)$ when e is close to zero. For Earth's orbit, $e \approx 0.017$, so that $r(\theta) \approx a$. (This is why it took astronomers so long to see the elliptical nature of Earth's orbit around the sun—the deviation of the orbit from a circle is so small that it requires *very* precise measurements to detect.)

D1. Recall from chapter 1 that the infinite limit of the ratio of successive Fibonacci numbers, $\frac{F_{n+1}}{F_n}$, is the golden ratio $\varphi = \frac{1+\sqrt{5}}{2}$. In this exercise we'll find an explicit formula for F_n in terms of φ and use it to prove that $\frac{F_{n+1}}{F_n} \to \varphi$.

(a) The golden ratio φ is defined to be the ratio of the total length $a + b$ of a line segment to the length a of its larger chunk when that ratio is equal to the ratio of the length a of the larger chunk to the length b of the shorter chunk:

$$\varphi = \frac{a+b}{a} = \frac{a}{b}$$

This shows that $1 + \frac{1}{\varphi} = \varphi$, or that φ is one solution to the quadratic equation $x^2 = x + 1$. Show that the other solution is $\tau = \frac{1-\sqrt{5}}{2}$. Then, show that $\tau = -\frac{1}{\varphi}$.

(b) Show that $x^3 = F_3 x + F_2$ by writing $x^3 = x \cdot x^2$ and using $x^2 = x + 1$. Show that $x^4 = F_4 x + F_3$ using a similar approach and the Fibonacci recurrence relation (see example 4.3).

(c) In general, $x^n = F_n x + F_{n-1}$, $n \geq 2$. (We get all these equations by multiplying $x^2 = x + 1$ by x and substituting in $x^2 = x + 1$.) We know that φ and τ solve $x^2 = x + 1$, so they must also solve every $x^n = F_n x + F_{n-1}$, $n \geq 2$. Therefore,

$$\varphi^n = F_n \varphi + F_{n-1}, \quad \tau^n = F_n \tau + F_{n-1}.$$

Subtract the latter equation from the former, and use the results of (a) to show that

$$F_n = \frac{1}{\sqrt{5}} (\varphi^n - \tau^n),$$

valid for $n \geq 1$. This is **Binet's formula**.

(d) Recalling from (a) that $\tau = -\frac{1}{\varphi} = -\varphi^{-1}$, substitute this into the F_n from (c) to show that

$$F_n = \frac{1}{\sqrt{5}} (\varphi^n - (-1)^n \varphi^{-n}).$$

(e) Finally, use the formula from (d) to show that $\displaystyle\lim_{n \to \infty} \frac{F_{n+1}}{F_n} = \varphi$.

D2. Consider the geometric sequence (ar^n), and let $f(x) = ar^x$.

(a) Let $r > 0$. Use the end behavior of the exponential function f, along with Alternating Series Remainder Theorem 4.2, to show $a_n = ar^n$ converges to 0 if $0 < r < 1$, to a if $r = 1$, and diverges if $r > 1$.

(b) Let $r < 0$. Write $r = -s$, so that $s > 0$. The geometric sequence now has the form $a_n = a(-s)^n = (-1)^n as^n$. Explain how adapting the results of (a) to the exponential function $g(x) = as^x$ shows that $a = ar^n$ converges to 0 if $-1 < r < 0$, and diverges if $r \leq 1$.

D3. This exercise proves the divergence test, theorem 4.6, using a proof by contradiction. Suppose that $\sum a_n$ converges and sums to S, and that $a_n \nrightarrow 0$ (i.e., the infinite limit of a_n is not zero).

(a) Use (4.9) to show that $a_n = S_n - S_{n-1}$.

(b) Use the result from (a) to show that $\lim\limits_{n \to \infty} (S_n - S_{n-1}) \neq 0$.

(c) From definition 4.6 we know that in a convergent series $S_n \to S$, and so $S_{n-1} \to S$ as well. Explain how this implies $(S_n - S_{n-1}) \to 0$. But from (b), $(S_n - S_{n-1}) \nrightarrow 0$, a contradiction.

D4. This exercise proves the contrapositive of the divergence test: If $\sum a_n$ converges, then $a_n \to 0$. Suppose that $\sum a_n$ converges and that its sum is S. Let S_n denote the nth partial sum of the series. Explain why $\lim\limits_{n \to \infty} S_n = S = \lim\limits_{n \to \infty} S_{n-1}$. Then, explain how to use this and (4.9) to show that $a_n \to 0$.

D5. Let x be a real number. For what x-values is

$$\sum_{n=0}^{\infty} e^{nx} = \frac{1}{1 - e^x}?$$

D6. A general telescoping series has the form

$$\sum_{n=1}^{\infty} (b_n - b_{n+k}), \quad \text{where } k \geq 1.$$

(a) Write out the first $k+1$ terms of the series to see that it takes that long to find a pair of bs that cancel.

(b) Explain why

$$\sum_{n=1}^{N} b_{n+k} = \sum_{n=k+1}^{N+k} b_n.$$

(It may help to write out both series.)

(c) Use your result from (b) to show that the Nth partial sum of the series

$$S_N = \sum_{n=1}^{N} (b_n - b_{n+k})$$

can be expressed as

$$S_N = \sum_{n=1}^{k} b_n - \sum_{n=N+1}^{N+k} b_n .$$

It follows that, if the limit as $N \to \infty$ of the highlighted term is zero, then the series converges, and its sum is $\sum_{n=1}^{k} b_n$.

D7. Find the error—and correct it—in the following claim: The series

$$\sum_{n=1}^{\infty} \left(\frac{1}{n} - \frac{1}{n+1} \right)$$

diverges because

$$\sum_{n=1}^{\infty} \left(\frac{1}{n} - \frac{1}{n+1} \right) = \sum_{n=1}^{\infty} \frac{1}{n} - \sum_{n=1}^{\infty} \frac{1}{n+1}$$

and both series on the right-hand side diverge.

D8. This exercise proves the absolute convergence test, theorem 4.14. Let $\sum a_n$ be a series, and suppose that $\sum |a_n|$ converges.

(a) Explain why $0 \leq a_n + |a_n| \leq 2|a_n|$.

(b) What theorem implies that $\sum 2|a_n|$ converges?

(c) What theorem, when used together with the results of (a) and (b), implies that the series $\sum (a_n + |a_n|)$ converges?

(d) What theorem allows you to say that

$$\sum (a_n + |a_n|) - \sum |a_n| = \sum a_n,$$

and that this last series converges?

D9. Early attempts to make sense of infinite series ran into the *rearrangement problem*: In some *convergent* infinite series, rearranging the order of the terms changes the series's sum. As an illustration, consider the alternating harmonic series $\sum \frac{(-1)^{n+1}}{n}$.

(a) We've shown via the alternating series test that the alternating harmonic series converges. Let

S denote its sum:

$$S = 1 - \frac{1}{2} + \frac{1}{3} - \frac{1}{4} + \cdots.$$

Write out the rearrangement where every positive term is followed by the next two negative terms.

(b) In the series obtained in (a), add the first two terms, the fourth and fifth terms, the seventh and eighth terms, etc., to obtain the new series

$$\frac{1}{2} - \frac{1}{4} + \frac{1}{6} - \cdots.$$

Show that this series's sum is $\frac{S}{2}$. Conclusion: Rearranging the alternating series's terms has halved the sum of the originally ordered series.

In general, if $\sum a_n$ is *conditionally* convergent, then one can find a rearrangement of the series that sums to *whatever real number we'd like!* (This is part of the **Riemann Rearrangement Theorem**.) However, if $\sum a_n$ is *absolutely* convergent, then every rearrangement of the series yields the same sum.

D10. Let $\sum a_n$ be a series, with $a_n \geq 0$, and let $L = \lim_{n \to \infty} |a_n|^{1/n}$. The **root test** says that $\sum a_n$ converges absolutely if $L < 1$, and diverges if $L > 1$ or $L = \infty$. Use the test to determine whether these series converge or diverge:

$$\sum_{n=1}^{\infty} \left(\frac{n^2}{3n^2 + 1} \right)^n, \qquad \sum_{n=1}^{\infty} \frac{n^n}{2^n}.$$

5   Connections between Integration and Series

Chapter Preview. We've now answered the three Big Questions that drove the development of Calculus 2. And in doing so, we've developed several new integration techniques and have applied them to calculate arc length, surface area, and volumes of surfaces of revolution and of solids with known cross sections. We've also built up the theory of infinite series and Taylor/Maclaurin power series and used them to approximate values of functions and sum infinite series. Pretty impressive work! But these two main pillars of Calculus 2—integration and infinite series—have yet to meet each other in this book. In this chapter we'll get them dancing. We'll waltz through several interconnections between the previous three chapters and use the results to deepen our answers to Calculus 2's Big Questions.

5.1   Partial Fractions and Telescoping Series

Near the beginning of chapter 4 we studied telescoping series (definition 4.8):

$$(b_1 - b_2) + (b_2 - b_3) + (b_3 - b_4) + \cdots .$$

Recall from theorem 4.8 that in these series the Nth partial sum

$$S_N = b_1 - b_{N+1},$$

from which we can determine these series' convergence by evaluating the limit as $N \to \infty$ of this S_N. But not every series that telescopes (i.e., has terms that cancel in a predictable pattern) comes already in the "$\sum(b_n - b_{n+1})$" form above. We saw an example of this in the series $\sum(\frac{1}{n} - \frac{1}{n+2})$ of (4.15). More complex examples come even more disguised, like $\sum \frac{1}{n^2-1}$. Let me show you how to use a section 2.8 technique—partial fractions—to help uncover such series' telescoping nature.

EXAMPLE 5.1   Show that $\displaystyle\sum_{n=1}^{\infty} \frac{8}{(n+1)(n+2)}$ converges, and find its sum.

Solution The partial fractions decomposition here is

$$\frac{8}{(n+1)(n+2)} = \frac{A}{n+1} + \frac{B}{n+2} \qquad \text{Using the guidelines in box 2.2 in section 2.8.}$$

$$8 = A(n+2) + B(n+1). \qquad \text{Multiplying the equation by } (n+1)(n+2)$$

Setting $n = -2$ yields $8 = -B \Rightarrow B = -8$; setting $n = -1$ yields $8 = A$. Thus,

$$\frac{8}{(n+1)(n+2)} = \frac{8}{n+1} - \frac{8}{n+2}$$

$$\Rightarrow \quad \sum_{n=1}^{\infty} \frac{8}{(n+1)(n+2)} = 8 \sum_{n=1}^{\infty} \left(\frac{1}{n+1} - \frac{1}{n+2} \right).$$

Writing out S_N now helps us recognize this as a telescoping series:

$$S_N = 8 \left[\left(\frac{1}{2} - \frac{1}{3} \right) + \left(\frac{1}{3} - \frac{1}{4} \right) + \left(\frac{1}{4} - \frac{1}{5} \right) + \cdots + \left(\frac{1}{N+1} - \frac{1}{N+2} \right) \right]$$

$$= 8 \left(\frac{1}{2} - \frac{1}{N+2} \right) = 4 - \frac{8}{N+2}.$$

And since $\frac{1}{N+2} \to 0$ as $N \to \infty$, the series converges and its sum is 4:

$$\lim_{N \to \infty} S_N = 4 - \lim_{N \to \infty} \frac{8}{N+2} = 4. \qquad \blacksquare$$

EXAMPLE 5.2   Show that $\displaystyle\sum_{n=2}^{\infty} \frac{1}{n^2 - 1}$ converges, and find its sum.

Solution   Since $n^2 - 1 = (n-1)(n+1)$, the partial fractions expansion is

$$\frac{1}{n^2 - 1} = \frac{A}{n-1} + \frac{B}{n+1} \qquad \text{Using the guidelines in box 2.2 in section 2.8.}$$

$$1 = A(n+1) + B(n-1). \quad \text{Multiplying the equation by } (n-1)(n+1)$$

Setting $n = 1$ yields $1 = 2A \Rightarrow A = \frac{1}{2}$; setting $n = -1$ yields $1 = -2B \Rightarrow B = -\frac{1}{2}$. Thus,

$$\frac{1}{n^2 - 1} = \frac{\frac{1}{2}}{n-1} - \frac{\frac{1}{2}}{n+1} \quad \Rightarrow \quad \sum_{n=2}^{\infty} \frac{1}{n^2 - 1} = \frac{1}{2} \sum_{n=2}^{\infty} \left(\frac{1}{n-1} - \frac{1}{n+1} \right).$$

Writing out S_N again helps us recognize this as a telescoping series:

$$S_N = \frac{1}{2} \left[\left(1 - \frac{1}{3} \right) + \left(\frac{1}{2} - \frac{1}{4} \right) + \left(\frac{1}{3} - \frac{1}{5} \right) + \cdots + \left(\frac{1}{N-2} - \frac{1}{N} \right) \right.$$

$$\left. + \left(\frac{1}{N-1} - \frac{1}{N+1} \right) \right]$$

$$= \frac{1}{2} \left(1 + \frac{1}{2} - \frac{1}{N} - \frac{1}{N+1} \right).$$

And since

$$\lim_{N \to \infty} S_N = \frac{1}{2} \lim_{N \to \infty} \left(1 + \frac{1}{2} - \frac{1}{N} - \frac{1}{N+1} \right) = \frac{1}{2} \left(1 + \frac{1}{2} \right) = \frac{3}{4},$$

we conclude that the series converges and sums to 3/4. $\qquad \blacksquare$

Related Exercises   1–4.

5.2 Power/Taylor Series and Integration

Further into chapter 4 we learned about power series and, in particular, how to differentiate them (theorem 4.19).

Recall that this differentiation is done term by term:

$$\text{If} \quad S(x) = \sum_{n=0}^{\infty} a_n(x-c)^n, \quad \text{then}$$

$$S'(x) = \sum_{n=0}^{\infty} a_n \cdot n(x-c)^{n-1} = \sum_{n=0}^{\infty} na_n(x-c)^{n-1}.$$

In other words, to *differentiate* $S(x)$ we *differentiate* the $(x-c)^n$. You might therefore guess that to *integrate* $S(x)$ we'd *integrate* the $(x-c)^n$. That's exactly what the next theorem says: Integrating a power series is likewise done term-wise.

THEOREM 5.1: INTEGRATION OF POWER SERIES. Let

$$S(x) = \sum_{n=0}^{\infty} a_n(x-c)^n = a_0 + a_1(x-c) + a_2(x-c)^2 + \cdots$$

be a power series with radius of convergence $R > 0$. Then, inside the corresponding interval of convergence $(c - R, c + R)$, f is integrable, and

$$\int S(x)\,dx = C + \sum_{n=0}^{\infty} a_n \frac{(x-c)^{n+1}}{n+1}$$

$$= C + a_0(x-c) + a_1\frac{(x-c)^2}{2} + a_2\frac{(x-c)^3}{3} + \cdots .$$

Moreover, the radius of convergence of this series is also R. Convergence at the endpoints $x = c \pm R$ may not be the same as that of S's.

In a nutshell, this theorem says that for a power series with radius of convergence R: (1) we can integrate it, and its integral, $\int S(x)\,dx$, also has radius of convergence R; (2) the integration is done term by term. Note also that we can express the theorem's result as

$$\int S(x)\,dx = \int \left[\sum_{n=0}^{\infty} a_n(x-c)^n \right] dx$$

$$= \sum_{n=0}^{\infty} a_n \left[\int (x-c)^n \, dx \right] = C + \sum_{n=0}^{\infty} a_n \frac{(x-c)^{n+1}}{n+1}.$$

Notice that, after the second equals sign, the integral $\int$ swapped places with the summation $\sum$. That is, the integral of the sum became the sum of the integral.[1]

EXAMPLE 5.3 Recall the Maclaurin expansion (from the table of common Maclaurin expansions, figure 4.9)

---

[1] A similar swap happened in the Differentiation of Power Series Theorem 4.19, where the derivative of the sum became the sum of the derivative.

$$e^x = \sum_{n=0}^{\infty} \frac{x^n}{n!} = 1 + x + \frac{x^2}{2!} + \frac{x^3}{3!} + \cdots, \qquad -\infty < x < \infty.$$

(a) Calculate the Maclaurin expansion of $\sqrt[3]{x}e^x$, and show that it converges absolutely for $-\infty < x < \infty$.

(b) One can show that the series in (a) converges to $\sqrt[3]{x}e^x$ for $-\infty < x < \infty$. Use this information to help expand $\int \sqrt[3]{x}e^x \, dx$ in a power series valid for all x.

(c) Use the expansion in (b) to express $\int_0^1 x^{1/3} e^x \, dx$ as an infinite series.

Solution

(a) By theorem 4.20, the Maclaurin series of $\sqrt[3]{x}e^x$ is the product of $x^{1/3}$ with the Maclaurin expansion of e^x:

$$M.S.[x^{1/3}e^x] = x^{1/3} \sum_{n=0}^{\infty} \frac{x^n}{n!}$$

$$= \sum_{n=0}^{\infty} \frac{x^{n+1/3}}{n!} \qquad \text{Since } x^{1/3}x^n = x^{n+1/3}$$

$$= x^{1/3} + x^{4/3} + \frac{x^{7/3}}{2!} + \cdots.$$

The same theorem also tells us that this series converges absolutely for (in this case) all x.

(b) We're told that this power series converges to $x^{1/3}e^x$ with radius of convergence $R = \infty$. Therefore, the Power Series Integration Theorem applies, and we can integrate the series term by term:

$$\int x^{1/3}e^x \, dx = C + \frac{x^{4/3}}{4/3} + \frac{x^{7/3}}{7/3} + \frac{x^{10/3}}{2!(10/3)} + \cdots$$

$$= \sum_{n=0}^{\infty} \frac{x^{n+\frac{4}{3}}}{\left(n+\frac{4}{3}\right)n!} + C.$$

Furthermore, this series converges to $\int x^{1/3}e^x \, dx$ for all x-values.

(c) Let $F(x)$ be the gray highlighted series in (b). By the Evaluation Theorem,

$$\int_0^1 x^{1/3}e^x \, dx = [F(x)]_0^1 = F(1) - F(0).$$

Translation: The answer to the integral is the difference "$F(1) - F(0)$," which we get by plugging in $x = 1$ and $x = 0$ into the gray highlighted series above and subtracting the results:

$$\sum_{n=0}^{\infty} \frac{(1)^{n+\frac{4}{3}}}{\left(n+\frac{4}{3}\right)n!} - \sum_{n=0}^{\infty} \frac{(0)^{n+\frac{4}{3}}}{\left(n+\frac{4}{3}\right)n!} = \sum_{n=0}^{\infty} \frac{1}{\left(n+\frac{4}{3}\right)n!}. \qquad \blacksquare$$

Related Exercises   5(a)–7(a) ((a) only).

A couple of comments on this example's results:

- *None* of the integration techniques we learned in chapter 2 work to evaluate $\int x^{1/3} e^x \, dx$. Yet in part (b) we were able to find this antiderivative (albeit in power series form). Takeaway: Theorem 5.1 gives us another integration technique—very cool!

- Note that the definite integral in (c) produced an infinite series. That entire infinite series is *the* value of the definite integral, so a partial sum of that series is an *approximation* to that value. Takeaway: Theorem 5.1 gives us another route to approximating definite integrals, one separate from the Riemann sum and Trapezoidal Rule methods we learned in sections 2.2–2.3; let's explore that next.

Using Taylor Series to Approximate Definite Integrals

In section 2.4 we developed theorems for approximating definite integrals to within a desired accuracy using Riemann sums or the Trapezoidal Rule. Since part (b) of example 5.3 above produced an infinite series answer to a definite integral, and since we studied techniques for approximating the sum of an infinite series in chapter 4, we can now put these together to develop a new technique for approximating definite integrals to within a desired accuracy. Let's learn how that works via an example.

EXAMPLE 5.4 Consider the definite integral $\int_0^1 e^{-x^3} \, dx$.

(a) Use an appropriate Taylor series to expand this definite integral in a power series, and determine the series's radius of convergence.

(b) Using the series from (a), determine the number of terms required to approximate the value of the definite integral with an error less than 0.001.

Solution

(a) First we need the Maclaurin series for e^{-x^3}. We'll find this using theorem 4.20 again by substituting $-x^3$ for x in the Maclaurin series for e^x (from example 5.3):

$$M.S.[e^{-x^3}] = \sum_{n=0}^{\infty} \frac{(-x^3)^n}{n!} = \sum_{n=0}^{\infty} \frac{(-1)^n x^{3n}}{n!}.$$

This series converges absolutely for all x (by theorem 4.20), so that $R = \infty$. Furthermore, one can show that it converges to e^{-x^3} for all x.* The Power Series Integration Theorem therefore applies and tells us that

* This uses an argument very similar to that in example 4.35.

$$\int_0^1 e^{-x^3} \, dx = \int_0^1 \left[\sum_{n=0}^{\infty} \frac{(-1)^n x^{3n}}{n!} \right] dx$$

$$= \sum_{n=0}^{\infty} \left[\frac{(-1)^n}{n!} \int_0^1 x^{3n} \, dx \right]$$

$$= \sum_{n=0}^{\infty} \frac{(-1)^n}{n!} \left[\frac{x^{3n+1}}{3n+1} \right]_0^1 \qquad \text{Evaluation Theorem}$$

$$= \sum_{n=0}^{\infty} \frac{(-1)^n}{(3n+1)n!} \qquad \text{Since} \quad \frac{1^{3n+1}}{3n+1} - \frac{0^{3n+1}}{3n+1} = \frac{1}{3n+1}$$

$$= 1 - \frac{1}{4} + \frac{1}{14} - \cdots + (-1)^n \frac{1}{(3n+1)n!} + \cdots.$$

Furthermore, the theorem tells us that this series converges for all x, too.

(b) The infinite series we've obtained is an alternating series with $b_n = \frac{1}{(3n+1)n!}$. Since $(3n+1)n!$ gets larger as n increases, b_n is a decreasing sequence. And since $b_n \to 0$ as $n \to \infty$, Alternating Series Remainder Theorem 4.16 applies and tells us that

$$\left| \int_0^1 e^{-x^3}\, dx - S_N \right| \leq \frac{1}{[3(N+1)+1](N+1)!}.$$

(The right-hand side here is b_{N+1}.) We'd now like the right-hand side of this inequality to be less than $0.001 = \frac{1}{1,000}$. By trial and error, we find that $N \geq 4$ works. Conclusion:

$$\int_0^1 e^{-x^3}\, dx \approx S_4 = \sum_{n=0}^{4} \frac{(-1)^n}{(3n+1)n!} = \frac{2,941}{3,640} \approx 0.80797,$$

and we're guaranteed that this estimate is within 0.001 of the definite integral's value. ∎

Related Exercises   5(b)–7(b) ((b) only), A1, D1–D2.

APPLICATIONS Applied exercise A1 shows you how to use Maclaurin series to approximate the period of a pendulum, which can help you tell time by counting the number of full swings of the pendulum.

EXPLORATIONS Exercise D1 explores Leibniz's formula for π, which he arrived at by integrating a particular Maclaurin series and using the results of this section. Exercise D2 guides you through the derivation of Taylor's Theorem (theorem 4.17) via repeated integration by parts.

5.3   Improper Integrals

We've been making heavy use of the Evaluation Theorem in this book. But this theorem assumes that the integrand f is continuous on the interval of integration, and that the limits of integration are finite. In this section we'll extend that theorem to the cases when f has discontinuities inside the interval of integration and/or when one (or both) of the endpoints of integration is infinite. As a bonus, in the next section we'll leverage our results to develop a new comparison test for series that uses integration. Let's begin with a couple of definitions.

DEFINITION 5.1: INFINITE DISCONTINUITY. Let f be a function defined near $x = c$ except possibly at $x = c$. If

$$\lim_{x \to c^{\pm}} f(x) = \infty \qquad \text{or} \qquad \lim_{x \to c^{\pm}} f(x) = -\infty,$$

then we say f has an **infinite discontinuity at c**.

You may recognize the limit conditions above from Calculus 1: They are precisely the ones that make $x = c$ into a vertical asymptote of f. Takeaway: An infinite discontinuity at $x = c$ is equivalent to a vertical asymptote at $x = c$.

We're now ready to study integration in the presence of these types of discontinuities (and possibly infinite limits of integration).

DEFINITION 5.2: IMPROPER INTEGRAL. Let f be a function, and let a and b be real numbers. The following integrals of f are all termed **improper integrals**:

1. $\displaystyle\int_{a}^{\pm\infty} f(x)\, dx, \quad \int_{\pm\infty}^{b} f(x)\, dx, \quad \int_{-\infty}^{\infty} f(x)\, dx.$

2. $\displaystyle\int_{a}^{b} f(x)\, dx$ if f has a finite number of infinite discontinuities in the interval $[a, b]$.

3. $\displaystyle\int_{a}^{b} f(x)\, dx$ if f is discontinuous at a, b, or both.

One implication of this: $\displaystyle\int_{a}^{b} f(x)\, dx$ is *not* improper when f is continuous on $[a, b]$ *and* both a and b are finite. In that context we're in Evaluation Theorem land. What we're discussing in this section is when one or both of those conditions fails.

EXAMPLE 5.5   Which of the integrals below are improper? Justify your answer.

(a) $\displaystyle\int_{0}^{1} \frac{1}{2x - 3}\, dx$ (b) $\displaystyle\int_{1}^{\infty} e^{-2x}\, dx$ (c) $\displaystyle\int_{0}^{1} \frac{\sin x}{x^2 + 1}\, dx$

(d) $\displaystyle\int_{0}^{\frac{\pi}{3}} \csc x\, dx$

Solution   Only (b) and (d) are improper. In (b) it's because one of the limits of integration is infinite. In (d) it's because $f(x) = \csc x$ has an infinite discontinuity at the left endpoint of integration, $x = 0$, since $\csc 0 = \frac{1}{\sin 0}$ and $\sin 0 = 0$. The other two integrals don't have an infinite limit of integration, don't have infinite discontinuities *in the interval of integration*, and don't have discontinuities at the endpoints of the integration interval.   ∎

Related Exercises   8(a)–11(a) ((a) only).

Now that we can recognize improper integrals, let's discuss how to evaluate them.

DEFINITION 5.3: IMPROPER INTEGRALS WITH INFINITE INTEGRATION LIMITS. Let f be a function, and let a, b, and c be real numbers.

1. If f is continuous on $[a, \infty)$, then we define

$$\int_a^\infty f(x)\, dx = \lim_{b \to \infty} \int_a^b f(x)\, dx,$$

provided the limit exists.

2. If f is continuous on $(-\infty, b]$, then we define

$$\int_{-\infty}^b f(x)\, dx = \lim_{a \to -\infty} \int_a^b f(x)\, dx,$$

provided the limit exists.

3. If f is continuous on the interval $(-\infty, \infty)$, then we define

$$\int_{-\infty}^\infty f(x)\, dx = \lim_{a \to -\infty} \int_a^c f(x)\, dx + \lim_{b \to \infty} \int_c^b f(x)\, dx,$$

where c is any real number, provided both limits exist.

In each case, if the associated limits exist we say that the improper integral **converges**. If the limits don't exist—in case 3, if at least one of the limits doesn't exist—then we say the improper integral **diverges**.

The "converges/diverges" terminology might remind you of sequences and series. That's intentional, because this definition does for integration what partial sums did for infinite series. To see that, we can compare, for example, the definition in case 1 above with the limit of an infinite series's partial sums from (4.8):

$$\sum_{n=1}^\infty a_n = \lim_{N \to \infty} S_N, \qquad \int_a^\infty f(x)\, dx = \lim_{b \to \infty} \int_a^b f(x)\, dx.$$

These are both manifestations of the Calculus 2 workflow (figure 1.3): Make sense of a quantity involving infinity (left-hand sides of the equations) by taking the infinite limit of its finite counterpart (right-hand sides of the equations). Practically speaking, this means that evaluating improper integrals reduces eventually to evaluating a limit. And that brings us back to the Growth Order Theorem in chapter 4, and to limit evaluation techniques like the limit laws (appendix C) and L'Hôpital's Rule (appendix F).

EXAMPLE 5.6   Does the improper integral $\displaystyle\int_{-\infty}^\infty xe^{-x^2}\, dx$ converge or diverge?

Solution The integrand $f(x) = xe^{-x^2}$ is continuous, so by the third part of definition 5.3,

$$\int_{-\infty}^\infty xe^{-x^2}\, dx = \lim_{a \to -\infty} \int_a^0 xe^{-x^2}\, dx + \lim_{b \to \infty} \int_0^b xe^{-x^2}\, dx.$$

(I chose $c = 0$ for simplicity, but you can choose any real number, since f is continuous.) We can find the antiderivative of f via the u-substitution $u = -x^2$,

$du = -2x\,dx$:

$$\int xe^{-x^2}\,dx = -\frac{1}{2}\int e^u\,du = -\frac{1}{2}e^u + C = -\frac{1}{2}e^{-x^2} + C = -\frac{1}{2e^{x^2}} + C.$$

Thus,

$$\int_{-\infty}^{\infty} xe^{-x^2}\,dx = \lim_{a \to -\infty}\left[-\frac{1}{2e^{x^2}}\right]_a^0 + \lim_{b \to \infty}\left[-\frac{1}{2e^{x^2}}\right]_0^b$$

$$= -\frac{1}{2}\lim_{a \to -\infty}\left[1 - \frac{1}{e^{a^2}}\right] - \frac{1}{2}\lim_{b \to \infty}\left[\frac{1}{e^{b^2}} - 1\right].$$

Since $1/e^{x^2} \to 0$ as $x \to \pm\infty$, we finally get

$$\int_{-\infty}^{\infty} xe^{-x^2}\,dx = -\frac{1}{2}(1) - \frac{1}{2}(-1) = -\frac{1}{2} + \frac{1}{2} = 0. \qquad \blacksquare$$

EXAMPLE 5.7   Does the improper integral $\displaystyle\int_{-\infty}^{0} xe^{-4x}\,dx$ converge or diverge?

Solution The integrand $f(x) = xe^{-4x}$ is continuous, so by the first part of definition 5.3,

$$\int_{-\infty}^{0} xe^{-4x}\,dx = \lim_{a \to -\infty}\int_a^0 xe^{-4x}\,dx.$$

We can find the antiderivative of f via integration by parts (section 2.5) and then a u-substitution. For the integration-by-parts portion, let's choose $u = x$, $dv = e^{-4x}\,dx$,* so that $v = -\frac{1}{4}e^{-4x}$ and $du = dx$. This yields

* Recall the ILATE guidelines from chapter 2 (figure 2.10).

$$\int xe^{-4x}\,dx = -\frac{1}{4}xe^{-4x} - \int -\frac{1}{4}e^{-4x}\,dx$$

$$= -\frac{1}{4}xe^{-4x} + \int \frac{1}{4}e^{-4x}\,dx.$$

The last integral can be calculated via the u-substitution $u = -4x$, $du = -4\,dx$, yielding

$$\int xe^{-4x}\,dx = -\frac{1}{4}xe^{-4x} - \frac{1}{16}e^{-4x} + C.$$

Thus,

$$\int_{-\infty}^{0} xe^{-4x}\,dx = \lim_{a \to -\infty}\int_a^0 xe^{-4x}\,dx$$

$$= \lim_{a \to -\infty}\left[-\frac{1}{4}xe^{-4x}\Big|_a^0 - \frac{1}{16}e^{-4x}\Big|_a^0\right].$$

Simplifying this yields

$$\lim_{a \to -\infty}\left[\frac{e^{-4a}(4a + 1)}{16} - \frac{1}{16}\right].$$

As $a \to -\infty$, $e^{-4a}(4a + 1)$ tends to $\infty(-\infty) = -\infty$, so the limit doesn't exist, and therefore the integral diverges. $\qquad \blacksquare$

Related Exercises   8(b)–11(b) ((b) only), A2–A3, D3.

APPLICATIONS Applied exercise A2 explores the average speed of molecules in an ideal gas, which turns out to be modeled by an improper integral. Applied exercise A3 discusses probability theory, the mean value of a "random variable," and the waiting time to speak to a customer service representative.

EXPLORATIONS Exercise D3 explores the *gamma function*, which both relates to factorials and shows up in the **Riemann hypothesis** (see the last footnote in section 4.4).

The next example revisits our work on volumes of revolution in chapter 3 and investigates the volume of a solid that has an infinitely long tail. Surprisingly, as we'll show, its volume is *finite*.

EXAMPLE 5.8   Let $f(x) = \frac{1}{x}$. Find the volume of the solid generated by revolving the region under the graph of f and bounded by $x \geq 1$ about the x-axis.

Solution Let's first work with the plane region under the graph of $y = 1/x$ and bounded by $x = 1$ and $x = b > 1$. Recalling the guidance from figure 3.17, the disk method (theorem 3.5) says that the volume of the solid obtained by revolving that region about the x-axis is

$$V_b = \pi \int_1^b \left(\frac{1}{x} \right)^2 dx = \pi \int_1^b \frac{1}{x^2} dx = \pi \left[-\frac{1}{x} \right]_1^b = \pi \left[-\frac{1}{b} + 1 \right].$$

The volume V of the solid described in the prompt is the limit as $b \to \infty$ of V_b. Taking this limit yields

$$V = \pi \lim_{b \to \infty} \left[-\frac{1}{b} + 1 \right] = \pi. \qquad \text{Since } \frac{1}{b} \to 0 \text{ as } b \to \infty \qquad \blacksquare$$

Related Exercises   12–16, D4.

EXPLORATIONS The solid produced in the example above looks like a trumpet with an infinitely long tip. One intriguing ramification of our solution is that *we can fill this infinitely long trumpet with a finite volume of water.* Exercise D4 shows, however, that we'd need an *infinite* amount of paint to paint that trumpet. Working with infinities never ceases to amuse!

Evaluating Integrals Containing (Possibly Infinite) Discontinuities

Let's wrap up the section by mirroring what we just did for the case in which f has (possibly infinite) discontinuities.

DEFINITION 5.4: IMPROPER INTEGRALS WITH INFINITE DISCONTINU-ITIES. Let f be a function, and let a, b, and c be real numbers.

1. If f is continuous on $[a, b)$ but has a discontinuity b, then we define

$$\int_a^b f(x)\, dx = \lim_{z \to b^-} \int_a^z f(x)\, dx,$$

provided the limit exists.

2. If f is continuous on $(a, b]$ but has a discontinuity a, then we define

$$\int_a^b f(x)\, dx = \lim_{z \to a^+} \int_z^b f(x)\, dx,$$

provided the limit exists.

3. If f is continuous on $[a, b]$ except for some c in (a, b) at which f has a discontinuity, then we define

$$\int_a^b f(x)\, dx = \lim_{z \to c^-} \int_a^z f(x)\, dx + \lim_{z \to c^+} \int_z^b f(x)\, dx,$$

where $z \neq c$ is any real number in $[a, b]$, provided both limits exist.

In each case, if the associated limits exist we say that the improper integral **converges**. If the limits don't exist—in case 3, if at least one of the limits doesn't exist—then we say the improper integral **diverges**.

EXAMPLE 5.9   Does the improper integral $\displaystyle\int_0^1 \frac{1}{(x-1)^2}\, dx$ converge or diverge?

Solution The integrand here has an infinite discontinuity at $x = 1$, which is inside the interval of integration $[0, 1]$, and is continuous at all other x-values in that interval. We're therefore in the first case of definition 5.4. Thus,

$$\int_0^1 \frac{1}{(x-1)^2}\, dx = \lim_{z \to 1^-} \int_0^z \frac{1}{(x-1)^2}\, dx.$$

Let's use the u-substitution $u = x - 1$, $du = dx$ to find the antiderivative of $\frac{1}{(x-1)^2} = (x-2)^{-2}$:

$$\int (x-1)^{-2}\, dx = \int u^{-2}\, du = -u^{-1} + C = -(x-1)^{-1} + C = -\frac{1}{x-1} + C.$$

Using now the Evaluation Theorem,

$$\lim_{z \to 1^-} \int_0^z \frac{1}{(x-1)^2}\, dx = \lim_{z \to 1^-} \left[-\frac{1}{x-1} \right]_0^z = \lim_{z \to 1^-} \left[-\frac{1}{z-1} - 1 \right].$$

But since $-\frac{1}{z-1} \to \infty$ as $z \to 1^-$, we conclude that the integral diverges.   ∎

EXAMPLE 5.10   Explain what's wrong with the following calculation. Then, correct the calculation.

$$\int_{-1}^1 \frac{1}{x}\, dx = [\ln |x|]_{-1}^1 = \ln 1 - \ln 1 = 0.$$

Solution The Evaluation Theorem is being used, but that requires the integrand to be continuous on the interval of integration, and $1/x$ isn't—it has an infinite discontinuity at $x = 0$, which is inside the interval of integration $[-1, 1]$. Conclusion: The integral is an improper integral, in particular, one of the case 3 integrals of definition 5.4. Thus,

$$\int_{-1}^{1} \frac{1}{x}\, dx = \lim_{z \to 0^-} \int_{-1}^{z} \frac{1}{x}\, dx + \lim_{z \to 0^+} \int_{z}^{1} \frac{1}{x}\, dx.$$

If even one of these integrals diverges, then the original integral will diverge. Looking at the second integral on the right-hand side,

$$\lim_{z \to 0^+} \int_{z}^{1} \frac{1}{x}\, dx = \lim_{z \to 0^+} \left[\ln |x|\right]_{z}^{1} = \ln 1 - \lim_{z \to 0^+} \ln z = - \lim_{z \to 0^+} \ln z.$$

Since $\ln z \to -\infty$ as $z \to 0^+$, this integral diverges, and therefore so does $\int_{-1}^{1} \frac{1}{x}\, dx$. ■

Related Exercises 17–20.

5.4 The Integral Test for Series

For our final act, let's end the chapter with a discussion of perhaps the ultimate combination of all the major topics we've studied thus far (as we'll soon see): the integral test. I mentioned this test in a margin note to theorem 4.7, where I promised to show you how it derives the convergence of p-series theorem 4.7. One of the exercises I'll soon recommend guides you through that derivation. But first, let's work our way up to the integral test itself.

To begin, let f be a function with domain $[1, \infty)$ that's also continuous on that domain. Suppose further that $f(x)$ is positive and decreasing, illustrated in figure 5.1.

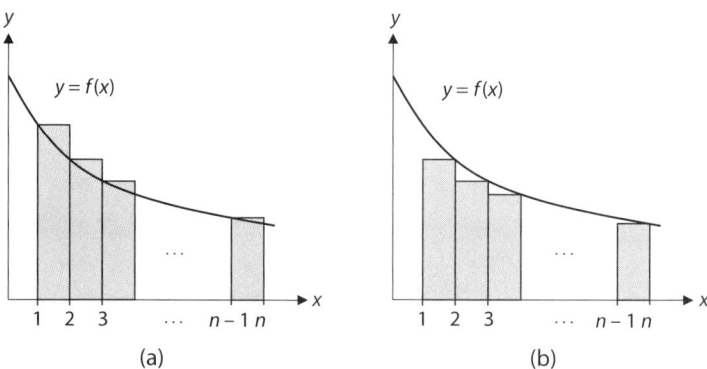

Figure 5.1: The left-hand (a) and right-hand (b) equipartition Riemann sum approximations (with $\Delta x = 1$) to the area under the curve $y = f(x)$.

Let's put on our chapter 2 hat and use Riemann sums to approximate the integral $\int_{1}^{n} f(x)\, dx$, where $n > 1$ is an integer. We'll first equipartition $[1, n]$ into $n - 1$ subintervals of equal width $\Delta x = 1$ (as in figure 5.1). The corresponding left- and right-hand equipartition Riemann sum approximations L_{n-1} and R_{n-1} are then*

$$L_{n-1} = [f(1) + f(2) + \cdots + f(n-1)](1) = \sum_{i=1}^{n-1} f(i),$$

$$R_{n-1} = [f(2) + f(3) + \cdots + f(n)](1) = \sum_{i=2}^{n} f(i).$$

We see this visually in figure 5.1: L_{n-1} is the total area of the shaded rectangles in figure (a); R_{n-1} is the total area of the shaded rectangles in figure (b). Now, since f is decreasing, recall from (2.13) that the right-hand Riemann sum underestimates the area under f, while the left-hand Riemann sum over overestimates it (again, we also see that from figure 5.1):

$$R_{n-1} \leq \int_{1}^{n} f(x)\, dx \leq L_{n-1}.$$

In terms of the nth partial sum $S_n = \sum_{i=1}^{n} f(i)$, this inequality becomes

$$S_n - f(1) \leq \int_{1}^{n} f(x)\, dx \leq S_{n-1}. \tag{5.1}$$

Suppose now that $\int_{1}^{\infty} f(x)\, dx$ converges to L. We know from the previous section that this happens when

$$\lim_{n \to \infty} \int_{1}^{n} f(x)\, dx = L.$$

In the context of the first inequality in (5.1), this implies that $S_n - f(1) \leq L$, or $S_n \leq L + f(1)$. Therefore, every partial sum S_n of the series $\sum f(i)$ is bounded above by $L + f(1)$. Furthermore, since f is a positive function, the values in the sequence of partial sums (S_n) always increase.* Thus, the graph of the sequence (S_n) approaches the horizontal asymptote $y = L$ as $n \to \infty$. Conclusion: (S_n) converges,[2] and therefore from definition 4.6 for infinite series convergence, the infinite series $\sum_{i=1}^{\infty} f(i)$ converges, too. We've just proven the convergence portion of the theorem below.

* For example, $S_2 = f(1) + f(2)$ is larger than $S_1 = f(1)$.

THEOREM 5.2: THE INTEGRAL TEST. Let f be a function with domain $[1, \infty)$ that is also continuous on that domain. Suppose that $f(x)$ is positive and decreasing. Then

$$\sum_{n=1}^{\infty} f(n) \qquad \text{and} \qquad \int_{1}^{\infty} f(x)\, dx$$

either both converge or both diverge.

A few quick comments on this theorem:

- What we just did used a lot of the main concepts developed in this book, so it's completely understandable if you didn't 100% follow the exposition. *That's okay*—you don't need to understand a theorem's derivation/proof to use it. (If you'd still like to understand what we did above, I recommend pausing after each claim made to think about it, and consulting the appropriate formula or section referenced.)

- Our derivation treated only the case when the improper integral $\int_{1}^{\infty} f(x)\, dx$ converges. Exercise D5 guides you through the case when it diverges.

[2] This also follows from the Monotone Convergence Theorem, which we won't discuss here.

- In practice, to use the integral test we take the infinite series, convert its a_n expression into a function $f(x)$, and then verify the test's hypotheses. Examples of the conversion step:

$$\sum_{n=1}^{\infty} \frac{1}{n^3+n^2} \rightarrow \int_1^{\infty} \frac{1}{x^3+x^2}\,dx, \qquad \sum_{n=1}^{\infty} \frac{(-1)^n}{n^3+n^2} \rightarrow \int_1^{\infty} \frac{(-1)^x}{x^3+x^2}\,dx.$$

The first series *can* be investigated via the integral test—because $f(x) = \frac{1}{x^3+x^2}$ is continuous, positive, and decreasing on $[1,\infty)$—but the second one *cannot*, because $\frac{(-1)^x}{x^3+x^2}$ is not always positive on $[1,\infty)$.

- Assuming the test's hypotheses are met, another potential hurdle is the integration step. Thus, *the integral test is a good option when evaluating the improper integral is relatively easy.* This depends in part on the complexity of the a_n expression and on your integration skills.

- Finally, the integral test can be generalized to work if f is only decreasing for $x \geq k$, where k is some positive number (i.e., if f is eventually a decreasing function).

EXAMPLE 5.11 Determine whether the integral test can be applied. If so, use it to determine whether the series converges or diverges.

(a) $\displaystyle\sum_{n=2}^{\infty} \frac{1}{n \ln n}$ (b) $\displaystyle\sum_{n=1}^{\infty} \sin n$

Solution

(a) Here $f(x) = \frac{1}{x \ln x}$, which is continuous and positive on $[2,\infty)$. To see if f is decreasing there, let's calculate f',

$$f(x) = (x \ln x)^{-1} \quad \Longrightarrow \quad f'(x) = -(x \ln x)^{-2}\left[\ln x + \frac{x}{x}\right],$$

using the Chain Rule first and then the Product Rule. This simplifies to

$$f'(x) = -\frac{1+\ln x}{(x \ln x)^2}.$$

The denominator is never negative. On the interval $[2,\infty)$ the function $1 + \ln x$ is always positive. Thus $f'(x) < 0$ on $[2,\infty)$, and so f is decreasing on $[2,\infty)$.[*] Conclusion: The integral test applies. We now investigate the improper integral cousin of the series:

* Recall from Calculus 1 that if $f'(x) < 0$ on an interval I, then f is decreasing on I.

$$\int_2^{\infty} \frac{1}{x \ln x}\,dx = \lim_{b \to \infty} \int_2^b \frac{1}{x \ln x}\,dx.$$

Employing now the u-substitution $u = \ln x$, $du = \frac{1}{x}\,dx$, to calculate the antiderivative yields

$$\int \frac{1}{x \ln x}\,dx = \int \frac{1}{u}\,du = \ln|u| + C = \ln|\ln x| + C.$$

Thus,

$$\lim_{b \to \infty} \int_2^b \frac{1}{x \ln x}\,dx = \lim_{b \to \infty} \left[\ln|\ln x|\right]_2^b$$

$$= \lim_{b\to\infty} [\ln|\ln b| - \ln|\ln 2|]$$

$$= \lim_{b\to\infty} [\ln|\ln b|] - \ln(\ln 2).$$

But $\ln b \to \infty$ as $b \to \infty$, and so this limit doesn't exist—the improper integral diverges. We conclude from the integral test that the infinite series in (a) diverges too.

(b) Here $f(x) = \sin x$. This isn't positive (or decreasing) on $[1, \infty)$, so we cannot use the integral test. ∎

Related Exercises 21–26, D6.

The conclusion from the example we just did echoes the overall conclusion we reached in chapter 4 about convergence tests: For any particular series, one test may be inapplicable or yield inconclusive information, while another may be applicable and yield a definitive conclusion. Returning, then, to the order of choosing tests, shown in box 4.2, I'd put the integral test last in the list of tests to try. That's because the test requires integrating—which can throw you into possibly needing to employ one of the integration techniques we've discussed—and also requires the integrand to be positive and decreasing. This ends up being too much work or too many requirements in most cases. Takeaway: Consider multiple convergence tests when examining a series to see which one will work best with the least amount of effort.

Approximating the Sum of Series Using the Integral Test

In section 4.8 we derived Alternating Series Remainder Theorem 4.16, which allowed us to estimate the sum of a convergent alternating series with nonincreasing terms to any desired accuracy, thereby connecting both the Approximation Question and the Infinite Sum Question from chapter 1. The integral test helps us do something similar, as we'll now see.

Suppose that the integral test's hypotheses are satisfied, and that $\sum f(n)$ converges to S. The remainder $E_N = S - S_N$ in approximating S with its Nth partial sum S_N is

$$E_N = \sum_{n=N+1}^{\infty} f(n) \leq \int_N^{\infty} f(x)\, dx,$$

since E_N is the sum of all the areas of the rectangles in figure 5.2 (the continuation of figure 5.1(b)), which is less than the area under the graph of $y = f(x)$ for $x \geq N$. Similarly,

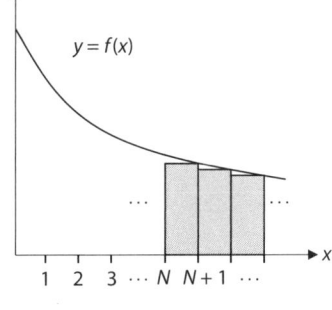

Figure 5.2

$$\int_{N+1}^{\infty} f(x)\, dx \leq E_N,$$

which follows from the analogous continuation of figure 5.1(a). We've proven the following theorem.

THEOREM 5.3: INTEGRAL TEST REMAINDER THEOREM. Let f be a function with domain $[1, \infty)$ that is also continuous on that domain. Suppose that $f(x)$ is positive and decreasing, and that $\sum f(n)$ converges. Let S be its sum, S_N its Nth partial sum, and let $E_N = S - S_N$ be the remainder in approximating S with S_N. Then,

$$\int_{N+1}^{\infty} f(x)\, dx \leq E_N \leq \int_{N}^{\infty} f(x)\, dx. \tag{5.2}$$

This theorem is the analogue of Alternating Series Remainder Theorem 4.16, except for series whose terms are positive and decreasing a_n expressions. And, like the bounds (4.21) on the alternating series' sum implied by that theorem, (5.2) implies the following bounds for the sum S of the series $\sum f(n)$:

* To get this, substitute $E_N = S - S_N$ into (5.2) and add S_N to all sides of the inequality.

$$S_N + \int_{N+1}^{\infty} f(x)\, dx \leq \sum_{n=1}^{\infty} f(n) \leq S_N + \int_{N}^{\infty} f(x)\, dx.^{\star} \tag{5.3}$$

EXAMPLE 5.12 Consider the convergent p-series $\sum_{n=1}^{\infty} \dfrac{1}{n^2}$. Let S denote its sum.

(a) Using the fact that $S_5 = \sum_{n=1}^{5} \dfrac{1}{n^2} = \dfrac{5{,}269}{3{,}600}$, find lower and upper bounds for the error in using S_5 to approximate S.

(b) Find lower and upper bounds for the sum of the series if only the fifth partial sum is used to approximate S.

(c) Determine the minimum number of terms required to approximate S with an error of less than 0.001.

Solution

(a) Let $f(x) = \frac{1}{x^2}$. This function is continuous, positive, and decreasing for $x \geq 1$. Furthermore, $\sum f(n) = \sum \frac{1}{n^2}$ converges (p-series with $p > 1$). Thus, the integral test remainder theorem applies. From (5.2), the error $E_5 = S - S_5$ satisfies

$$\int_{6}^{\infty} \frac{1}{x^2}\, dx \leq E_5 \leq \int_{5}^{\infty} \frac{1}{x^2}\, dx.$$

Now, for $a > 0$,

$$\int_{a}^{\infty} \frac{1}{x^2}\, dx = \lim_{b \to \infty} \int_{a}^{b} \frac{1}{x^2}\, dx = \lim_{b \to \infty} \left[-\frac{1}{x} \right]_{a}^{b} = \lim_{b \to \infty} \left[\frac{1}{a} - \frac{1}{b} \right] = \frac{1}{a}.$$

Using this in the inequality for E_5 yields

$$\frac{1}{6} \leq E_5 \leq \frac{1}{5} \qquad \Longleftrightarrow \qquad 0.1\overline{6} \leq E_5 \leq 0.2.$$

(b) Using (5.3),

$$S_5 + \int_{6}^{\infty} \frac{1}{x^2}\, dx \leq \sum_{n=1}^{\infty} \frac{1}{n^2} \leq S_5 + \int_{5}^{\infty} \frac{1}{x^2}\, dx.$$

Using the S_5 value from (a),

$$\frac{5,269}{3,600} + \frac{1}{6} \le \sum_{n=1}^{\infty} \frac{1}{n^2} \le \frac{5,269}{3,600} + \frac{1}{5}.$$

This generates the interval $1.6302\overline{7} \le S \le 1.9636\overline{1}$.

(c) Using only the right-hand side of (5.2),

$$E_N \le \int_N^{\infty} \frac{1}{x^2}\, dx = \frac{1}{N},$$

using our work from part (a). Setting $\frac{1}{N} \le 0.001 = \frac{1}{1,000}$ yields $N \ge 1,000.$ ∎

Related Exercises   27–29.

5.5   Parting Thoughts

I foreshadowed at the end of chapter 4 that this chapter would be a capstone experience. I hope that's been your experience, now that you've worked through it. We made several new connections between sequences, series, power series, integration, and improper integrals. Furthermore, these connections altogether addressed all three of chapter 1's Big Questions, and did so in new ways. Not bad for such a brief chapter!

The final comment I'll leave you with—for now—is a congratulations. I mean it— you've now worked through five chapters' worth of Calculus 2, including (hopefully) many, many exercises. That took diligence; it took persistence; it took grit. And *you* did it. What's left of the book are this chapter's exercises, the epilogue, and the appendixes. I particularly recommend working out the applied and theoretical exercises in this chapter. Among the gems in there are applications to probability, a proof of Taylor's Theorem, an introduction to the gamma function, and a discussion of the **Riemann zeta function** and its connection with p-series and prime numbers (exercise D7). After that, I have a few more words of encouragement for you in the epilogue, along with a preview of what could be next for you in your mathematics adventure. But for now, put this book down and go out for a treat. You've earned it!

Chapter 5   Exercises

1–4: Find a formula for the series' Nth partial sum S_N, and then find its sum S.

1. $\displaystyle\sum_{n=1}^{\infty} \frac{2}{(2n-1)(2n+1)}$

2. $\displaystyle\sum_{n=1}^{\infty} \frac{4}{n(n+4)}$

3. $\displaystyle\sum_{n=1}^{\infty} \frac{7}{n^2 + 3n + 2}$

4. $\displaystyle\sum_{n=1}^{\infty} \frac{6n+3}{n^2(n+1)^2}$

5–7: (a) Use an appropriate Taylor series to expand the indefinite integral in a power series (set $C = 0$), and determine the series's radius of convergence.

(b) Using the series from (a), determine the number of terms required to approximate the value of the definite integral with an error less than the specified tolerance E.

5. $\displaystyle\int \frac{1}{1+x^3}\,dx, \int_0^1 \frac{1}{1+x^3}\,dx, E = 0.001$

6. $\displaystyle\int \cos(x^2)\,dx, \int_0^1 \cos(x^2)\,dx, E = 0.001$

7. $\displaystyle\int e^{-x^2/2}\,dx, \int_0^1 e^{-x^2/2}\,dx, E = 0.0005$

8–11: (a) Determine whether the integral is an improper integral, and if so, explain why. (b) If the integral is improper, does it converge or diverge?

8. $\displaystyle\int_0^\infty \frac{1}{x^2+1}\,dx$ 9. $\displaystyle\int_0^\infty 2e^{-x}\cos x\,dx$

10. $\displaystyle\int_e^\infty \frac{1}{x\ln^2 x}\,dx$ 11. $\displaystyle\int_1^\infty \frac{e^{1-\sqrt{x}}}{\sqrt{x}}\,dx$

12. Find the volume of the solid obtained by revolving the area under the graph of $f(x) = \frac{1}{(x+1)^{7/2}}$ for $x \geq 0$ about the x-axis.

13. Find (a) the area under the graph of $f(x) = e^{-3x}$ for $x \geq 0$ and (b) the volume of the solid obtained by revolving that area about the x-axis.

14. Let $f(x) = e^{-x}$. (a) Write out the integral representing the volume of the solid obtained by revolving f about the y-axis for $x \geq 0$ using the shell method. (b) Evaluate your integral from (a).

15. Let $f(x) = \frac{9}{x}$. (a) Write out the integral representing the arc length of f for $x \geq 1$. (b) Explain why the arc length in (a) is infinite *without* evaluating the integral.

16. Find the area under the graph of $f(x) = \frac{x}{\sqrt{4-x^2}}$ for $0 \leq x \leq 2$.

17–20: (a) Determine whether the integral is an improper integral, and if so, explain why. (b) If the integral is improper, does it converge or diverge?

17. $\displaystyle\int_0^1 \frac{1}{2\sqrt{x}}\,dx$ 18. $\displaystyle\int_0^1 \frac{1}{\sqrt{1-x^2}}\,dx$

19. $\displaystyle\int_0^2 \frac{1}{4-x^2}\,dx$ 20. $\displaystyle\int_0^{\pi/2} \frac{\sin x}{2\sqrt{1-\cos x}}\,dx$

21–26: Use the integral test to determine whether the series converges or diverges.

21. $\displaystyle\sum_{n=1}^\infty \frac{2\ln n}{n^3}$ 22. $\displaystyle\sum_{n=5}^\infty \frac{3}{n\sqrt{\ln n}}$

23. $\displaystyle\sum_{n=1}^\infty \frac{\arctan n}{n^2+1}$ 24. $\displaystyle\sum_{n=1}^\infty \frac{5}{(2n+3)^3}$

25. $\displaystyle\sum_{n=1}^\infty \frac{4n}{2n^2+5}$ 26. $\displaystyle\sum_{n=3}^\infty \frac{4n}{(4n+5)^{3/2}}$

27–29: (a) Use the integral test remainder theorem to find the maximum error in approximating the sum of the convergent series with the indicated partial sum. (b) Use the integral test remainder theorem to determine the minimum number of terms required to approximate the sum of the series with an error of less than 0.001.

27. $\displaystyle\sum_{n=1}^\infty \frac{1}{n^4}, S_3 = \frac{1{,}393}{1{,}296}$

28. $\displaystyle\sum_{n=1}^\infty ne^{-n^2}, S_3 = \frac{3+2e^5+e^8}{e^9}$

29. $\displaystyle\sum_{n=1}^\infty \frac{1}{n^2+1}, S_4 = \frac{73}{85}$

A1. Measuring time using a pendulum Consider a pendulum of length ℓ, and denote by $\theta > 0$ the initial angle of the pendulum with respect to a vertical line before it's released from rest (see diagram below).

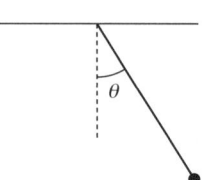

The time to complete one full swing of the pendulum (the **period**) is

$$T = \frac{4}{\omega} \int_0^{\pi/2} \frac{1}{\sqrt{1 - k^2 \sin^2 x}} \, dx,$$

where $\omega = \sqrt{g/\ell}$, g is the acceleration of gravity, and $k^2 = \sin^2(\theta/2)$.

(a) The integral above is hard to calculate. Identify the appropriate Maclaurin series that allows us to write $\frac{1}{\sqrt{1 - k^2 \sin^2 x}} \approx 1 + \frac{k^2 \sin^2 x}{2}$ for small values of $k \sin x$.

(b) Use the result from part (a) to show that $T \approx \frac{2\pi}{\omega} + \frac{\pi}{2} \frac{k^2}{\omega}$. When θ (the initial angle) is small, since $\sin \theta \approx \theta$ for small θ (from the Maclaurin series for $\sin x$), this approximation reduces further to $T \approx \frac{2\pi}{\omega} = 2\pi \sqrt{\frac{\ell}{g}}$. With $\theta = 3°$ and $\ell = 1$ meter, $T \approx 2$ seconds. This helps us measure time by counting the number of full swings of the pendulum.

A2. Speed of molecules in an ideal gas
In an **ideal gas** molecules move around in random trajectories and have little to no interaction with each other. The average speed v_a of molecules in such a gas is

$$v_a = \frac{4k^{3/2}}{\sqrt{\pi}} \int_0^\infty v^3 e^{-kv^2} \, dv,$$

where $c > 0$ is a constant. Show that $v_a = \frac{2}{\sqrt{k\pi}}$.

A3. Applications to probability
"Random" events are not really random. There is always some probability that the event will occur. A **probability density function** (pdf) f for a **random variable** X encodes such probabilities. Specifically, if X has pdf f, then the probability that a measurement of X will return a value between a and b is

$$P(a \leq X \leq b) = \int_a^b f(x) \, dx.$$

Probability values lie between 0 (the event never happens) to 1 (the event always happens). And since all probabilities must add to 1, we require that

$$\int_A^B f(x) \, dx = 1,$$

where A and B are the smallest and largest, respectively, possible values of X. (It's possible that $A = -\infty$ and $B = \infty$.) Often we are also interested in the **mean value** (the average) of X. This is defined as

$$\overline{X} = \int_A^B x f(x) \, dx.$$

(a) An **exponential probability density** $f(x) = \frac{e^{-x/t}}{t}$, where $x > 0$ and $t > 0$, is sometimes used to model waiting times. With $A = 0$ and $B = \infty$, verify that $\int_A^B f(x) \, dx = 1$.

(b) Show that $\overline{X} = t$ for the exponential density in part (a).

(c) A customer calls in to a company and is put on hold. Assuming that the waiting time T is a random variable with an exponential probability density function with mean 10 minutes, what is the probability that the customer will wait more than 20 minutes?

D1. Leibniz's formula for π
Use the fact that

$$\int_0^1 \frac{1}{1 + x^2} \, dx = [\arctan x]_0^1 = \frac{\pi}{4}$$

and the Maclaurin series for $\frac{1}{1+x^2}$ to show that

$$\frac{\pi}{4} = \sum_{n=0}^{\infty} (-1)^n \frac{1}{2n + 1}.$$

Discuss the convergence of the series.

D2. Deriving Taylor's Theorem (theorem 4.17)
Let f be differentiable up to order $n + 1$ on $[a, b]$.

(a) Use the Evaluation Theorem to show that

$$f(x) = f(a) + \int_a^x f'(t) \, dt.$$

(b) Use integration by parts to show that

$$f(x) = f(a) + (x - a)f'(a) + \int_a^x (x - t)f''(t) \, dt.$$

(c) Show that, after n uses of integration by parts,

$$f(x) = T_n(x) + \int_a^x \frac{f^{(n+1)}(t)}{n!} (x-t)^n \, dt.$$

One can use the Mean Value Theorem for integrals to show that the highlighted term equals $\frac{f^{(n+1)}(z)}{(n+1)!}(x-a)^{n+1}$ for some z in between x and a. This recovers (4.29).

D3. Let $r \neq 0$ be a real number that is not a negative integer (e.g., -3). The **gamma function** $\Gamma(r) = \int_0^\infty x^{r-1} e^{-x} \, dx$.

(a) Let $r = n$, a positive integer. Use integration by parts to show that $\Gamma(n+1) = n\Gamma(n)$.

(b) Apply the result from (a) over and over again to conclude that $\Gamma(n+1) = n!$.

D4. Let S be the surface area of the infinitely long trumpet from example 5.8. Show, via (3.16), that the surface area of this solid is

$$S = 2\pi \lim_{b \to \infty} \int_1^b \frac{1}{x} \sqrt{1 + \frac{1}{x^4}} \, dx.$$

Then, explain how the fact that

$$\frac{1}{x}\sqrt{1 + \frac{1}{x^4}} > \frac{1}{x}$$

implies that S is infinite.

D5. To prove the other half of the integral test, suppose that the test's hypotheses are satisfied but that $\int_1^\infty f(x) \, dx$ diverges.

(a) Explain why $\int_1^n f(x) \, dx$ tends to ∞ as $n \to \infty$.

(b) Use the result from (a) and (5.1) to conclude that $\sum_{n=1}^\infty f(n)$ diverges.

D6. This exercise derives the p-series convergence theorem 4.7, from the integral test.

(a) Show that $\int_1^\infty \frac{1}{x} \, dx$ diverges. Then, explain how this helps you conclude that $\sum_{n=1}^\infty \frac{1}{n}$ diverges.

(b) Let $p \neq 1$ be a positive real number. Show that

$$\int_1^\infty \frac{1}{x^p} \, dx = -\frac{1}{1-p} + \lim_{b \to \infty} \frac{b^{1-p}}{1-p}.$$

(c) Explain how the result from (b) implies that $\sum_{n=1}^\infty \frac{1}{n^p}$ diverges for $0 < p < 1$ and converges for $p > 1$.

D7. The Riemann zeta function In the last footnote in section 4.4 I briefly mentioned the Riemann hypothesis and its connection to prime numbers. This exercise will probe that connection. For x a real number, with $|x| \geq 1$, the **Riemann zeta function** over the real numbers, $\zeta(x)$, is defined as

$$\zeta(x) = \sum_{n=1}^\infty \frac{1}{n^x}.$$

(Note that this is a p-series when $x \geq 0$.) Euler proved the remarkable result that for $x > 1$,

$$\zeta(x) = \left(\frac{1}{1-2^{-x}}\right)\left(\frac{1}{1-3^{-x}}\right)\left(\frac{1}{1-5^{-x}}\right) \cdots,$$

where the product goes on forever and each successive term is of the form $\frac{1}{1-p^{-x}}$, where p is the next prime number.

(a) The **Bernoulli numbers** B_n are a sequence of special numbers occurring in the powers series

$$\frac{x}{e^x - 1} = \sum_{n=0}^\infty \frac{B_n x^n}{n!}.$$

Calculate the first four terms in the Maclaurin series expansion of $\frac{x}{e^x-1}$, and extract from that the Bernoulli numbers B_0 through B_4.

(b) Let $m \geq 1$ be an integer. Remarkably,

$$\zeta(2m) = \frac{(2\pi)^{2m}|B_{2m}|}{2(2m)!}.$$

Use this formula, along with your results from (a), to calculate the sums of the p-series

$$\sum_{n=1}^\infty \frac{1}{n^2}, \quad \sum_{n=1}^\infty \frac{1}{n^4}.$$

(c) The formula in part (b) helps sum even-powered p-series. For $r > 1$ (not necessarily an even number), it turns out that

$$\zeta(r) = \frac{1}{\Gamma(r)} \int_0^\infty \frac{x^{r-1}}{e^x - 1}\, dx,$$

where $\Gamma(r)$ is the gamma function from exercise D3. Use this to write out an integral formula for the sum of the p-series $\displaystyle\sum_{n=1}^\infty \frac{1}{n^3}$. (One can then use the methods developed in this book to approximate that integral, and therefore the sum.)

Epilogue

Let me be the first to congratulate you for working through this book. Calculus is a subject that is viewed by many as difficult, abstract, and inaccessible. I sincerely hope this book gave you the opposite experience. (And if so, tell your friends!)

We started our journey by studying the three Big Questions that drove the development of Calculus 2. We then tackled them by switching to a dynamics mindset and employing the Calculus 2 workflow. This is the Way, or *dao*, of Calculus 2. And here's a secret: *Every subject in mathematics has its own* dao, *its own way of doing things*—the *dao* of differential equations, the *dao* of linear algebra, etc. In fact, mathematics as a discipline has its own *dao*—its own way of contemplating and making sense of things and transforming them into new concepts and knowledge. I encourage you to embrace this *dao* framework and to be on the lookout for the guiding principles (e.g., "shift to a dynamics mindset") and practices (e.g., the Calculus 2 workflow) embedded in your next mathematics adventure.

With that in mind, let me point you in a few possible future directions, now that you've studied Calculus 2.

- **Multivariable calculus.** All the calculus covered in Calculus 1 and 2 uses single-variable functions, like $f(x)$. In multivariable calculus you study *multi*variable functions, like $g(x, y, z)$. You're already familiar with these—the temperature T of the air surrounding you right now, for example, is a function of your three-dimensional spatial location plus time: $T = T(x, y, z, t)$. Quantifying how such multivariable functions change leads to "partial derivatives" in multivariable calculus. Other core calculus concepts, such as integration and Taylor series, also generalize to multiple variables. Multivariable calculus, therefore, is an excellent follow-up to Calculus 2, and one whose *dao* is very similar to Calculus 2's.

- **Differential equations**. In many, *many* places in mathematics (and its real-world applications) a function is related to its derivative (or higher-order derivatives) by an equation. Example: For $y = e^x$, we know that $y' = e^x$. In other words, $y' = y$. This is a *differential equation*, an equation involving a function and its derivatives. Differential equations, as a subject, therefore studies such relationships and develops methods to solve such equations. This subject tends to be more applied than multivariable calculus. That's because differential equations are used throughout the sciences—social, life, and physical—to model and make sense of the world around us. The *dao* of differential equations is different from that of the calculus *daos* (which all involve a dynamics mindset plus a "limit" workflow). The dynamics mindset is retained, but the workflows are different.

- **Linear algebra.** For a break from calculus, consider linear algebra. We've solved some systems of equations in this book, mainly in connection with solving for the constants in a system of linear equations generated from a partial fractions decomposition. A big chunk of linear algebra studies when such systems of linear equations have solutions, how many solutions they have, and how we can find them using, for example, matrices. Organizing all those results then leads to some beautiful theories (e.g., vector spaces) that have come to underlie much of the modern mathematics that comes after linear algebra. Linear algebra also has *many* applications to real-world contexts, including to economics, demography, computer graphics, physics, and beyond. The *dao* of linear algebra is a completely new *dao* (relative to the calculus *daos*). In brief, it shifts the perspective from calculus's dynamics mindset to what one could call a structural perspective: How is a set of objects organized? Are there "ingredient members" of the set that, when combined in the right ways, generate all members of the set (like how north, south, east, and west generate every direction one could walk on a plane)?

There are even more follow-up math courses to consider (e.g., statistics), but I'll stop here. Thank you again for trusting me to be your Calculus 2 tour guide. I hope you enjoyed the book, and that you continue studying mathematics.

Acknowledgments

My name is on the cover of this book, but in truth it was a collaborative effort. Without the support of my family—Zoraida, Emilia, and Alicia, in particular—this book would not exist. I'm also grateful to the entire team at Princeton University Press, and in particular to my editor, Diana Gillooly. Your collective work on this book has produced a resource for students that costs a fraction of what a calculus textbook does yet teaches the same mathematics and looks just as professional. Thank you to the reviewers, students and faculty alike, who took the time to provide valuable feedback on early drafts of the book. And finally, thank *you*. Ultimately, all the time and energy we collectively put into realizing this book were directed toward one goal: *to help you learn Calculus 2*. Thank you for choosing this book as your guide on that adventure.

Appendixes A–D: Precalculus and Calculus Review

Appendix A: Algebra and Geometry Review

1. Number Systems

- *Natural numbers* ($\mathbb{N}$): $1, 2, 3, \ldots$

- *Integers* ($\mathbb{Z}$): $0, \pm 1, \pm 2, \ldots$

- *Rational numbers* ($\mathbb{Q}$): $\dfrac{p}{q}$, where p and $q \neq 0$ are integers

- *Real numbers* ($\mathbb{R}$): integers plus irrational numbers (like π)

2. Algebra

Arithmetic Operations

$$\frac{a}{b} \pm \frac{c}{d} = \frac{ad \pm bc}{bd} \qquad \frac{1}{\frac{c}{d}} = \frac{d}{c}$$

$$\frac{a}{b} \cdot \frac{c}{d} = \frac{ac}{bd} \qquad a(b \pm c) = ab \pm ac$$

Interval Notation

$$(a, b) = \{a < x < b\}$$
$$[a, b) = \{a \leq x < b\}$$
$$[a, b] = \{a \leq x \leq b\}$$
$$(a, b] = \{a < x \leq b\}$$

Rules of Exponents

$$x^m x^n = x^{m+n} \qquad \left(\frac{x}{y}\right)^n = \frac{x^n}{y^n}$$
$$\frac{x^m}{x^n} = x^{m-n}$$
$$(x^m)^n = x^{mn} \qquad x^{1/n} = \sqrt[n]{x}$$
$$\sqrt[n]{x^m} = x^{m/n}$$
$$x^{-n} = \frac{1}{x^n} \qquad \sqrt[n]{xy} = \sqrt[n]{x}\sqrt[n]{y}$$
$$(xy)^n = x^n y^n \qquad \frac{\sqrt[n]{x}}{\sqrt[n]{y}} = \sqrt[n]{\frac{x}{y}}$$

Factoring, Expanding

$$x^2 - y^2 = (x + y)(x - y)$$
$$x^3 + y^3 = (x + y)\left(x^2 - xy + y^2\right)$$
$$x^3 - y^3 = (x - y)\left(x^2 + xy + y^2\right)$$
$$(x + y)^2 = x^2 + 2xy + y^2$$
$$(x - y)^2 = x^2 - 2xy + y^2$$
$$(x + y)^3 = x^3 + 3x^2 y + 3xy^2 + y^3$$
$$(x - y)^3 = x^3 - 3x^2 y + 3xy^2 - y^3$$

Quadratic Formula

The solutions to $ax^2 + bx + c = 0$ are

$$x = \frac{-b \pm \sqrt{b^2 - 4ac}}{2a}.$$

3. Geometry

Areas and Volumes

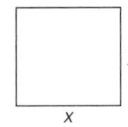  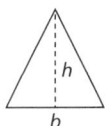

$$\text{Area} = xy \qquad \text{Area} = \pi r^2 \qquad \text{Area} = \tfrac{1}{2}bh$$

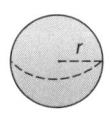 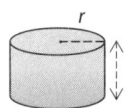 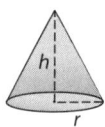

$$V = \tfrac{4}{3}\pi r^3 \qquad V = \pi r^2 h \qquad V = \tfrac{1}{3}\pi r^2 h$$

Distance Formula

The distance d between two points (x_1, y_1) and (x_2, y_2) in the plane is

$$d = \sqrt{(x_2 - x_1)^2 + (y_2 - y_1)^2}.$$

Lines

- Slope of line through (x_1, y_1) and (x_2, y_2):
$$m = \frac{\Delta y}{\Delta x} = \frac{y_2 - y_1}{x_2 - x_1}$$

- Point-slope equation of line through (x_1, y_1) with slope m: $y - y_1 = m(x - x_1)$

- Slope-intercept equation of line with slope m and y-intercept b: $y = mx + b$

Circles and Ellipses

- Equation of circle with center (h, k) and radius r:

$$(x - h)^2 + (y - k)^2 = r^2$$

- Equation of ellipse with center (h, k) and major axis parallel to the x-axis:

$$\frac{(x - h)^2}{a^2} + \frac{(y - k)^2}{b^2} = 1,$$

where $a > b$, the length of the major axis is $2a$, and the length of the minor axis is $2b$. (If the major axis is parallel to the y-axis, swap a^2 and b^2 in the equation, and require $b > a$.)

4. Trigonometry

Angles and Right Triangles

- *Definition of a radian*: An angle of "1 radian" is the central angle θ of a circular arc whose length s is equal to the radius r of the circle.

- More generally, for θ in radians, $s = r\theta$. We call the perimeter of the circle the circumference C of the circle, and $C = 2\pi r$.

- *Angle conversions*: π radians $= 180°$, $1° = \frac{\pi}{180}$ radians, 1 radian $= \frac{180°}{\pi}$.

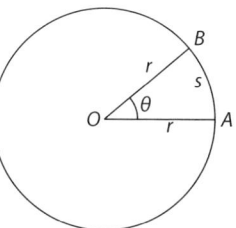

Trigonometry and Right Triangles, Trigonometric Values

Relative to the right triangle below,

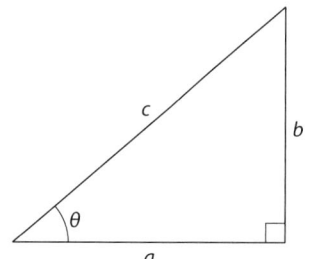

$$\sin \theta = \frac{b}{c} \qquad \csc \theta = \frac{c}{b}$$

$$\cos \theta = \frac{a}{c} \qquad \sec \theta = \frac{c}{a}$$

$$\tan \theta = \frac{b}{a} \qquad \cot \theta = \frac{a}{b}$$

θ	$\cos \theta$	$\sin \theta$	$\tan \theta$
0 (0°)	1	0	0
$\frac{\pi}{6}$ (30°)	$\frac{\sqrt{3}}{2}$	$\frac{1}{2}$	$\frac{\sqrt{3}}{3}$
$\frac{\pi}{4}$ (45°)	$\frac{\sqrt{2}}{2}$	$\frac{\sqrt{2}}{2}$	1
$\frac{\pi}{3}$ (60°)	$\frac{1}{2}$	$\frac{\sqrt{3}}{2}$	$\sqrt{3}$
$\frac{\pi}{2}$ (90°)	0	1	undefined

Appendix B: **Precalculus (Functions) Review**

Basic Concepts

- A **function** f is a rule that assigns to each input x a unique output $y = f(x)$.
- The set of allowable inputs of f is its **domain**. The set of all outputs of f is its **range**.

Polynomial, Rational, and Power Functions

- A linear function has the form $f(x) = mx + b$, where m is the slope and $(0, b)$ the y-intercept.
- A quadratic function has the form $f(x) = ax^2 + bx + c$.
- A polynomial function has the form $f(x) = a_n x^n + a_{n-1} x^{n-1} + \cdots + a_1 x + a_0$, where $n \geq 0$ is an integer and is called the degree of the polynomial.
- A rational function as the form $f(x) = \frac{p(x)}{q(x)}$, where p and q are polynomials.
- A power function has the form $f(x) = ax^b$.

Trigonometric Functions

Relative to the figure below,

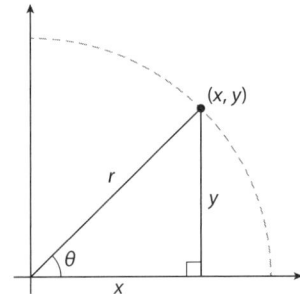

$$\sin \theta = \frac{y}{r} \qquad \csc \theta = \frac{r}{y}$$

$$\cos \theta = \frac{x}{r} \qquad \sec \theta = \frac{r}{x}$$

$$\tan \theta = \frac{y}{x} \qquad \cot \theta = \frac{x}{y}$$

Quotient Trigonometric Functions

$$\csc \theta = \frac{1}{\sin \theta} \qquad \sec \theta = \frac{1}{\cos \theta}$$

$$\tan \theta = \frac{\sin \theta}{\cos \theta} \qquad \cot \theta = \frac{1}{\tan \theta}$$

Basic Trigonometric Identities

$$\sin^2 \theta + \cos^2 \theta = 1 \qquad \sin(-\theta) = -\sin \theta$$

$$1 + \tan^2 \theta = \sec^2 \theta \qquad \cos(-\theta) = \cos \theta$$

$$1 + \cot^2 \theta = \csc^2 \theta \qquad \tan(-\theta) = -\tan \theta$$

$$\sin\left(\tfrac{\pi}{2} - \theta\right) = \cos \theta$$

$$\cos\left(\tfrac{\pi}{2} - \theta\right) = \sin \theta$$

$$\tan\left(\tfrac{\pi}{2} - \theta\right) = \cot \theta$$

Law of Sines

$$\frac{\sin A}{a} = \frac{\sin B}{b} = \frac{\sin C}{c}$$

Law of Cosines

$$a^2 = b^2 + c^2 - 2bc \cos A$$

Addition, Subtraction Identities

$$\sin(x + y) = \sin x \cos y + \cos x \sin y$$

$$\sin(x - y) = \sin x \cos y - \cos x \sin y$$

$$\cos(x + y) = \cos x \cos y - \sin x \sin y$$

$$\cos(x - y) = \cos x \cos y + \sin x \sin y$$

$$\tan(x + y) = \frac{\tan x + \tan y}{1 - \tan x \tan y}$$

$$\tan(x - y) = \frac{\tan x - \tan y}{1 + \tan x \tan y}$$

Double-Angle Identities

$$\sin 2x = 2 \sin x \cos x$$

$$\cos 2x = \cos^2 x - \sin^2 x$$

$$\tan 2x = \frac{2 \tan x}{1 - \tan^2 x}$$

$$\sin^2 x = \frac{1 - \cos 2x}{2}$$

$$\cos^2 x = \frac{1 + \cos 2x}{2}$$

Exponential and Logarithmic Functions

- An exponential function has the form $f(x) = ab^x$, where $a \neq 0$, $b > 0$, and $b \neq 1$. If $b > 1$ the graph of f is increasing; if $0 < b < 1$ it's decreasing. We call b the base of the exponential function and a the initial value (since $f(0) = a$).

- A logarithmic function has the form $f(x) = \log_b x$, where $b > 0$ and $b \neq 1$. If $b = 10$ we write $\log_{10} x = \log x$, called the common logarithm. If $b = e$ (Euler's number, $e \approx 2.71$) we write $\log_e x = \ln x$, called the natural logarithm.

- Exponential and logarithmic functions undo each other:

$$b^{\log_b c} = c, \quad \log_b(b^x) = x$$

- *The Rules of Logarithms*: For x and y positive real numbers, and r any real number,

$$\log_b(xy) = \log_b x + \log_b y,$$

$$\log_b\left(\frac{x}{y}\right) = \log_b x - \log_b y,$$

$$\log_b(x^r) = r\log_b x.$$

- *Change of base formula*:

$$\log_b c = \frac{\log_a c}{\log_a b}$$

Function Operations

Addition, subtraction:

$$(f + g)(x) = f(x) + g(x)$$

$$(f - g)(x) = f(x) - g(x)$$

Multiplication, division:

$$(fg)(x) = f(x)g(x)$$

$$\left(\frac{f}{g}\right)(x) = \frac{f(x)}{g(x)}$$

Composition:

$$(f \circ g)(x) = f(g(x))$$

Graphs of Functions

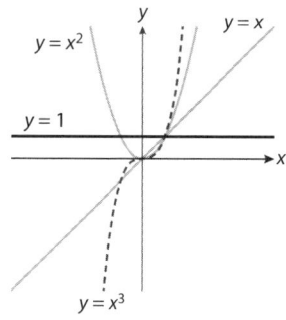

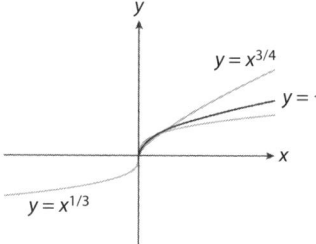

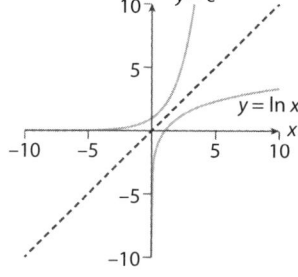

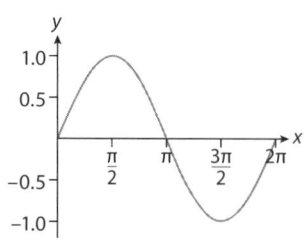

$y = \sin x$

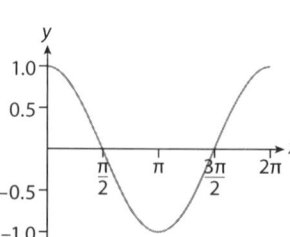

$y = \cos x$

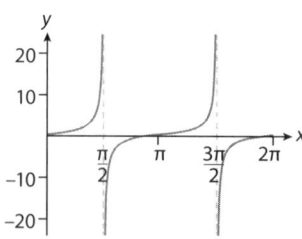

$y = \tan x$

Appendix C: Calculus Review I—Up to Differentiation

1. Limits

One-Sided Limits

Let c and L be real numbers.

- *Left-hand limit*: We write $\lim_{x \to c^-} f(x) = L$ when the values $f(x)$ can be made as close as desired to L by making x sufficiently close to (but never equal to) c, where $x < c$. As a shorthand, we also write $f(x) \to L$ as $x \to c^-$.

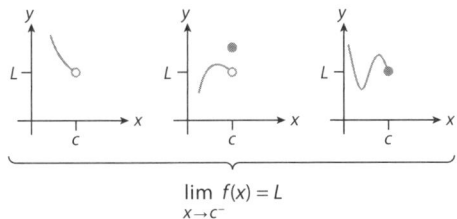

$$\lim_{x \to c^-} f(x) = L$$

- *Right-hand limit*: We write $\lim_{x \to c^+} f(x) = L$ when the values $f(x)$ can be made as close as desired to L by making x sufficiently close to (but never equal to) c, where $x > c$. As a shorthand, we also write $f(x) \to L$ as $x \to c^+$.

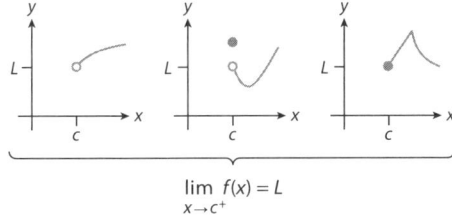

$$\lim_{x \to c^+} f(x) = L$$

Two-Sided Limits

Let c and L be real numbers. We write $\lim_{x \to c} f(x) = L$ when the values $f(x)$ can be made as close as desired to L by making x sufficiently close to (but never equal to) c. As a shorthand, we also write $f(x) \to L$ as $x \to c$.

Existence of Two-Sided Limit

The $\lim_{x \to c} f(x)$ exists only if

$$\lim_{x \to c^-} f(x) = L \quad \text{and} \quad \lim_{x \to c^+} f(x) = L.$$

If either (or both) of these conditions fails, the two-sided limit does not exist.

Nonexistence of Limits

- If $f(x) \to \pm\infty$ as $x \to c$, $x \to c^-$, or $x \to c^+$, then the associated limit does not exist. Furthermore, in this case $x = c$ is a vertical asymptote of the graph of f. Example: $f(x) = \frac{1}{x}$ as $x \to 0$, illustrated in figure (a) below

- If $f(x)$ oscillates wildly as $x \to c$, $x \to c^-$, or $x \to c^+$, with no discernible y-value as its limiting value, then the associated limit does not exist. Example: $g(x) = \sin\left(\frac{1}{x}\right)$ as $x \to 0$, illustrated in figure (b) below

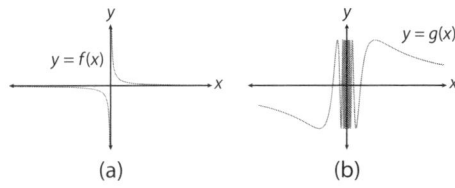

(a)   (b)

Continuity

- We say f is **continuous from the left at** $x = c$ if $\lim_{x \to c^-} f(x) = f(c)$.

- We say f is **continuous from the right at** $x = c$ if $\lim_{x \to c^+} f(x) = f(c)$.

- We say f is **continuous at** $x = c$ if $\lim_{x \to c} f(x) = f(c)$.

Criteria for Continuity

For f to be continuous at a number c in its domain, $f(c)$ must exist, $\lim_{x \to c} f(x)$ must exist, and $\lim_{x \to c} f(x)$ must equal $f(c)$.

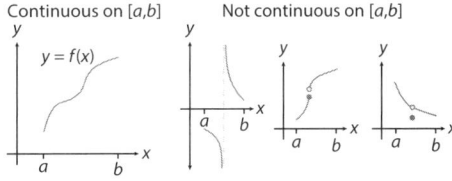

Continuous on $[a,b]$ Not continuous on $[a,b]$

- *Continuity on an interval*: We say that f is continuous on an interval I if f is continuous for all x in I. When f is continuous on $(-\infty, \infty)$, we call f continuous.

Families of Continuous Functions

The following function families are continuous at all points on their domain: polynomials; power functions; rational functions; exponential functions; $\sin x$, $\cos x$, $\tan x$, $\sec x$, $\csc x$, $\cot x$; and $\log_b x$.

Combining Continuous Functions

- If f and g are continuous at c, then so are $f \pm g$, fg, af (where a is a real number), and $\frac{f}{g}$ (provided $g(c) \neq 0$).

- If f is continuous at a, and $\lim_{x \to c} g(x) = a$, then

$$\lim_{x \to c} f(g(x)) = f\left(\lim_{x \to c} g(x)\right) = f(a).$$

Intermediate Value Theorem

Suppose that f is continuous on $[a, b]$ and that $f(a) \neq f(b)$. If z is any number between $f(a)$ and $f(b)$, then there exists a number c in (a, b) such that $f(c) = z$. (Translation: The graph of a continuous function passes through every point between the points $(a, f(a))$ and $(b, f(b))$.)

Evaluating Limits

The Limit Laws

Let "lim" be a stand-in for $\lim_{x \to c}$, $\lim_{x \to c^-}$, or $\lim_{x \to c^+}$. Suppose that f and g are functions, and that $\lim f(x)$ and $\lim g(x)$ both exist. Then:

1. $\lim [f(x) \pm g(x)] = \lim f(x) \pm \lim g(x)$.

2. $\lim [kf(x)] = k[\lim f(x)]$, where k is a real number.

3. $\lim [f(x)g(x)] = [\lim f(x)][\lim g(x)]$.

4. $\lim \dfrac{f(x)}{g(x)} = \dfrac{\lim f(x)}{\lim g(x)}$, provided $[\lim g(x)] \neq 0$.

5. For n a positive integer, $\lim \sqrt[n]{f(x)} = \sqrt[n]{\lim f(x)}$, provided $[\lim f(x)] \geq 0$ if n is even.

6. For n a positive integer, $\lim [f(x)]^n = [\lim f(x)]^n$.

7. Supposing f is continuous at $\lim g(x)$, $\lim f(g(x)) = f(\lim g(x))$.

Techniques

Factor. Example:

$$\lim_{x \to 1} \frac{x^2 - 1}{x - 1} = \lim_{x \to 1} \frac{(x-1)(x+1)}{x-1}$$
$$= \lim_{x \to 1} (x + 1) = 2$$

For rational functions, factor out x^n from the numerator and denominator, where n is the largest power of x in the denominator. Example:

$$\lim_{x \to \infty} \frac{x^2 + 7}{x^3 + x} = \lim_{x \to \infty} \frac{x^3 \left(\frac{1}{x} + \frac{7}{x^3}\right)}{x^3 \left(1 + \frac{1}{x^2}\right)}$$
$$= \lim_{x \to \infty} \frac{\frac{1}{x} + \frac{7}{x^3}}{1 + \frac{1}{x^2}} = 0$$

Rationalize. Example: $\lim_{x \to 0} \dfrac{\sqrt{1+x} - 1}{x}$. Since

$$\frac{\sqrt{1+x} - 1}{x} \cdot \frac{\sqrt{1+x} + 1}{\sqrt{1+x} + 1}$$
$$= \frac{(1+x) - 1}{x(\sqrt{1+x} + 1)}$$
$$= \frac{1}{\sqrt{1+x} + 1}, \tag{A.1}$$

then

$$\lim_{x \to 0} \frac{\sqrt{1+x} - 1}{x} = \lim_{x \to 0} \frac{1}{\sqrt{1+x} + 1} = \frac{1}{2}.$$

L'Hopita's Rule. This is covered in appendix F.

2. Differentiation

Foundational Concepts

- The **derivative function** $f'(x)$ of f, denoted by $f'(x)$, is defined by

$$f'(x) = \lim_{h \to 0} \frac{f(x+h) - f(x)}{h},$$

for all x-values for which the limit exists.

- *Leibniz notation:* $\frac{dy}{dx} = f'(x)$.

- $f'(a)$ is the slope of the line tangent to the graph of $y = f(x)$ at the point $(a, f(a))$.

- $f'(a)$ measures the instantaneous rate of change of $y = f(x)$ at $x = a$.

Differentiability

- We call f **differentiable at** $x = a$ if $f'(a)$ exists. If f is differentiable for all x inside an interval I, we say that f is **differentiable on** I.

- Three common situations for which $f'(a)$ does not exist are when, at $x = a$, the graph of f has a kink, a gap, or a vertical tangent. These are illustrated, left to right, in the figure below:

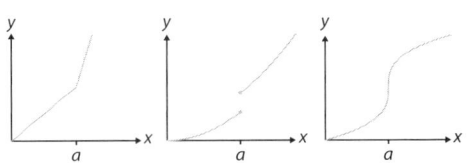

- *Theorem*: If f is differentiable at a, then f is continuous at a. If f is discontinuous at a, then f is not differentiable at a.

Foundational Derivatives

- $(1)' = 0$

- $(c)' = 0$ for any constant c

- $(x)' = 0$

- $(x^n)' = nx^{n-1}$ (Power Rule)

- $(\sin x)' = \cos x$

- $(\cos x)' = -\sin x$

- $(\tan x)' = \sec^2 x$

- $(\sec x)' = \sec x \tan x$

- $(\csc x)' = -\csc x \cot x$

- $(\cot x)' = -\csc^2 x$

- $(e^x)' = e^x$

- $(b^x)' = b^x \ln b$

- $(\ln x)' = \dfrac{1}{x}$

- $(\log_a x)' = \dfrac{1}{x(\ln a)}$

Differentiation Rules

Constant Multiple, Sum, and Difference Rules

$$\frac{d}{dx}[cf(x)] = cf'(x)$$

$$\frac{d}{dx}[f(x) + g(x)] = f'(x) + g'(x)$$

$$\frac{d}{dx}[f(x) - g(x)] = f'(x) - g'(x)$$

Product, Quotient, and Chain Rules

$$\frac{d}{dx}[f(x)g(x)] = f'(x)g(x) + f(x)g'(x)$$

$$\frac{d}{dx}\left[\frac{f(x)}{g(x)}\right] = \frac{f'(x)g(x) - f(x)g'(x)}{[g(x)]^2}$$

$$\frac{d}{dx}f(g(x)) = f'(g(x))g'(x)$$

3. Applications of Differentiation

Linearization

- The approximation
$$f(x) \approx f(a) + f'(a)(x - a) \text{ for } x \text{ near } a$$
is called the **linear approximation** of f at a.

- The linear function
$$L(x) = f(a) + f'(a)(x - a)$$
is called the **linearization** of f at a.

- *Linearization interpretation of the derivative*: A one-unit increase in the x-value a increases the y-value $f(x)$ by approximately $f'(a)$ (if $f'(a) > 0$), or decreases the y-value by approximately $f'(a)$ (if $f'(a) < 0$).

- *Differentials*: For $y = f(x)$, $dy = f'(x)\,dx$ is the **differential** of y. Here dy is the infinitesimal change in y resulting from the infinitesimal change dx in x.

The Increasing/Decreasing Test

Theorem: Let f be differentiable on (a, b). Then:

(a) If $f'(x) > 0$ for all x in (a, b), then f is increasing on that interval.

(b) If $f'(x) < 0$ for all x in (a, b), then f is decreasing on that interval.

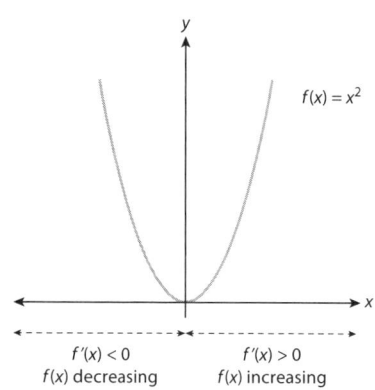

$f(x) = x^2$

$f'(x) < 0$ $f'(x) > 0$
$f(x)$ decreasing $f(x)$ increasing

Optimization

Critical Numbers, Values, and Points

Definition: Let f be a function and c be in the domain of f, where c is not an endpoint of the domain of f. We then say that:

(a) c is a **critical number of** f if $f'(c) = 0$ or $f'(c)$ does not exist;

(b) $f(c)$ is a **critical value of** f if c is a critical number of f; and

(c) $(c, f(c))$ is a **critical point of** f if c is a critical number of f.

Local Extrema

• *Definition*: Let f be a function and c in the domain of f. We then say that:

(a) f has a **local maximum at** c if $f(c) \geq f(x)$ for x near c; and

(b) f has a **local minimum at** c if $f(c) \leq f(x)$ for x near c.

In either case, we refer to $f(c)$ as "the" local extremum.

• Example: In the figure below, f has local minima at a, c, and e; it has local maxima at b and d.

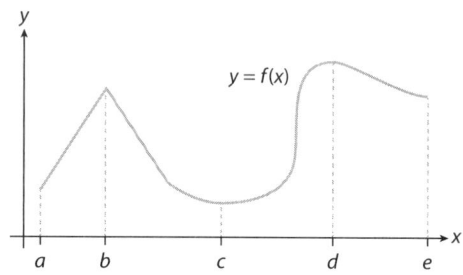

$y = f(x)$

The First Derivative Test

Theorem: Let f be a function and c a critical number of f.

(a) If f' changes sign from positive to negative as we cross $x = c$, then f has a local maximum at c.

(b) If f' changes sign from negative to positive as we cross $x = c$, then f has a local minimum at c.

(c) If f' does not change sign as we cross $x = c$, then there is no local extremum at c.

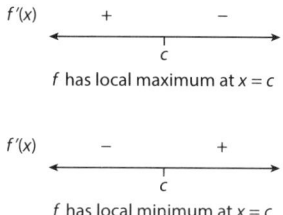

$f'(x)$ +   –
c
f has local maximum at $x = c$

$f'(x)$ –   +
c
f has local minimum at $x = c$

Absolute Extrema

Definition: Let f be a function defined on an interval I, and let c be a number in I. We then say that:

(a) f has an **absolute maximum at** c if $f(c) \geq f(x)$ for all x in I. We then call $f(c)$ the absolute maximum on I; and

(b) f has an **absolute minimum at** c if $f(c) \leq f(x)$ for all x in I. We then call $f(c)$ the absolute minimum on I.

When the interval I is all real numbers, we simply say that $f(c)$ is the absolute minimum (or absolute maximum)

The Extreme Value Theorem

Suppose f is a function continuous on $[a, b]$. Then f has both an absolute maximum and an absolute minimum on $[a, b]$.

Single Critical Number Test

Theorem: Suppose f is continuous on an interval I and has only one critical number c inside I. If c is a local maximum, then $f(c)$ is the absolute maximum of f on I. Similarly, if c is a local minimum, then $f(c)$ is the absolute minimum of f on I.

Finding the Absolute Extrema of a Continuous Function Defined on a Closed Interval $[a, b]$

1. Find the critical numbers in the interval (a, b).

2. Calculate the associated critical values, and also calculate $f(a)$ and $f(b)$.

3. The absolute maximum is the largest of the numbers in step 2; the absolute minimum is the smallest of the numbers in step 2.

Second- and Higher-Order Derivatives

* *Theorem*: Let f be twice differentiable on (a, b). Then:

(a) If $f''(x) > 0$ for all x in (a, b), then f' is increasing on that interval.

(b) If $f''(x) < 0$ for all x in (a, b), then f' is decreasing on that interval.

* *Concavity definition*: Let f be a function defined on an interval I. We say that:

(a) f is **concave up on** I if the graph of f is above the graph of its tangent lines on I; and

(b) f is **concave down on** I if the graph of f is below the graph of its tangent lines on I.

* Concave-up portions of a graph look U-shaped; concave down portions look upside-down U-shaped.

* *Theorem (concavity test)*: Let f be a function defined on an interval I. Then:

(a) If $f''(x) > 0$ for all x in I, then f is concave up on I.

(b) If $f''(x) < 0$ for all x in I, then f is concave down on I.

* *Inflection point definition*: We say f has an **inflection point** at $x = c$ if the graph of f changes concavity as we cross $x = c$.

* *Theorem (second derivative test)*: Suppose that f'' is continuous on an interval including c, and that $f'(c) = 0$. Then:

(a) If $f''(c) > 0$ then f has a local minimum at $x = c$.

(b) If $f''(c) < 0$ then f has a local maximum at $x = c$.

Appendix D: Calculus Review II—Integration

Foundational Concepts

- When $f(x) \geq 0$ on $[a, b]$ and is continuous on that interval, the **definite integral**

$$\int_a^b f(x)\, dx$$

measures the area between the graph of f and the x-axis, and bounded by $x = a$ and $x = b$.

- *The Fundamental Theorem of Calculus*: Suppose $f(x)$ is continuous on $[a, b]$, and define the function

$$A(t) = \int_a^t f(x)\, dx,$$

where $a \leq t \leq b$. Then $A(t)$ is continuous on $[a, b]$, differentiable on (a, b), and $A'(t) = f(t)$.

- *Antiderivatives definition*: Suppose $F'(x) = f(x)$. We then call F the **antiderivative** of f.

- *Indefinite integral definition*: F an antiderivative of f is equivalent to

$$\int f(x)\, dx = F(x) + C.$$

- *The Evaluation Theorem*: Suppose that f is continuous on $[a, b]$ and that F is an antiderivative of f (i.e., $F'(x) = f(x)$). Then,

$$\int_a^b f(x)\, dx = F(b) - F(a) = F(x)\big|_a^b.$$

Foundational Integrals

$$\int 1\, dx = x + C$$

$$\int a\, dx = ax + C \text{ for any constant } a$$

$$\int x\, dx = \frac{x^2}{2} + C$$

$$\int x^n\, dx = \frac{x^{n+1}}{n+1}, \text{ for } n \neq -1$$

$$\int \cos x\, dx = \sin x + C$$

$$\int \sin x\, dx = -\cos x + C$$

$$\int \tan x\, dx = \ln|\sec x| + C$$

$$\int \sec^2 x\, dx = \tan x + C$$

$$\int \sec x \tan x\, dx = \sec x + C$$

$$\int \csc x \cot x\, dx = -\csc x + C$$

$$\int \csc^2 x\, dx = -\cot x + C$$

$$\int e^x\, dx = e^x + C$$

$$\int b^x\, dx = \frac{b^x}{\ln b} + C$$

$$\int \frac{1}{x}\, dx = \ln|x| + C$$

$$\int \frac{1}{x(\ln a)}\, dx = \log_a x + C$$

$$\int \frac{1}{x^2 + 1} = \arctan x + C$$

Integration Properties

Constant Multiple, Sum, and Difference Rules

$$\int [cf(x)]\, dx = c \int f(x)\, dx$$

$$\int [f(x) \pm g(x)]\, dx = \int f(x)\, dx \pm \int g(x)\, dx$$

Definite Integral Properties

$$\int_a^c f(x)\, dx = \int_a^b f(x)\, dx + \int_b^c f(x)\, dx \int_a^b f(x)\, dx =$$
$$- \int_b^a f(x)\, dx$$

$$\int_a^a f(x)\, dx = 0$$

If $f(x) \leq g(x)$ for every x in $[a, b]$, then

$$\int_a^b f(x)\, dx \leq \int_a^b g(x)\, dx$$

Net Signed Area

If $f(x)$ has both positive and negative values inside the interval $[a, b]$, then

$$\int_a^b f(x)\, dx = A_+ - A_-,$$

where A_+ denotes the sum of all areas above the x-axis (and bounded by the graph of f and $x = a$ and $x = b$) and A_- the sum of all areas below the x-axis (and bounded by the graph of f and $x = a$ and $x = b$).

Integration by u-Substitution

To integrate $\int f(g(x)) g'(x)\, dx$, let $u = g(x)$ so that $du = g'(x)\, dx$. Substitute these into the integral to obtain an integral in terms of u. Integrate that, and then convert back to xs via $u = g(x)$. This technique is best used when the original integrand contains a composition of two functions.

Appendix E: Integration Basics

This appendix aims to provide a succinct introduction to the basics of integration theory. This way, a reader having forgotten the origins of the various concepts and formulas reviewed in appendix D can understand where all that comes from. The content in this appendix is an abridged version of chapter 5 in Calculus Simplified *[2]; section 5.1 of that book discusses the history of the "area under a curve problem," which drove the development of integration theory. In this appendix we pick up the story with the Fundamental Theorem of Calculus, and then move on to properties of the integral and finally to develop u-substitution, a very useful integration technique.*

E.1 The Fundamental Theorem of Calculus

Consider the problem of calculating the area between the graph of a continuous function $f(x)$ and the x-axis, and bounded by $x = a$ and $x = t$. In integral notation, we're looking for

$$A(t) = \int_a^t f(x)\, dx.$$

(If you've forgotten where this notation comes from, see the first few paragraphs of section 2.1.) Let's assume, for now, that $f(x) \geq 0$, that is, that it's a nonnegative function. (We'll generalize our results in a later section.) Visually, then, we're looking for the area of a region like that of the shaded region in figure 2.2. Calculus is a dynamics mindset—as we discussed in chapter 1—so let's build up to the Fundamental Theorem of Calculus by investigating the effect on $A(t)$ of a small change Δt in t:

$$A(t + \Delta t) - A(t) = \int_a^{t+\Delta t} f(x)\, dx - \int_a^t f(x)\, dx = \int_t^{t+\Delta t} f(x)\, dx. \qquad \text{(E.1)}$$

The leftmost plot in figure E.1 helps us understand what just happened in this equation. The first integral in the equation is the area under the graph of f and between $x = a$ and $x = t + \Delta t$ (the sum of the areas of the two shaded regions in the figure). The second integral in the equation is the area of only the darker-shaded region in the figure. Subtracting these two areas—as is done in (E.1)—yields the lighter shaded area in the figure, which in integral notation is the rightmost integral in (E.1).

Great. Let's now pretend that the lightly shaded region in the figure is a column of water (figure E.1, second plot). When we remove the top "lid" (the graph of f), the water settles down to some y-value. As I've illustrated in the figure, that y-value is the output of some x-value z: $y = f(z)$, where $t \leq z \leq t + \Delta t$.[1] Conclusion: The area of the lightly shaded region is the area of the *rectangle* of base length Δt and height

---

[1] This follows from the fact that f is continuous and the Intermediate Value Theorem.

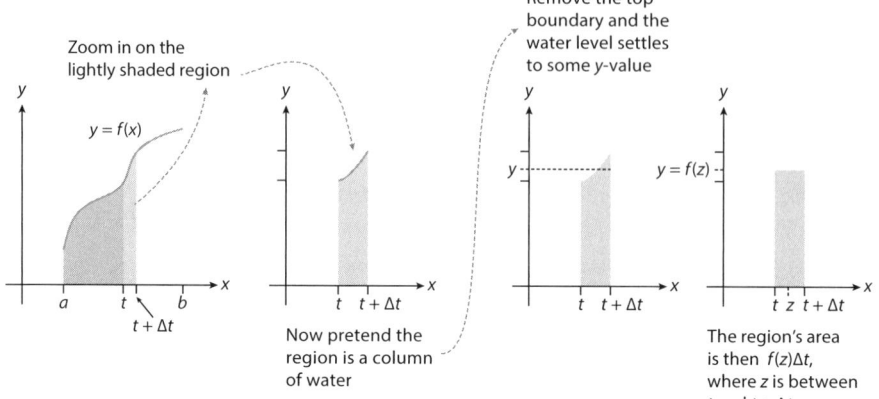

Figure E.1

$f(z)$ (figure E.1, last plot). So, (E.1) becomes

$$A(t + \Delta t) - A(t) = f(z)\Delta t. \tag{E.2}$$

Dividing by Δt and taking the limit of both sides yields

$$\lim_{\Delta t \to 0} \frac{A(t + \Delta t) - A(t)}{\Delta t} = \lim_{\Delta t \to 0} f(z). \tag{E.3}$$

The left-hand side is the limit definition of the derivative $A'(t)$; to calculate the right-hand side, we recall that $t \le z \le t + \Delta t$. Thus, as $\Delta t \to 0$, z approaches t. Conclusion:

$$A'(t) = f(t). \tag{E.4}$$

(We used an intuitive argument here; for a more formal argument, see *Calculus Simplified* [2], chapter 5, exercise 36.) In a later section I'll indicate how our argument can be modified to account for the possibility that $f(x) < 0$ for some x-values. So let me time travel a bit and give you this extended result now. It's what we today call the **Fundamental Theorem of Calculus**. Here is the formal statement of the theorem, published by Leibniz in 1693.

THEOREM E.1: THE FUNDAMENTAL THEOREM OF CALCULUS. Suppose $f(x)$ is continuous on $[a, b]$, and define the function $A(t)$ by

$$A(t) = \int_a^t f(x)\, dx, \tag{E.5}$$

where $a \le t \le b$. Then $A(t)$ is continuous on $[a, b]$, differentiable on (a, b), and $A'(t) = f(t)$.

EXAMPLE E.1   Consider $f(x) = 1$ on the interval $[0, 5]$, and let t be inside this interval.

(a) Show that

$$\int_0^t 1\,dx = t. \tag{E.6}$$

(b) Verify theorem E.1 in this setting.

Solution

(a) The integral $\int_0^t 1\,dx$ is the area of a rectangle with width t and height 1, which is $(t)(1) = t$.

(b) We know that $f(x) = 1$ is a continuous function (in particular, continuous on $[0, 5]$). We just calculated that $A(t) = t$; this too is a continuous function (in particular, continuous on $[0, 5]$). Moreover, since $A'(t) = 1$ (by the Power Rule), A is differentiable (in particular, differentiable on $(0, 5)$), and we see that $A'(t) = f(t)$. ∎

The Fundamental Theorem of Calculus connects integration to differentiation, another pillar of calculus. Figure E.2 visualizes this connection and helps you begin to appreciate why theorem E.1 is so fundamental. Assuming that f is continuous, the theorem's workflow is (1) integrate $f(x)$ to get $A(t)$, and (2) differentiate $A(t)$ to get $f(t)$. (Note that $f(t)$ and $f(x)$ are the same function.) The first big revelation: *Differentiation and integration undo each other!* Therefore, theorem E.1 ties together, in one simple equation, two of the pillars of calculus: differentiation and integration.

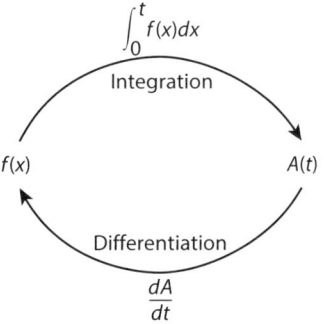

Figure E.2: Differentiation and integration are inverse processes.

Beyond the specific revelation that differentiation and integration undo each other, theorem E.1 implies that we can use *derivatives* to help us calculate definite *integrals*. (What a plot twist!) We'll explore that in the next section.

E.2   Antiderivatives and the Evaluation Theorem

The Fundamental Theorem of Calculus (theorem E.1) suggests that if f is continuous, then $f(t) = A'(t)$, the derivative of the definite integral in (E.5). To calculate that definite integral, then, all we have to do is "undo" that derivative. That process is aptly called **antidifferentiation**.

DEFINITION E.1: Suppose $F'(x) = f(x)$. We then call F the **antiderivative** of f.

In general, then, F is a function that *differentiates to* $f(x)$. Example: If $f(x) = 2x$, then one antiderivative is $F(x) = x^2$ (since $F'(x) = 2x = f(x)$). Another antiderivative is $F(x) = x^2 + 5$. The most general antiderivative is $F(x) = x^2 + C$, where C is any real number.

Let's now employ the antiderivatives viewpoint to extract from the Fundamental Theorem of Calculus a way to calculate $A(t)$. To begin, suppose F is an antiderivative of the function f in the theorem, so that $F'(t) = f(t)$. We also know that $A'(t) = f(t)$ (where A is defined by (E.5)). Therefore, $F'(t) = A'(t)$. One can show that this implies that $A(t) = F(t) - C.$[2] Substituting in (E.5) yields

$$\int_a^t f(x)\, dx = F(t) - C. \tag{E.7}$$

When $t = a$ we get $\int_a^a f(t)\, dt = F(a) - C$. But the integral on the left-hand side is the area under the graph of f and bounded by $x = a$ and $x = a$, which is zero. Thus, $0 = F(a) - C$, so that $C = F(a)$. Using this in (E.7) and writing $t = b$ yields the following corollary of theorem E.1.

THEOREM E.2: THE EVALUATION THEOREM. Suppose that f is continuous on $[a, b]$ and that F is an antiderivative of f (i.e., $F'(x) = f(x)$). Then,

$$\int_a^b f(x)\, dx = F(b) - F(a). \tag{E.8}$$

EXAMPLE E.2 (Video)   The Power Rule tells us that $\left(x^3\right)' = 3x^2$. Use this result to help you calculate

$$\int_0^1 (3x^2)\, dx.$$

Solution Here $f(x) = 3x^2$, which is continuous on $[0, 1]$. Moreover, $\left(x^3\right)' = 3x^2$ tells us that $F(x) = x^3$ is an antiderivative of f. It follows from theorem E.2 that

$$\int_0^1 (3x^2)\, dx = F(1) - F(0) = 1^3 - 0^3 = 1. \qquad\blacksquare$$

EXAMPLE E.3 (Video)   The Power Rule tells us that $\left(\sqrt{x}\right)' = \frac{1}{2\sqrt{x}}$. Use this result to help you calculate

$$\int_1^4 \left(\frac{1}{2\sqrt{x}}\right) dx.$$

Solution Here $f(x) = \frac{1}{2\sqrt{x}}$, which is continuous on $[1, 4]$. Moreover, $\left(\sqrt{x}\right)' = \frac{1}{2\sqrt{x}}$ tells us that $F(x) = \sqrt{x}$ is an antiderivative of f. It follows from theorem E.2 that

$$\int_1^4 \left(\frac{1}{2\sqrt{x}}\right) dx = F(4) - F(1) = \sqrt{4} - \sqrt{1} = 1. \qquad\blacksquare$$

Having now seen how we could use antiderivatives to calculate a definite integral, here are some things you should know.

[2] See *Calculus Simplified* [2], chapter 5, exercise 14 (with d replaced by F).

- Often we use the shorthand $F(x)\big|_a^b$ for $F(b) - F(a)$, so that (E.8) becomes

$$\int_a^b f(x)\, dx = F(x)\big|_a^b.$$

- Replacing x with t (or any other letter) in (E.8) changes nothing. For that reason we refer to x as a **dummy variable**.

- The Evaluation Theorem is also called the "Fundamental Theorem of Calculus, Part 2" in some textbooks.

The Evaluation Theorem has converted the problem of calculating the definite integral of a continuous function to the problem of finding an antiderivative of that function. These antiderivatives are not unique, as the following theorem shows.

THEOREM E.3: Suppose F is an antiderivative of f (i.e., $F' = f$). Then $F(x) + C$, where C is any constant, is also an antiderivative of f.

(The proof is very simple: $(F(x) + C)' = F'(x) = f(x)$.) We will therefore henceforth refer to, for example, $F(x) = x^2 + C$ as "the" antiderivative of $f(x) = 2x$. (This is technically an abuse of the English language because C could be any real number, so there are many different formulas for F, clashing with the usage of "the" in the sentence.) Finally, we can rid ourselves of these wordy statements via the following notation.

DEFINITION E.2: THE INDEFINITE INTEGRAL. Let F be an antiderivative of f, so that $F' = f$. We then write

$$\int f(x)\, dx = F(x) + C \qquad\qquad \text{(E.9)}$$

and call the left-hand side the **indefinite integral of** f.

Note the use of the symbol $\int$ here. But be careful to not confuse (E.9) with the definite integral: the definite integral produces a *number* (the area under the graph of f), while the indefinite integral produces a *function* (the most general antiderivative of f).

Since indefinite integrals are just new notation for antidifferentiation, the indefinite integral is just the reverse process of differentiation:

$$F'(x) = f(x) \quad \Longleftrightarrow \quad \int f(x)\, dx = F(x) + C. \qquad\qquad \text{(E.10)}$$

For example,

$$(x^2)' = 2x \quad \Longleftrightarrow \quad \int 2x\, dx = x^2 + C.$$

This condenses the wordier "$F(x) = x^2 + C$ is the antiderivative of $f(x) = 2x$."

The equivalence (E.10) gives us a wealth of antiderivatives; *simply take the differentiation results from calculus, read them in right-to-left order, and add in the indefinite integral sign and the "+C" in the right places.* For example, returning to the first sentence in examples E.2–E.3,

$$\int 3x^2 \, dx = x^3 + C, \qquad \int \left(\frac{1}{2\sqrt{x}} \right) dx = \sqrt{x} + C.$$

These results come from the Power Rule for differentiation:

$$(x^m)' = mx^{m-1},$$

so we can follow our prescription (E.10) to write down the integral version of the Power Rule:

$$(x^m)' = mx^{m-1} \quad \Longleftrightarrow \quad \int mx^{m-1} \, dx = x^m + C.$$

A more user-friendly formula is obtained by substituting in $m - 1 = n$ and solving for the indefinite integral of x^n. This yields the following theorem.

THEOREM E.4: THE INTEGRAL VERSION OF THE POWER RULE.

$$\int x^n \, dx = \frac{x^{n+1}}{n+1} + C, \quad n \neq -1 \tag{E.11}$$

Note the requirement that $n \neq -1$. (The integral of $\frac{1}{x}$ turns out to be a logarithm, as discussed later in this appendix.) One particularly tricky instance of this theorem is when $n = 0$. In that case, (E.11) yields

$$\int 1 \, dx = x + C. \tag{E.12}$$

EXAMPLE E.4 (Video)   Calculate $\int x^2 \, dx$.

Solution Setting $n = 2$ in (E.11) yields $\int x^2 \, dx = \frac{x^3}{3} + C.$ ■

EXAMPLE E.5 (Video)   Calculate $\int_0^1 x^2 \, dx$.

Solution We just calculated a family of antiderivatives for x^2 ($\frac{x^3}{3} + C$). We can choose any of these to use in the Evaluation Theorem. Choosing $C = 0$, we get that $F(x) = \frac{x^3}{3}$ is an antiderivative of $f(x) = x^2$. Therefore, according to the Evaluation Theorem:

$$\int_0^1 x^2 \, dx = \frac{x^3}{3} \bigg|_0^1 = \frac{1}{3}.$$ ■

EXAMPLE E.6 (Video)   Calculate $\int \frac{1}{x^2} \, dx$.

Solution Since $\frac{1}{x^2} = x^{-2}$, using (E.11) with $n = -2$ yields

$$\int \frac{1}{x^2} \, dx = \frac{x^{-1}}{-1} + C = -\frac{1}{x} + C.$$ ■

EXAMPLE E.7 (Video)   Calculate $\int \sqrt{x}\,dx$.

Solution Writing $\sqrt{x} = x^{1/2}$ and using (E.11) with $n = 1/2$ yields

$$\int \sqrt{x}\,dx = \frac{x^{3/2}}{\frac{3}{2}} + C = \frac{2x^{3/2}}{3} + C. \qquad \blacksquare$$

Example E.5 illustrates the fact that F can be *any* antiderivative of f when it comes to using the Evaluation Theorem. For this reason, we will always select the $C = 0$ antiderivative of f when using the Evaluation Theorem.[3]

We've learned how to integrate only one function at a time thus far. In the next section we'll learn how to integrate combinations of functions (e.g., a sum or difference of two functions).

E.3   Properties of Integrals

Like the limit laws from calculus, the indefinite and definite integrals satisfy various properties that help us calculate them. The first few mimic the first few derivative rules one typically learns in calculus: the Sum, Difference, and Constant Multiple Rules.

> **THEOREM E.5: PROPERTIES OF THE INTEGRAL.** Suppose f and g are continuous on $[a, b]$, and let c be a real number. Then:
>
> 1. **The Sum Rule:** $\displaystyle\int_a^b [f(x) + g(x)]\,dx = \int_a^b f(x)\,dx + \int_a^b g(x)\,dx$
>
> 2. **The Difference Rule:** $\displaystyle\int_a^b [f(x) - g(x)]\,dx = \int_a^b f(x)\,dx - \int_a^b g(x)\,dx$
>
> 3. **The Constant Multiple Rule:** $\displaystyle\int_a^b [cf(x)]\,dx = c\int_a^b f(x)\,dx$
>
> Moreover, the rules above also hold if the definite integral is replaced by an indefinite integral.

In addition to the rules above, the following additional rules hold for definite integrals.

> **THEOREM E.6: ADDITIONAL PROPERTIES OF THE DEFINITE INTEGRAL.** Suppose f and g are continuous on $[a, b]$, and let c be a real number. Then:
>
> 1. $\displaystyle\int_a^c f(x)\,dx = \int_a^b f(x)\,dx + \int_b^c f(x)\,dx.$

---

[3] Using any other antiderivative, like $F(x) + 7$, won't change the result in the Evaluation Theorem, since $[F(b) + 7] - [F(a) + 7] = F(b) - F(a)$, the same result as using $F(x)$ (i.e., with $C = 0$).

2. $\displaystyle\int_a^b f(x)\,dx = -\int_b^a f(x)\,dx.$

3. $\displaystyle\int_a^a f(x)\,dx = 0.$

4. If $f(x) \le g(x)$ for every x in $[a, b]$, then $\displaystyle\int_a^b f(x)\,dx \le \int_a^b g(x)\,dx.$

Property 1 tells us that we can split the calculation of the area under a curve into a sum of two different area calculations. (Importantly, while we think of b as being between a and c in that property, it need not be.) Property 2 tells us that swapping the limits of integration multiplies the original value of the definite integral by -1. Property 3 merely reflects the fact that the area under the graph of f between $x = a$ and $x = a$ is zero (a fact we've already used). Finally, property 4 says that if the graph of f is at or below the graph of g, then the area under the graph of f will be less than or equal to the area under the graph of g. Let's now illustrate these properties through a couple of examples.

EXAMPLE E.8 (Video)   Calculate $\displaystyle\int (x^2 - x)\,dx.$

Solution

$\displaystyle\int (x^2 - x)\,dx = \int x^2\,dx - \int x\,dx$ Indefinite integral version of the Difference Rule, theorem E.5

$\displaystyle\qquad\qquad\quad = \frac{x^3}{3} - \frac{x^2}{2} + C.$ Using (E.11)   ∎

EXAMPLE E.9 (Video)   Calculate $\displaystyle\int_0^9 (3\sqrt{x} + 9x^2)\,dx.$

Solution First, let's find the antiderivative of $f(x) = 3\sqrt{x} + 9x^2$:

$\displaystyle\int (3\sqrt{x} + 9x^2)\,dx = 3\int x^{1/2}\,dx + 9\int x^2\,dx$ Sum, Constant Multiple Rules; theorem E.5

$\displaystyle\qquad\qquad\qquad\quad = 3\left(\frac{2}{3}x^{3/2}\right) + 9\left(\frac{x^3}{3}\right) + C$ Using (E.11)

$\displaystyle\qquad\qquad\qquad\quad = 2x^{3/2} + 3x^3 + C.$ Simplifying

Selecting $C = 0$ and using that result in the Evaluation Theorem:

$$\int_0^9 (3\sqrt{x} + 9x^2)\,dx = \left[2x^{3/2} + 3x^3\right]_0^9 = 2,241.$$ ∎

We can now integrate many of the common combinations of functions we'll encounter—but not all; we'll return to this point in section E.6—but as we'll see in the next section, the Difference Rule in theorem E.5 will force us to reinterpret what quantity the definite integral yields.

E.4   Net Signed Area

Consider the integral $\int_0^1 (-1)\, dx$. Our understanding of the definite integral as yielding the area under the graph of $f(x)$ does not apply here, because the x-axis—which has been the *bottom* boundary of the area defined by all the integrals we've calculated thus far—is actually *above* the function $f(x) = -1$. So, we need to reinterpret what the definite integral means when the graph of f dips below the x-axis. That's where theorem E.5 comes in—the Constant Multiple Rule implies that

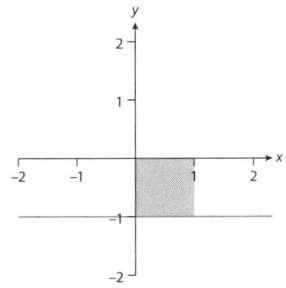

Figure   E.3: $f(x) = -1$ and $\int_0^1 (-1)\, dx$ (the negative of the area of the shaded region).

$$\int_0^1 (-1)\, dx = (-1) \int_0^1 1\, dx = -1,$$

since the second definite integral is 1. The first equation here literally tells us that $\int_0^1 (-1)\, dx$ is -1 times the area under the graph of $f(x) = 1$. This is why some interpret $\int_0^1 (-1)\, dx$ as "negative area." But such a thing doesn't exist, so I'll interpret $\int_0^1 (-1)\, dx = -1$ as "there's 1 unit of area below the x-axis," as illustrated in figure E.3.

What we've just done generalizes rather easily. To wit, if $f(x)$ has both positive and negative values inside the interval $[a, b]$, then

$$\int_a^b f(x)\, dx = A_+ - A_-, \tag{E.13}$$

where A_+ denotes the sum of all areas above the x-axis and A_- the sum of all areas below the x-axis. Thus, *in general the definite integral yields a **net signed area***. The "net" part of the phrase describes the subtraction present in (E.13); the "signed area" describes the possibility that the resulting number—which we previously thought of as the area under the curve—may be negative.

EXAMPLE E.10 (Video)   Calculate $\displaystyle\int_0^2 (x - 1)\, dx$ using the Evaluation Theorem and also (E.13).

Solution Using the Difference Rule (from theorem E.5), (E.11), and the Evaluation Theorem:

$$\int_0^2 (x - 1)\, dx = \int_0^2 x\, dx - \int_0^2 1\, dx$$

$$= \frac{x^2}{2}\bigg|_0^2 - x\big|_0^2 = 2 - 2 = 0.$$

Figure E.4 illustrates our answer. The region below the x-axis (the darker shaded region) has area $\frac{1}{2}$, so that $A_- = \frac{1}{2}$. The lighter-shaded region above the x-axis has area $\frac{1}{2}$ too, so that $A_+ = \frac{1}{2}$. Therefore, from (E.13),

$$\int_0^2 (x-1)\, dx = A_+ - A_- = \frac{1}{2} - \frac{1}{2} = 0. \quad \blacksquare$$

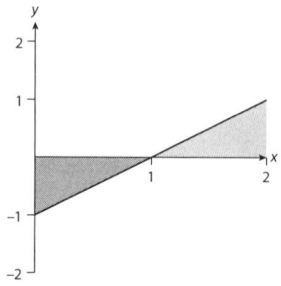

Figure E.4: The function $f(x) = x - 1$, along with the shaded region between f, the x-axis, and bounded by $[0, 1]$ (the darker-shaded region), and the similar region bounded by $[1, 2]$ (the lighter-shaded region).

Alright, we've now covered everything you need to know about the basics of integration. The next section applies all we've learned to exponentials, logarithms, and trigonometric functions to extend our results to those function families.

E.5   Integrating Transcendental Functions

Let's start with the rules for integrating exponential functions. From their differentiation rules, in view of (E.10) we immediately get the following integration rules.

THEOREM E.7:

$$\int b^x\, dx = \frac{b^x}{\ln b} + C, \qquad \int e^x\, dx = e^x + C. \qquad \text{(E.14)}$$

Let's now apply the same approach to finally calculate the integral of $1/x$ (which is not covered by (E.11)). We'll need the following slight generalization of the differentiation rule for $\ln x$:

$$\frac{d}{dx}(\ln |x|) = \frac{1}{x}. \qquad \text{(E.15)}$$

Using this in (E.10) yields the following integration rule.

THEOREM E.8:

$$\int \frac{1}{x}\, dx = \ln |x| + C. \qquad \text{(E.16)}$$

Let me illustrate the preceding two theorems through a few examples.

EXAMPLE E.11 (Video)   Calculate $\int 3e^x\, dx$.

Solution

$$\int 3e^x\, dx = 3\int e^x\, dx \qquad \text{Constant Multiple Rule; theorem E.5}$$

$$= 3e^x + C. \qquad \text{Using (E.14)} \qquad \blacksquare$$

EXAMPLE E.12 (Video)   Calculate $\int (x^2 + 2^x)\, dx$.

Solution

$$\int (x^2 + 2^x)\, dx = \int x^2\, dx + \int 2^x\, dx \qquad \text{Sum Rule; theorem E.5}$$

$$= \frac{1}{3}x^3 + \frac{2^x}{\ln 2} + C. \qquad \text{Using (E.11) and (E.14)} \qquad ■$$

EXAMPLE E.13 (Video)   Calculate $\int \dfrac{x^2 + 1}{2x}\, dx$.

Solution We first simplify: $f(x) = \frac{x^2+1}{2x} = \frac{x}{2} + \frac{1}{2x}$. Then,

$$\int \frac{x^2 + 1}{2x}\, dx = \frac{1}{2}\int x\, dx + \frac{1}{2}\int \frac{1}{x}\, dx \quad \text{Sum, Constant Multiple Rules; theorem E.5}$$

$$= \frac{x^2}{2} + \frac{1}{2}\ln |x| + C. \qquad \text{Using (E.11) and (E.16)} \qquad ■$$

Let's now pivot to discussing the integration of trigonometric functions. Applying (E.10) to the differentiation rules for $\cos x$, $\sin x$, and $\tan x$ yields the following theorem.

THEOREM E.9:

$$\int \cos x\, dx = \sin x + C, \quad \int \sin x\, dx = -\cos x + C, \quad \int \sec^2 x\, dx = \tan x + C.$$

Similarly, applying to (E.10) to the reciprocal trigonometric functions $\sec x$, $\csc x$, and $\cot x$ yields the following rules.

THEOREM E.10:

$$\int \sec x \tan x\, dx = \sec x + C, \quad \int \csc^2 x\, dx = -\cot x + C, \quad \int \csc x \cot x\, dx = -\csc x + C.$$

EXAMPLE E.14 (Video)   Calculate $\int_0^\pi (\sin x + \cos x)\, dx$.

Solution Let's first find the antiderivative of $f(x) = \sin x + \cos x$:

$$\int (\sin x + \cos x)\, dx = \int \sin x\, dx + \int \cos x\, dx \qquad \text{Sum Rule; theorem E.5}$$

$$= -\cos x + \sin x. \qquad \text{Using theorem E.9}$$

Then, using the Evaluation Theorem:

$$\int_0^\pi (\sin x + \cos x)\, dx = [-\cos x + \sin x]_0^\pi$$

$$= (-\cos \pi + \sin \pi) - (-\cos 0 + \sin 0)$$

$$= 1 - (-1) = 2. \qquad ■$$

E.6 The Substitution Rule

Thus far we've developed the rules for integrating sums, differences, and constant multiples of functions (along with a few specialized rules for integrating particular functions, like (E.11)). Let's now develop the integral rule analogue of the Chain Rule, what is sometimes called the "anti-Chain Rule."

To get there, let's apply the equivalence (E.10) to the Chain Rule:

$$\frac{d}{dx}[F(g(x))] = F'(g(x))g'(x) \quad \Longleftrightarrow \quad \int F'(g(x))g'(x)\,dx = F(g(x)) + C. \quad \text{(E.17)}$$

The integrand in the rightmost equation has too much going on. Let's make it look simpler by introducing $u = g(x)$. This yields

$$\int F'(g(x))g'(x)\,dx = \int F'(u)g'(x)\,dx. \quad \text{(E.18)}$$

Defining now the *differential du* to be $g'(x)\,dx$, (E.18) then becomes

$$\int F'(g(x))g'(x)\,dx = \int F'(u)\,du.$$

Finally, letting $F' = f$ yields the following theorem.

> **THEOREM E.11: THE SUBSTITUTION RULE.** Suppose f is continuous on an interval I, and $g(x)$ is differentiable and has range I. Then for $u = g(x)$,
>
> $$\int f(g(x))g'(x)\,dx = \int f(u)\,du. \quad \text{(E.19)}$$

You'll often hear this technique referred to as "u-substitution." Being the "anti-Chain Rule," this integration technique works best (though not exclusively) on integrands that contain composite functions. In such cases, choosing u to be the "inner" function in the composition works well. The following examples illustrate this.

EXAMPLE E.15 (Video)   Calculate $\int 2x(x^2 + 1)^{100}\,dx$.

Solution The integrand contains the composite function $(x^2 + 1)^{100}$; it's "inner" function is $x^2 + 1$. So, let's try setting $u = g(x)$:

$$u = x^2 + 1 \quad \Longrightarrow \quad du = 2x\,dx.$$

Substituting these into the integral yields

$$\int 2x(x^2 + 1)^{100}\,dx = \int u^{100}\,du.$$

Using (E.11) with $n = 100$, this integrates to $\frac{u^{101}}{101} + C$. That's not the end of the calculation, though, since we should end up with a function in the same variable we started with. So, we substitute $u = x^2 + 1$ back in to get

$$\int 2x(x^2 + 1)^{100}\,dx = \frac{(x^2 + 1)^{101}}{101} + C. \qquad \blacksquare$$

EXAMPLE E.16 (Video)   Calculate $\int_0^2 x(x^2 + 4)^3 \, dx$.

Solution Here $u = x^2 + 4$ seems to be the logical choice (it's the "inside" function of the composition $(x^2 + 4)^5$). Letting $u = x^2 + 4$, $du = 2x \, dx$. Dividing both sides by 2 yields $\frac{1}{2} du = x \, dx$. Then, (E.19) and (E.11) yield

$$\int x(x^2 + 4)^3 \, dx = \int u^3 \left(\frac{1}{2} du\right) = \frac{1}{2} \int u^3 \, du = \frac{u^4}{8} + C = \frac{(x^2 + 4)^4}{8} + C.$$

(E.20)

We've now found an antiderivative of $x(x^2 + 4)^3$. The Evaluation Theorem E.2 then implies that

$$\int_0^2 x(x^2 + 4)^3 \, dx = \frac{(x^2 + 4)^4}{8} \Big|_0^2 = 480. \qquad \blacksquare$$

EXAMPLE E.17 (Video)   Calculate $\int \frac{x}{\sqrt{1 + x^2}} \, dx$.

Solution For reasons similar to those in example E.16, the logical choice here is $u = x^2 + 1$. Then $du = 2x \, dx$. Dividing both sides by 2 yields $\frac{1}{2} du = x \, dx$. Then, (E.19) and (E.11) yield

$$\int \frac{x}{\sqrt{1 + x^2}} \, dx = \frac{1}{2} \int u^{-1/2} \, du = u^{1/2} + C = \sqrt{1 + x^2} + C. \qquad \blacksquare$$

EXAMPLE E.18 (Video)   Calculate $\int \sqrt{x + 1} \, dx$.

Solution The only viable choice for u is $x + 1$. Letting $u = x + 1$, we have $du = 1 \, dx$; employing (E.19) then yields

$$\int \sqrt{x + 1} \, dx = \int 1 \cdot \sqrt{x + 1} \, dx = \int \sqrt{u} \, du = \frac{2u^{3/2}}{3} + C = \frac{2(x + 1)^{3/2}}{3} + C. \qquad \blacksquare$$

EXAMPLE E.19 (Video)   Calculate $\int x^5 \sqrt{1 + x^2} \, dx$.

Solution This is the most challenging example yet. But hopefully your gut should tell you to choose $u = 1 + x^2$. Then $du = 2x \, dx$ and $\frac{1}{2} du = x \, dx$. Substituting in what we know thus far yields

$$\frac{1}{2} \int x^4 \sqrt{u} \, du.$$

(I siphoned off one x from x^5 to use $x \, dx = \frac{1}{2} du$.) We now need to relate x to u to complete the substitution. But since $u = 1 + x^2$, then $x^2 = u - 1$, and so $x^4 = (u - 1)^2$. Thus,

$$\frac{1}{2} \int x^4 \sqrt{u} \, du = \frac{1}{2} \int \sqrt{u}(u - 1)^2 \, du$$

$$= \frac{1}{2} \int \sqrt{u}(u^2 - 2u + 1) \, du$$

$$= \frac{1}{2} \int \left[u^{5/2} - 2u^{3/2} + u^{1/2}\right] \, du.$$

The properties of integrals (from theorem E.6) and (E.11) yield

$$\frac{1}{2} \int \left[u^{5/2} - 2u^{3/2} + u^{1/2} \right] du = \frac{1}{2} \left(\frac{2u^{7/2}}{7} - \frac{2u^{5/2}}{5} + \frac{2u^{3/2}}{3} \right) + C.$$

(Each integral generates its own arbitrary constant, but these can be added together to form another arbitrary constant, which is the C in the equation.) Substituting back in $u = 1 + x^2$ finally yields

$$\int x^5 \sqrt{1 + x^2} \, dx = \frac{(1 + x^2)^{7/2}}{7} - \frac{(1 + x^2)^{5/2}}{5} + \frac{(1 + x^2)^{3/2}}{3} + C. \qquad \blacksquare$$

EXAMPLE E.20 (Video)   Calculate the integrals below.

(a) $\displaystyle\int_0^1 2xe^{x^2} \, dx$ (b) $\displaystyle\int \frac{1}{x+1} \, dx$ (c) $\displaystyle\int \frac{x^2 + 2x + 1}{x^2 + 1} \, dx$

Solution

(a) Let $u = x^2$, so that $du = 2x \, dx$. Equations (E.19) and (E.14) then yield

$$\int 2xe^{x^2} \, dx = \int e^u \, du = e^u + C = e^{x^2} + C.$$

Therefore,

$$\int_0^1 2xe^{x^2} \, dx = e^{x^2} \Big|_0^1 = e - 1.$$

(b) Let $u = x + 1$, so that $du = dx$. Equations (E.19) and (E.16) then yield

$$\int \frac{1}{x+1} \, dx = \int \frac{1}{u} \, du = \ln|u| + C = \ln|x+1| + C.$$

(c) Let's first simplify the function:

$$\frac{x^2 + 2x + 1}{x^2 + 1} = 1 + \frac{2x}{x^2 + 1}.$$

Then, by part 1 of theorem E.6,

$$\int \frac{x^2 + 2x + 1}{x^2 + 1} \, dx = \int 1 \, dx + \int \frac{2x}{x^2 + 1} \, dx.$$

The first integral yields $x + C_1$ (from (E.12)). To calculate the second, let $u = x^2 + 1$, so that $du = 2x \, dx$. Equations (E.19) and (E.16) then yield

$$\int \frac{2x}{x^2 + 1} \, dx = \int \frac{1}{u} \, du = \ln|u| + C_2 = \ln|x^2 + 1| + C_2.$$

We conclude that

$$\int \frac{x^2 + 2x + 1}{x^2 + 1} \, dx = x + \ln(x^2 + 1) + C.$$

(We don't need the absolute value around $x^2 + 1$ since that quantity is always positive. Also, I added C_1 and C_2 to produce C.) $\blacksquare$

EXAMPLE E.21 (Video)   Calculate the integrals:

(a) $\int \tan x \, dx$ (b) $\int \cot x \, dx$

Solution

(a) Since $\tan x = \frac{\sin x}{\cos x}$, letting $u = \cos x$ we have $du = -\sin x \, dx$, so that (E.19) yields

$$\int \tan x \, dx = \int \frac{\sin x}{\cos x} \, dx = -\int \frac{1}{u} \, du.$$

Here we need (E.16); we conclude that

$$\int \tan x \, dx = -\ln|\cos x| + C = \ln|\sec x| + C.$$

(b) Since $\cot x = \frac{\cos x}{\sin x}$, letting $u = \sin x$ we have $du = \cos x \, dx$, so that (E.19) yields

$$\int \cot x \, dx = \int \frac{\cos x}{\sin x} \, dx = \int \frac{1}{u} \, du.$$

Here we need (E.16) again; we conclude that

$$\int \cot x \, dx = \ln|\sin x| + C. \qquad \blacksquare$$

EXAMPLE E.22 (Video)   Calculate the integrals below.

(a) $\int x^2 \cos(x^3) \, dx$ (b) $\int \sec^2(2x) \, dx$ (c) $\int_0^{\pi/4} \cos(2x) \, dx$

Solution

(a) Let $u = x^3$. Then $du = 3x^2 \, dx$, and (E.19) along with theorem E.9 yields

$$\int x^2 \cos(x^3) \, dx = \frac{1}{3} \int \cos u \, du = \frac{1}{3} \sin u + C = \frac{1}{3} \sin(x^3) + C.$$

(b) Letting $u = 2x$, we have $du = 2 \, dx$. Then, (E.19) along with theorem E.10 yields

$$\int \sec^2(2x) \, dx = \frac{1}{2} \int \sec^2 u \, du = \frac{1}{2} \tan u + C = \frac{1}{2} \tan(2x) + C.$$

(c) Using the substitution $u = 2x$, $du = 2 \, dx$ yields

$$\int \cos(2x) \, dx = \frac{1}{2} \sin(2x) + C.$$

Setting $C = 0$ and applying (E.8) then yields

$$\int_0^{\pi/4} \cos(2x) \, dx = \frac{1}{2} \sin(2x)\Big|_0^{\pi/4} = \frac{1}{2}\left(\sin \frac{\pi}{2} - 0\right) = \frac{1}{2}. \qquad \blacksquare$$

A few parting words of wisdom on u-substitution.

- *u-Substitution should be used only when the integrand contains a composite function.* (This reflects the technique's origin in the Chain Rule, which is used to differentiate composite functions.) You should then try letting $u = g(x)$, where $g(x)$ is the "inner" function in the composition.

- The substitution $u = g(x)$ converts $f(g(x))$ to $f(u)$; nothing difficult there. However, the remaining part of the integral, namely, $g'(x)\,dx$, also gets transformed—into du. Therefore, the complete substitution is

$$u = g(x), \quad du = g'(x)\,dx.$$

- Once you've transformed the integral to one involving u's and (hopefully) calculated the resulting integral, don't forget to transform variables back to x (using $u = g(x)$).

One last bit about u-substitution. Though we've used the technique thus far only to help us calculate indefinite integrals, it works just as well for calculating definite integrals. Let me illustrate what I mean by returning to example E.16. Since $u = x^2 + 4$ in that example, the upper limit of integration $x = 2$ becomes $u = 8$; the lower limit of integration $x = 0$ becomes $u = 4$. Thus,

$$\int_0^2 x(x^2 + 4)^3\,dx = \frac{1}{2} \int_4^8 u^3\,du = \frac{1}{2} \left[\frac{u^4}{4} \right]_4^8 = 480,$$

the same answer we obtained.

We've now covered the basics of integration. We've focused a lot on finding antiderivatives and using the Evaluation Theorem to evaluate definite integrals. But it's not always possible to find those precious antiderivatives (sometimes the integrands are just too complex). Our best recourse in such situations is to approximate the definite integral. Sections 2.2–2.4 discuss how to do that using sums of areas of rectangles or trapezoids. The remainder of chapter 2 then returns to the antiderivative question and develops additional techniques for finding antiderivatives.

Appendix F: L'Hôpital's Rule

L'Hôpital's Rule is one method for evaluating limits yielding indeterminate forms. Before reviewing it, let's first define "indeterminate form."

DEFINITION F.1: INDETERMINATE FORMS OF TYPE 0/0 OR ∞/∞. Let "lim" be a stand-in for $\lim\limits_{x \to c}$, $\lim\limits_{x \to c^+}$, $\lim\limits_{x \to c^-}$, $\lim\limits_{x \to \infty}$, or $\lim\limits_{x \to -\infty}$, and suppose that f and g are functions. If

$$\lim f(x) = 0 \text{ and } \lim g(x) = 0 \quad \text{or} \quad \lim f(x) = \pm\infty \text{ and } \lim g(x) = \pm\infty,$$

then we say that $\lim \frac{f(x)}{g(x)}$ is **indeterminate of type** 0/0 (in the former case) or ∞/∞ (in the latter case).

The rationale for using the word "indeterminate" here is that limits approaching 0/0 or ∞/∞ don't have a determined value—a single value—like most other limits do. For example,

$$\lim_{x \to 0} \frac{x}{x} = 1, \qquad \lim_{x \to 0} \frac{2x}{x} = 2, \qquad \lim_{x \to \infty} \frac{x}{x} = 1, \qquad \lim_{x \to \infty} \frac{2x}{x} = 2.$$

In the first two examples, both limits approach 0/0 yet yield different answers. In the next two examples, both limits approach ∞/∞ yet again yield two different answers.

Now, although the limits above yield indeterminate forms, we can easily evaluate them (using, e.g., the fact that $x/x = 1$ for $x \neq 0$). In Calculus 1 you learned a variety of techniques for evaluating more complex limits yielding indeterminate forms (e.g., "unrationalization"); appendix C reviews those techniques. But sometimes none of those approaches work. Example:

$$\lim_{x \to \infty} \frac{\ln x^2}{x}.$$

None of those Calculus 1 techniques help evaluate this limit, which yields the indeterminate form ∞/∞. But the following theorem does.

THEOREM F.1: L'HÔPITAL'S RULE. Let "lim" be a stand-in for $\lim\limits_{x \to c}$, $\lim\limits_{x \to c^+}$, $\lim\limits_{x \to c^-}$, $\lim\limits_{x \to \infty}$, or $\lim\limits_{x \to -\infty}$, and suppose that f and g are functions. If $\lim \frac{f(x)}{g(x)}$ is indeterminate of type 0/0 or ∞/∞, then

$$\lim \frac{f(x)}{g(x)} = \lim \frac{f'(x)}{g'(x)},$$

provided the limit on the right-hand side exists or yields $\pm\infty$.

Translation: If we get the indeterminate form 0/0 or ∞/∞ for a limit, we can try evaluating $\lim \frac{f'(x)}{g'(x)}$, the limit of the ratio of derivatives, and if we get a number or $\pm\infty$, that's the limit. Note that evaluating this new limit requires differentiation; appendix C reviews that content if you need a refresher. Finally, to not confuse simplifications of expressions with applications of L'Hôpital's Rule, hereafter I'll use an equals sign with an "L" on top, $\overset{L}{=}$, to indicate when I've applied the rule.

EXAMPLE F.1 (Video)   Evaluate the limit.

(a) $\lim\limits_{x\to 0} \dfrac{e^{3x}-1}{x}$ (b) $\lim\limits_{x\to\infty} \dfrac{\ln x^2}{x}$ (c) $\lim\limits_{x\to\infty} \dfrac{x^2}{e^x}$

Solution

(a) This limit yields 0/0. Applying L'Hôpital's Rule: $\lim\limits_{x\to 0} \dfrac{e^{3x}-1}{x} \overset{L}{=} \lim\limits_{x\to 0} \dfrac{3e^{3x}}{1} = 3$.

(b) This one yields ∞/∞. Using $\ln x^2 = 2\ln x$, L'Hôpital's Rule then yields

$$\lim_{x\to\infty} \frac{2\ln x}{x} \overset{L}{=} \lim_{x\to\infty} \frac{\frac{2}{x}}{1} = 0.$$

(c) This one yields ∞/∞. Applying L'Hôpital's Rule: $\lim\limits_{x\to\infty} \dfrac{x^2}{e^x} \overset{L}{=} \lim\limits_{x\to\infty} \dfrac{2x}{e^x}$. This limit is still indeterminate. Applying L'Hôpital's Rule to this new one (yep, we can do it again), $\lim\limits_{x\to\infty} \dfrac{2x}{e^x} \overset{L}{=} \lim\limits_{x\to\infty} \dfrac{2}{e^x} = 0$.   ∎

Related Exercises   1–6.

The indeterminate forms 0/0 and ∞/∞ are the ones we'll run into most often. But there are many other indeterminate forms. These include $0\cdot\infty$, $\infty-\infty$, 1^∞, ∞^0, and 0^0. As before, in each case we can come up with limits that approach these outputs yet yield different values. In the following subsections we'll extend L'Hôpital's Rule to help us evaluate limits yielding these new indeterminate forms.

The Indeterminate Forms $0\cdot\infty$ and $\infty-\infty$

For the indeterminate forms $0\cdot\infty$ and $\infty-\infty$, our strategy is to convert the limit into 0/0 or ∞/∞ form via algebraic manipulation. In the $0\cdot\infty$ case we typically do this by using the fact that

$$f(x)g(x) = \frac{f(x)}{\frac{1}{g(x)}}.$$

Example:

$$-x\ln x = -\frac{\ln x}{\frac{1}{x}}.$$

As $x\to 0^+$, the left-hand side approaches $0\cdot\infty$, while the right-hand side approaches ∞/∞. We can therefore apply L'Hôpital's Rule to that representation of the function.

In the case of $\infty-\infty$, we look for ways to combine the parts that are separately tending to ∞ to reduce the new expression to one of the previously studied indeterminate forms. Example: $\sqrt{x^2+1}-x$. As $x\to\infty$, this tends to $\infty-\infty$. But since (for $x\ge 0$)

$$\sqrt{x^2+1}-x = \sqrt{x^2\left(1+\frac{1}{x^2}\right)} - x$$

$$= x\sqrt{1 + \frac{1}{x^2}} - x$$

$$= x\left[\sqrt{1 + \frac{1}{x^2}} - 1\right],$$

as $x \to \infty$ this new expression tends to $0 \cdot \infty$. From here, we can use the $x = \frac{1}{1/x}$ approach to convert the expression into one that tends to $0/0$ and then use L'Hôpital's Rule.

EXAMPLE F.2   Evaluate the limit.

(a) $\lim\limits_{x \to \infty} e^{-x}\sqrt{x}$ (b) $\lim\limits_{x \to 1^+} \left(\frac{1}{\ln x} - \frac{1}{x - 1}\right)$

Solution

(a) As $x \to \infty$, $e^{-x}\sqrt{x} \to 0 \cdot \infty$. We can convert this into an ∞/∞ indeterminate form by recognizing that $e^{-x}\sqrt{x} = \frac{\sqrt{x}}{e^x}$. Applying L'Hôpital's Rule yields

$$\lim_{x \to \infty} \frac{\sqrt{x}}{e^x} \overset{L}{=} \lim_{x \to \infty} \frac{\frac{1}{2\sqrt{x}}}{e^x} = \lim_{x \to \infty} \frac{1}{2e^x\sqrt{x}} = 0.$$

(b) As $x \to 1^+$, the function in parentheses tends to $\infty - \infty$. Now,

$$\frac{1}{\ln x} - \frac{1}{x - 1} = \frac{x - 1 - \ln x}{(x - 1)\ln x}.$$

Noting that

$$\frac{\frac{d}{dx}(x - 1 - \ln x)}{\frac{d}{dx}[(x - 1)\ln x]} = \frac{1 - 1/x}{(x - 1)(1/x) + \ln x} = \frac{x - 1}{x - 1 + x\ln x},$$

L'Hôpital's Rule yields

$$\lim_{x \to 1^+} \left(\frac{1}{\ln x} - \frac{1}{x - 1}\right) \overset{L}{=} \lim_{x \to 1^+} \frac{x - 1}{x - 1 + x\ln x}.$$

But this new fraction again tends to an indeterminate form (in this case $0/0$) as $x \to 1^+$. So, let's try applying L'Hôpital's Rule again. Noting that

$$\frac{\frac{d}{dx}(x - 1)}{\frac{d}{dx}(x - 1 + x\ln x)} = \frac{1}{2 + \ln x},$$

we finally see that

$$\lim_{x \to 1^+} \frac{x - 1}{x - 1 + x\ln x} \overset{L}{=} \lim_{x \to 1^+} \frac{1}{2 + \ln x} = \frac{1}{2}. \qquad \blacksquare$$

Related Exercises   7–10.

The Indeterminate Forms 0^0, ∞^0, and 1^∞

For the indeterminate forms 0^0, ∞^0, and 1^∞, we convert them to previously studied indeterminate forms by writing

$$f(x)^{g(x)} = e^{\ln f(x)^{g(x)}} = e^{g(x)\ln f(x)}. \tag{F.1}$$

This converts all three of the "power" indeterminate forms 0^0, ∞^0, and 1^∞ into the "product" form, $0 \cdot \infty$, arising from $g(x) \ln f(x)$. If we're successful in evaluating the limit of $g(x) \ln f(x)$—say we get L—we then have to remember that our final answer will be e^L.

EXAMPLE F.3 (Video)   Evaluate $\lim\limits_{x \to 0^+} (\sin x)^x$.

Solution This limit yields the indeterminate form 0^0. With $f(x) = \sin x$ and $g(x) = x$, (F.1) converts $(\sin x)^x$ to

$$e^{x \ln(\sin x)}.$$

As $x \to 0^+$, $x \ln(\sin x) \to 0 \cdot (-\infty)$. Writing $x \ln(\sin x) = \frac{\ln(\sin x)}{\frac{1}{x}}$ converts this to $-\infty/\infty$. So, *now* we can apply L'Hôpital's Rule:

$$\lim_{x \to 0^+} \frac{\ln(\sin x)}{\frac{1}{x}} \overset{L}{=} \lim_{x \to 0^+} \frac{\frac{\cos x}{\sin x}}{-\frac{1}{x^2}} = \lim_{x \to 0^+} \frac{-x^2}{\tan x}.$$

This new limit yields the indeterminate form $0/0$. So, one more application of L'Hôpital's Rule:

$$\lim_{x \to 0^+} \frac{-x^2}{\tan x} \overset{L}{=} \lim_{x \to 0^+} \frac{-2x}{\sec^2 x} = \frac{0}{1} = 0.$$

This is the L-value I talked about earlier. So, the final answer is $e^L = e^0 = 1$.   ∎

Related Exercises   11–14.

Additional Tips, Tricks, and Takeaways

- First things first: *We can only apply L'Hôpital's Rule if the limit yields an indeterminate form*. In particular, if after an application of L'Hôpital's Rule the limit doesn't yield an indeterminate form, we *cannot* apply L'Hôpital's Rule again. (We'd use other limit evaluation techniques in that case.)

- Next, note that sometimes it pays to simplify the function *before* and/or *after* an application of L'Hôpital's Rule. We did that in a couple of instances in the examples above, and it helped.

- Finally, note that sometimes getting functions into the 0/0 or ∞/∞ forms requires some creativity. Often dividing by the reciprocal does the trick.

With all this in mind, try exercises 15–24. Some of those require L'Hôpital's Rule, some don't. This type of mixed practice will help you further distinguish when—and when not—to use L'Hôpital's Rule.

Appendix F   Exercises

1–6: (a) Does L'Hôpital's Rule apply to the limit? Justify your answer. (b) Evaluate the limit.

1. $\lim\limits_{x\to 0} \dfrac{\sin x}{2x}$ **2.** $\lim\limits_{x\to 0^+} \dfrac{\cos x}{\sin x}$ **3.** $\lim\limits_{x\to 0} \dfrac{1 - e^{2x}}{2x}$

4. $\lim\limits_{x\to\infty} \dfrac{\sin x}{x}$ **5.** $\lim\limits_{x\to\infty} \dfrac{\ln(1 + e^x)}{x}$

6. $\lim\limits_{x\to\infty} \dfrac{3^x}{x^3}$

7–14: Evaluate the limit using L'Hôpital's Rule.

7. $\lim\limits_{x\to 0} \left[\dfrac{1}{\ln(1 + x)} - \dfrac{1}{x} \right]$ **8.** $\lim\limits_{x\to 0} (\csc x - \cot x)$

9. $\lim\limits_{x\to 1} \csc(\pi x) \cdot \ln x$

10. $\lim\limits_{x\to 0^+} \cot x \cdot \ln[1 + \sin(2x)]$

11. $\lim\limits_{x\to 0^+} x^x$ **12.** $\lim\limits_{x\to\infty} x^{1/x}$

13. $\lim\limits_{x\to 0^+} (\sin x)^{\tan x}$

14. $\lim\limits_{x\to 0^+} (\tan x)^x$

15–24: Evaluate the limit.

15. $\lim\limits_{x\to 0} \dfrac{5^x - 4^x}{x}$ **16.** $\lim\limits_{x\to 1} \dfrac{x^3 - 1}{x + 2}$

17. $\lim\limits_{x\to 0^+} \dfrac{\tan x}{3x^2}$ **18.** $\lim\limits_{x\to 0^+} \dfrac{\cos x}{\sqrt{x}}$

19. $\lim\limits_{x\to 0} \dfrac{e^{5x} - 1}{\tan x}$

20. $\lim\limits_{x\to 0^+} x^3 \cot x$ **21.** $\lim\limits_{x\to 0^+} (e^x + x)^{1/x}$

22. $\lim\limits_{x\to\infty} (1 + x)^{1/x}$ **23.** $\lim\limits_{x\to 1^+} (\ln x)^{x-1}$

24. $\lim\limits_{x\to 0} \dfrac{e^x}{x^2}$

Answers to Exercises

Chapter 2 Exercises

1. 5.12   **2.** ≈1.18   **3.** ≈0.77

4. ≈2.57   **5.** ≈1.73   **6.** ≈1.17

7. $R_n = 26 + \frac{10}{n}$ **8.** $R_n = \frac{14}{3} + \frac{6}{n} + \frac{4}{3n^2}$

9. $R_n = \frac{5}{4} + \frac{5}{2n} + \frac{5}{4n^2}$

10. Setting $\frac{e(\ln 2)}{n} \leq 0.01$ yields $n \geq 189$

11. Setting $\frac{1}{24n^2} \leq 0.01$ yields $n \geq 3$

12. Setting $\frac{9}{16n^2} \leq 0.01$ yields $n \geq 8$

13. $\frac{1}{3}x^3 \ln x - \frac{1}{9}x^3 + C$

14. $-5x \cos x + 5 \sin x + C$

15. $x \arcsin x + \sqrt{1 - x^2} + C$

16. $2 - \frac{5}{e}$

17. $e^{x^2}(x^2 - 1) + C$ **18.** -2π

19. $24 - \frac{65}{e}$

20. $5(x^4 - 12x^2 + 24) \sin x - x(x^4 - 20x^2 + 120) \cos x + C$

21. 16   **22.** $[4 + \cos(2x)] \tan^3 x \sec^2 x + C$

23. $2 \sin(x) - \frac{1}{2} \cos(2x) + C$

24. $-5 \cos^3(x) + 3 \cos^5(x) + C$

25. $7 \sin^5(x) - 5 \sin^7(x) + C$

26. $\frac{35 \sin^3(x)}{3} - 14 \sin^5(x) + 5 \sin^7(x) + C$

27. 6π **28.** $\frac{15\pi}{16}$ **29.** $\frac{\sqrt{x^2 - 9}}{x} + C$

30. $(x^2 + 162)\sqrt{x^2 - 81} + C$

31. $9 \arcsin(x/3) - x\sqrt{9 - x^2} + C$

32. $\frac{1}{\sqrt{9 - x^2}} + C$

33. $7 \ln|2 - x| - 6 \ln|1 - x| + C$

34. $\ln\left|\frac{x}{x+4}\right| + C$

35. $-\frac{4}{x-2} + \ln\left|\frac{2-x}{x+2}\right| + C$

36. $-\frac{30}{x-3} + \ln\left|\frac{x-3}{x+3}\right| + C$

37. $\frac{1}{2}\left[\ln|x + 1| + \sqrt{2} \arctan\left(\frac{x-1}{\sqrt{2}}\right)\right] + C$

38. $\ln(x^2 + 4) - 3 \arctan(x/2) + 2 \arctan(x) + C$

39. $5e^4 - 1$ **40.** $8 \cos^{10} x - 10 \cos^8 x + C$

41. $15 \ln|x - 1| + 3 \ln|x + 1| + C$

42. $2 \ln 2 - 1$ **43.** $\text{arcsec}(x/2) + C$

44. $\ln|x + 2| + \frac{8x + 13}{2(x+2)^2} + C$

45. $[(x^2 + 1) \arctan x] - x + C$

46. $4x + \sin(4x) + C$

47. $\ln\left|\frac{(x-1)(x+1)}{x^2+1}\right| + C$

48. $(x^4 \ln x)[(8 \ln x) - 4] + x^4 + C$

A1. Here $\Delta x = \frac{4-0}{4} = 1$, and $x_i = 0 + i\Delta x = i$. Using this in (2.6) yields $L_4 = [c(0) + c(1) + c(2) + c(3)](1)$. I'll use the following values here: $c(0) = 0$, $c(1) = 0.375$, $c(2) = 0.55$, and $c(3) = 0.35$. This yields $L_4 = 1.275$ mg/L per second. Therefore, $F \approx \frac{5}{1.275} \approx 3.92$ L/second.

A2. (a) $P(75 \leq X \leq 85) = \frac{1}{5\sqrt{2\pi}} \int_{75}^{85} e^{-(x-80)^2/50} \, dx$

(b) Here $\Delta x = \frac{85-75}{5} = 2$, and $x_i = a + i\Delta x = 75 + 2i$. Thus, $L_5 = [n(75) + n(77) + \cdots + n(83)](1)$, where $n(x) = \frac{1}{5\sqrt{2\pi}} e^{-(x-80)^2/50}$. This yields $L_5 \approx 0.676$.

A3. (a) Here $\Delta x = \frac{100-0}{10} = 10$, and $x_i = 10i$. With $f(x) = s(x)/100{,}000$, (2.11) becomes

$T_{10} = \frac{1}{2}[f(0) + 2f(10) + \cdots + 2f(90) + f(100)](10)$.
Using the table data yields $T_{10} = 76.968$ years.

(b) From $\frac{1}{100,000} \int_0^\omega x(-s'(x))\,dx$, use $u = x$, $dv = -s'(x)\,dx$ (and so $v = -s(x)$). Recall also that $s(\omega) = 0$.

(c) From $\frac{1}{100,000} \int_0^\omega x^2(-s'(x))\,dx$, use $u = x^2$, $dv = -s'(x)\,dx$ (and so $v = -s(x)$). Recall also that $s(\omega) = 0$.

(d) Let's estimate the integral in v first. With $\Delta x = \frac{100-0}{10} = 10$, $x_i = 10i$. Setting $f(x) = 2xs(x)/100,000$ in (2.5) gives $R_{10} = [f(10) + f(20) + \cdots + f(100)](10)$. Using the table data yields $R_{10} = 6{,}197.414$ (years)^2. Then $v = 6{,}197.414 - (76.968)^2 \approx 273.34$ (years)^2. Finally, $d = \sqrt{v} \approx 16.53$ years.

A4. Using integration by parts (with $u = r$, $dv = e^{-r}\,dr$) on $200\pi \int_0^x re^{-r}\,dr$ yields $P(x) = 200\pi [1 - e^{-x}(x+1)]$.

A5. Use the trig substitution $x = z\tan\theta$, then use the Evaluation Theorem.

A6. Here $Q = 2\pi \int_0^R \frac{r}{(r^2+4)^{3/2}}\,dr$. Using the trig substitution $x = 2\tan\theta$ yields $Q = \pi - \frac{2\pi}{\sqrt{R^2+4}}$.

D1. (a) Subtract L_n from each side of the inequality.

(b) $R_n - L_n = [f(x_n) - f(x_0)]\Delta x = [f(b) - f(a)]\frac{b-a}{n}$

(c) Using (b) in (a) yields $0 \leq EL_n \leq \frac{(b-a)[f(b)-f(a)]}{n} = \frac{(b-a)|f(b)-f(a)|}{n}$.

D3. $x[(\ln x)^2 - 2(\ln x) + 2]$

D4. $u = x^3$; writing $x^3 = x \cdot x^2$ and using $u = x^2$, $dv = x\sin(x^2)\,dx$ works.

Chapter 3 Exercises

1.

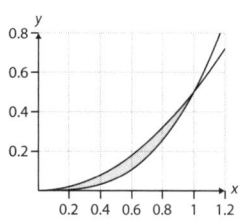

area: 1/24

2.

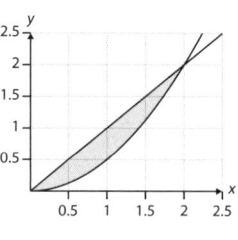

area: 2/3

3.

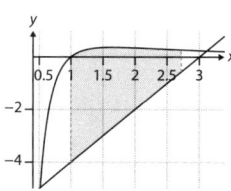

area: $e(6 - e) - 3 - \frac{4}{e}$

4.

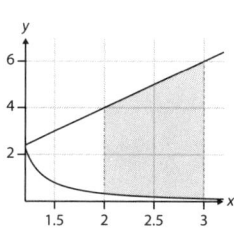

area: $5 + \ln\frac{2}{3}$

5.

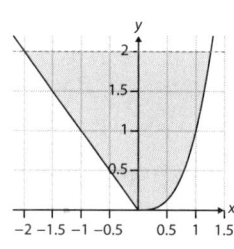

area: $2 + \frac{3}{\sqrt[3]{4}}$

6.

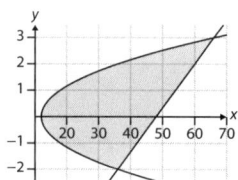

The graphs intersect at $y = -2$ and $y = 3$; area: 125.

7.

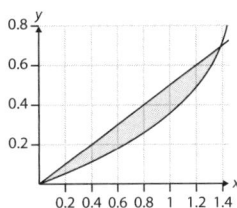

The graphs intersect at $y = 0$ and $y = \ln 2$; area: $2 - (\ln 2)^2 - \ln 4$.

8.

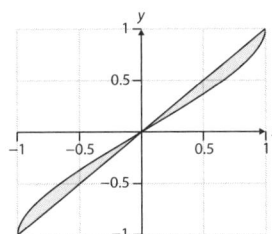

The graphs intersect at $y = -1$, $y = 0$, and $y = 1$; area: $\frac{4\sqrt{2}-2}{3} - 1$.

9. (a) 144; (b) 36π; (c) $36\sqrt{3}$

10. (a) $\frac{1}{30}$; (b) $\frac{\pi}{120}$; (c) $\frac{\sqrt{3}}{120}$

11. (a) $\frac{1}{5}$; (b) $\frac{\pi}{20}$; (c) $\frac{\sqrt{3}}{20}$

12. (a) $\frac{\pi}{2}$; (b) $\frac{\pi^2}{8}$; (c) $\frac{\pi\sqrt{3}}{8}$

13. (a) $\frac{e^2-1}{4}$; (b) $\frac{\pi(e^2-1)}{16}$; (c) $\frac{\sqrt{3}(e^2-1)}{16}$

14. (a) $\frac{1}{3}$; (b) $\frac{\pi}{12}$; (c) $\frac{\sqrt{3}}{12}$

15. (a) $\frac{1}{30}$; (b) $\frac{\pi}{120}$; (c) $\frac{\sqrt{3}}{120}$

16. (a) $\frac{1}{30}$; (b) $\frac{\pi}{120}$; (c) $\frac{\sqrt{3}}{120}$

17. (a) $\frac{9}{70}$; (b) $\frac{9\pi}{280}$; (c) $\frac{9\sqrt{3}}{280}$

18. (a) $\frac{11\pi}{3}$; (b) $\frac{11\pi}{3}$

19. (a) $\frac{9\sqrt[3]{9}\pi}{5}$; (b) $\frac{81\pi\sqrt[3]{3}}{14}$

20. 2π **21.** πe^2 **22.** $\frac{\pi^2}{4}$

23. $\pi(e-2)$ **24.** $\frac{\pi}{4}$

25. $\pi\left(4\ln 2 - \frac{3}{2}\right)$ **26.** $\frac{2\pi^2+3\sqrt{3}\pi}{24}$

27. $\frac{4\pi}{3}$ **28.** $\frac{56\pi}{15}$ **29.** $\frac{\pi}{2}$ **30.** $\frac{16\pi}{3}$

31. $2\pi(17\sqrt{17}-1)$ **32.** $2\pi^2$

33. $\frac{\pi}{3}$ **34.** $\frac{\pi}{2}$ **35.** $\frac{3\pi}{5}$

36. $\sqrt{26}$ **37.** $\frac{5\sqrt{5}-1}{6}$ **38.** $\frac{\ln(2+\sqrt{3})}{2}$

39. $\frac{2}{27}(11\sqrt{22}-4)$ **40.** $45\pi\sqrt{26}$

41. $\frac{1}{6}(17\sqrt{17}-27)$ **42.** $\frac{37\sqrt{37}-1}{54}$

43. 12π **44.** (a) $\frac{56\pi}{15}$; (b) $\frac{72\pi}{5}$

45. (a) $\frac{7\pi^2}{2}$; (b) $\frac{17\pi^2}{2}-12\pi$

46. $\frac{\pi}{2}$ **47.** $\frac{56\pi}{15}$

48. $\pi^2 - \frac{2\pi}{3}$ **49.** $\frac{\pi(7\pi-16)}{4}$

50. $\frac{2\pi}{5}$ **51.** $\pi\left(\frac{8\sqrt{2}-3}{12}\right)$

52. $\frac{\pi}{2}$ **53.** $\frac{5\pi}{12}$ **A1.** $G = \frac{1}{3}$

A2. $V = \pi \int_{-1}^{1}[(2+\sqrt{1-x^2})^2 - (2-\sqrt{1-x^2})^2]\,dx$

A3. (a) The shell method, because to use the washer method we'd need to solve $y = \sin(x^2)$ for x.

(b) Using the shell method: $V = 2\pi \int_0^{\sqrt{\pi}} x\sin(x^2)\,dx = 2\pi$.

A4. $2\sinh(1) = e - \frac{1}{e}$

D2. (a) Use the facts that $S_1 = \pi r_1(\ell_1 + \ell_2)$ and $S_2 = \pi r_2\ell_2$.

Chapter 4 Exercises

1. $a_n = \cos(n\pi)$, $n \geq 1$

2. $a_n = 4n+1$, $n \geq 0$

3. $a_n = \left(\frac{n}{n+1}\right)^n$, $n \geq 2$

4. $a_n = \frac{(-1)^{n+1}}{2^{n+1}}$, $n \geq 0$

5. $a_n = 2e^n$, $n \geq 2$

6. $a_n = \frac{(-1)^{n+1}n^2}{n+1}$, $n \geq 2$

7. $a_2 = 4$, $a_3 = 16$ **8.** $a_2 = 9$, $a_3 = 17$

9. $\frac{(3!)(4!)}{6\cdot5\cdot4!} = \frac{3!}{6\cdot5} = \frac{6}{6\cdot5} = \frac{1}{5}$

10. $\frac{n!}{(n+1)n!} = \frac{1}{n+1}$

11. $\frac{(n-1)!}{(n+1)n(n-1)!} = \frac{1}{n(n+1)}$

12. $\frac{(2n+2)(2n+1)(2n)!}{(2n)!} = (2n+2)(2n+1)$

13. (c)   **14.** (b)   **15.** (a)

16. Converges to 0   **17.** Converges to 1

18. Converges to 0   **19.** Converges to 0

20. Diverges   **21.** Converges to π

22. Converges to $\ln 2$ **23.** Converges to $\frac{1}{2}$

24. Diverges

25. Converges to 0   **26.** Converges to 0

27. Converges to 0   **28.** Converges to 0

29. Converges to 1   **30.** Diverges

31. Converges to 0   **32.** Converges to 10

33. Converges to e

34. $S_1 = 2, S_2 = 10, S_3 = 28$

35. $S_1 = \frac{2}{5}, S_2 = \frac{7}{10}, S_3 = \frac{159}{170}$

36. No, because $S_N \to \infty$ as $N \to \infty$

37. Yes, to $\ln 3$: $S_N = \ln \frac{3N}{N+5} \to \ln 3$ as $N \to \infty$.

38. No, since $a_n \to 0$

39. Yes; $a_n \to \infty$, so the series diverges.

40. Yes; $a_n \to 1$, so the series diverges.

41. No; $|a_n| \to 0$, and so $a_n \to 0$. (theorem 4.4)

42. No, since $a_n \to 0$ (by theorem 4.5)

43. Yes; a_n eventually oscillates between -1 and $+1$, so $a_n \not\to 0$ and thus the series diverges.

44. Diverges ($S_N = 1 - e^{N+1}$)   **45.** Converges to 1

46. Converges to $\frac{1}{\sqrt{2}}$ **47.** Converges to e^{-1}

48. Converges to $\frac{1}{1-4/5} = 5$

49. Converges to $\frac{2}{1-(-1/4)} = \frac{8}{5}$

50. Diverges ($|r| = 1.1 > 1$)

51. Diverges ($|r| = \pi > 1$)

52. Converges to $\frac{2}{1-(-11/13)} = \frac{26}{24}$

53. Converges to $\frac{2}{1-2/9} = \frac{18}{7}$ (use $3^{2n} = (3^2)^n = 9^n$)

54. $\sum_{n=0}^{\infty} \frac{5}{10} \left(\frac{1}{10}\right)^n = \frac{5}{9}$

55. $3 + \sum_{n=0}^{\infty} \frac{14}{100} \left(\frac{1}{100}\right)^n = 3 + \frac{14}{99} = \frac{311}{99}$

56. yes to both; both geometric series with $r = 4/5$ and $r = -1/4$

57. no; $r = \pi$ geometric series diverges

58. yes to both; both geometric series with $r = -11/13$ and $r = 2/9$

59. yes to both; geometric series with $r = \frac{1}{e}$ and p-series with $p = e > 1$

60. Converges (use $3^n + 2 > 3^n$)

61. Diverges (use $3^n + 2 < 5 \cdot 3^n$)

62. Converges (use $n^5 + 10 > n^5$)

63. Diverges (use $\sqrt{n} - 1 < \sqrt{n}$)

64. Diverges (use $b_n = \frac{5^n}{3^n}$)

65. Diverges (use $b_n = \frac{1}{n}$)

66. Converges (use $b_n = \frac{1}{7^n}$)

67. Converges (use $b_n = \frac{1}{n^2}$)

68. (b) conditionally; limit comparison

69. (b) $\sum |a_n| = \sum \frac{1}{\sqrt{n}}$, which is a divergent p-series

70. (b) absolutely; limit comparison

71. (b) conditionally; limit comparison

72. (b) conditionally; .limit comparison

73. (b) absolutely; limit comparison

74. Converges absolutely   **75.** Diverges

76. Converges absolutely

77. Diverges   **78.** Converges absolutely

79. Converges absolutely

80. (b) From (4.21): $-\frac{31}{36} - \frac{1}{4^2} \le S \le -\frac{31}{36} + \frac{1}{4^2}$

(c) Solving $b_{N+1} = \frac{1}{(N+1)^2} < 0.001$ yields $N \ge 31$.

81. (b) From (4.21): $-\frac{197}{216} - \frac{1}{4^3} \le S \le -\frac{197}{216} + \frac{1}{4^3}$

(c) Solving $b_{N+1} = \frac{1}{(N+1)^3} < 0.001$ yields $N \ge 9$.

82. (a) $M_1(x) = 1$, $M_2(x) = 1 - \frac{x^2}{2!}$, $M_3(x) = 1 - \frac{x^2}{2!}$
(Note: $M_4(x) = 1 - \frac{x^2}{2!} + \frac{x^4}{4!}$)

(b) $M_{2n}(x) = 1 - \frac{x^2}{2!} + \frac{x^4}{4!} - \cdots + (-1)^n \frac{x^{2n}}{(2n)!}$,
$M_{2n+1}(x) = M_{2n}(x)$

(c) $T_1(x) = \cos 1 - \sin(1)(x-1)$, $T_2(x) = \cos 1 - \sin(1)(x-1) - \frac{\cos(1)}{2!}(x-1)^2$, $T_3(x) = \cos 1 - \sin(1)(x-1) - \frac{\cos(1)}{2!}(x-1)^2 + \frac{\sin(1)}{3!}(x-1)^3$

83. (a) $M_1(x) = x$, $M_2(x) = x + x^2$,
$M_3(x) = x + x^2 + \frac{x^3}{2!}$

(b) $M_n(x) = x + x^2 + \frac{x^3}{2!} + \cdots + \frac{x^n}{(n-1)!}$

(c) $T_1(x) = e + 2e(x-1)$,

$T_2(x) = e + 2e(x-1) + \frac{3e}{2!}(x-1)^2$,

$T_3(x) = e + 2e(x-1) + \frac{3e}{2!}(x-1)^2 + \frac{4e}{3!}(x-1)^3$

84. (a) $M_1(x) = 1 - x$, $M_2(x) = 1 - x + x^2$,
$M_3(x) = 1 - x + x^2 - x^3$

(b) $M_n(x) = 1 - x + x^2 - x^3 + \cdots + (-1)^n x^n$

(c) $T_1(x) = \frac{1}{2} - \frac{x-1}{2^2}$, $T_2(x) = \frac{1}{2} - \frac{x-1}{2^2} + \frac{(x-1)^2}{2^3}$,
$T_3(x) = \frac{1}{2} - \frac{x-1}{2^2} + \frac{(x-1)^2}{2^3} - \frac{(x-1)^3}{2^4}$

85. (a) $M_1(x) = x$, $M_2(x) = x$, $M_3(x) = x - \frac{x^3}{3}$

(b) $M_{2n+1}(x) = x - \frac{x^3}{3} + \cdots + (-1)^n \frac{x^{2n+1}}{2n+1}$,
$M_{2n+2}(x) = M_{2n+1}(x)$

(c) $T_1(x) = \frac{\pi}{4} + \frac{x-1}{2}$, $T_2(x) = \frac{\pi}{4} + \frac{x-1}{2} - \frac{(x-1)^2}{2\cdot 2!}$,
$T_3(x) = \frac{\pi}{4} + \frac{x-1}{2} - \frac{(x-1)^2}{2\cdot 2!} + \frac{(x-1)^3}{2\cdot 3!}$

86. (a) $M_1(x) = \frac{\pi x}{2}$, $M_2(x) = \frac{\pi x}{2}$, $M_3(x) = \frac{\pi x}{2} - \frac{\pi^3 x^3}{2^3 \cdot 3!}$

(b) $M_{2n+1}(x) = \frac{\pi x}{2} - \frac{\pi^3 x^3}{2^3 \cdot 3!} + \cdots + (-1)^n \frac{\pi^{2n+1} x^{2n+1}}{2^{2n+1} \cdot (2n+1)!}$, $M_{2n+2}(x) = M_{2n+1}(x)$

(c) $T_1(x) = 1$, $T_2(x) = 1 - \frac{\pi^2 (x-1)^2}{2^2 \cdot 2!}$,
$T_3(x) = 1 - \frac{\pi^2 (x-1)^2}{2^2 \cdot 2!}$
(Note: $T_4(x) = 1 - \frac{\pi^2 (x-1)^2}{2^2 \cdot 2!} + \frac{\pi^4 (x-1)^4}{2^4 \cdot 4!}$)

87. (a) $M_1(x) = x$, $M_2(x) = x - \frac{x^2}{2}$,
$M_3(x) = x - \frac{x^2}{2} + \frac{x^3}{3}$

(b) $M_n(x) = x - \frac{x^2}{2} + \frac{x^3}{3} + \cdots + (-1)^n \frac{x^n}{n}$

(c) $T_1(x) = \ln 2 + \frac{x-1}{2}$, $T_2(x) = \ln 2 + \frac{x-1}{2} - \frac{(x-1)^2}{2^2 \cdot 2}$, $T_3(x) = \ln 2 + \frac{x-1}{2} - \frac{(x-1)^2}{2^2 \cdot 2} + \frac{(x-1)^3}{2^3 \cdot 3}$

88. (a) Since $f^{(5)}(x) = -\sin x$ and $|f^{(5)}(x)| \le 1$,
$|R_4(x)| \le \frac{(0.3)^5}{5!} \approx 0.000021$.

(b) Solving $\frac{(0.3)^{n+1}}{(n+1)!} \le 0.001$ by trial and error yields $n+1 \ge 4$, so $n \ge 3$.

89. (a) Since $f^{(5)}(x) = e^x$ and $|f^{(5)}(x)| \le e$ for $0 \le x \le 1$, $|R_5(x)| \le \frac{e(1)^6}{6!} \approx 0.0038$.

(b) Solving $\frac{e(1)^{n+1}}{(n+1)!} \le 0.001$ by trial and error yields $n+1 \ge 7$, so $n \ge 6$.

90. (a) $R = \frac{1}{2}$ (b) $-\frac{1}{2} \le x \le \frac{1}{2}$

91. (a) $R = 5$ (b) $-5 < x < 5$

92. (a) $R = 1$ (b) $2 \le x < 4$

93. (a) $R = 3$ (b) $-1 < x < 5$

94. $\displaystyle\sum_{n=0}^{\infty} (3x)^n$, $|x| < \frac{1}{3}$

95. $\displaystyle\sum_{n=0}^{\infty} (-1)^n x^{2n+4}$, $|x| < 1$

96.   Differentiating $\displaystyle\frac{1}{1-x} = \sum_{n=0}^{\infty} x^n$ yields $\displaystyle\frac{1}{(1-x)^2} = \sum_{n=0}^{\infty} n x^{n-1}$, which converges on $|x| < 1$.

97. Because $\dfrac{1}{1+x^2} = \sum_{n=0}^{\infty}(-x^2)^n = \sum_{n=0}^{\infty}(-1)^n x^{2n}$, differentiating this yields $-\dfrac{2x}{(1+x^2)^2} = \sum_{n=0}^{\infty} 2n(-1)^n x^{2n-1}$, which converges on $|x| < 1$. From the series laws, multiplying this by $-\frac{1}{2}$ yields $\dfrac{x}{(1+x^2)^2} = \sum_{n=0}^{\infty} n(-1)^{n+1} x^{2n-1}$, which converges on $|x| < 1$.

98. $M.S.[e^{-x}] = \sum_{n=0}^{\infty}(-1)^n \dfrac{x^n}{n!}$, $(-\infty, \infty)$ (via the ratio test).

99. The function here is $f(x) = \cosh x$, the "hyperbolic cosine" function; $M.S.[\cosh x] = \sum_{n=0}^{\infty} \dfrac{x^{2n}}{(2n)!}$, $(-\infty, \infty)$ (via the ratio test).

100. The function here is $f(x) = \sinh x$, the "hyperbolic sine" function; $M.S.[\sinh x] = \sum_{n=0}^{\infty} \dfrac{x^{2n+1}}{(2n+1)!}$, $(-\infty, \infty)$ (via the ratio test).

101. $M.S.[\arctan(x)] = \sum_{n=0}^{\infty}(-1)^n \dfrac{x^{2n+1}}{2n+1}$, $|x| \le 1$ (via the ratio test).

102. $M.S.[\ln(1+x^3)] = \sum_{n=1}^{\infty}(-1)^{n+1} \dfrac{x^{3n}}{n}$

103. $M.S.[\cos x^{3/2}] = \sum_{n=0}^{\infty}(-1)^n \dfrac{x^{3n}}{(2n)!}$

104. $M.S.[\ln \frac{1+x}{1-x}] = \sum_{n=0}^{\infty} \dfrac{2x^{2n+1}}{2n+1}$, using $\ln \frac{1+x}{1-x} = \ln(1+x) - \ln(1-x)$

106. Use the facts that all derivatives of $\sinh x$ are either $\sinh x$ or $\cosh x$, and that $|\sinh x| < |\cosh x|$.

107. (a) $\sqrt[4]{1+x} = 1 + \frac{x}{4} - \frac{3x^2}{32} + \frac{7x^3}{128} + \cdots$

(b) $\sqrt[4]{1.1} \approx 1.024$

108. (a) $(1+x)^{-2/3} = 1 - \frac{2x}{3} + \frac{5x^2}{9} - \frac{40x^3}{81} + \cdots$

(b) $(1.1)^{-2/3} \approx 0.938$

109. $\arctan(1) = \frac{\pi}{2}$ **110.** e^2

111. $1 - e^{-1}$ **112.** 1   **113.** 2

A1. (a) $y_1 = 5$ feet, $y_2 = 2.5$ feet, $y_3 = 1.25$ feet

(b) $y_n = 10(0.5)^n$ feet, which generates a geometric sequence with $a = 5$ feet and $r = 0.5$

(c) $y_1 = 5$ ft, $y_{n+1} = 0.5 y_n$ ft, $n \ge 1$

(d) $\lim_{n \to \infty} y_n = 0$ feet, so the basketball eventually comes to rest on the floor

A2. (a) $T_0 = \$10$, $T_1 = \$10 + \$30 = \$40$, $T_2 = \$10 + 2(\$30) = \$70$

(b) $T_n = 10 + 30n$ dollars, which generates an arithmetic sequence with $b = \$10$ and $m = \$30$/month.

(c) $T_0 = \$10$, $T_{n+1} = (30 + T_n)\$$, $n \ge 0$

(d) $\lim_{n \to \infty} T_n = \infty$

A3. (a) $P_0 = 309.3$ million, $P_1 = (1.0075)(309.3) \approx 311.62$ million, $P_2 = (1.0075)^2(309.3) \approx 313.96$ million

(b) $P_n = 309.3(1.0075)^n$ million, which generates a geometric sequence with $a = 309.3$ million and $r = 1.0075$

(c) $P_0 = 309.3$ million, $P_{n+1} = (1.0075)P_n\$$, $n \ge 0$

(d) $\lim_{n \to \infty} P_n = \infty$

A4. (a) $G_0 = 100 \cdot \frac{100}{C_0} = 100 \cdot \frac{100}{100} = \$100$, $G_1 = 100 \cdot \frac{100}{C_1} = 100 \cdot \frac{100}{100(1.03)} = 100\frac{100}{103} \approx \$97.09$, $G_2 = 100 \cdot \frac{100}{C_2} = 100 \cdot \frac{100}{100(1.03)^2} \approx \$94.26$.

(b) $G_n = \frac{100}{(1.03)^n} = 100(1.03)^{-n}\$$, which generates a geometric sequence with $a = \$100$ and $r = (1.03)^{-1} \approx 0.97$.

(c) $G_0 = \$100$, $G_{n+1} = \frac{G_n}{1.03}\$$, $n \ge 0$

(d) $\lim_{n \to \infty} G_n = 0$, meaning that eventually the 1984 $100 buys nothing.

A5. (c) $\lim_{t\to\infty} M_n = M_0 e^r$. This is the balance n years after opening the account. Here $t\to\infty$ leads to "continuous compounding."

A6. (a) The n central angles of the n inscribed triangles must add to 2π, so $n\theta = 2\pi$, and so $\theta = 2\pi/n$.

(b) The area of the triangle pictured in the figure is $A = \frac{1}{2}rh$, where h is the altitude. But $\sin\theta = h/r$, so $h = r\sin\theta$, and thus $A = \frac{1}{2}r^2 \sin\theta = \frac{1}{2}r^2 \sin\left(\frac{2\pi}{n}\right)$. There are n such inscribed triangles so the total area is $A_n = \frac{1}{2}nr^2 \sin\left(\frac{2\pi}{n}\right)$.

(c) Substituting $n = \frac{2\pi}{m}$ into A_n yields A_m.

(d) Writing $A(x) = \pi r^2 \left(\frac{\sin x}{x}\right)$, since $A(x) \to \pi r^2$ as $x\to\infty$, the theorem implies that $A_m \to \pi r^2$ as $m\to\infty$. Interpretation: As we cram more isosceles triangles into the circle, the total area of those triangles approaches the area of the circle.

A7. (a) $f_n = 440 \cdot 2^{\frac{n}{12}}$ Hz, a geometric sequence with $a = 440$ and $r = 2^{1/12}$.

(b) We have: $f_{12n} = 440 \cdot 2^{\frac{12n}{12}} = 440 \cdot 2^n = 2^n f_0$ Hz. Interpretation: The frequency of the musical note n octaves above A above middle C (i.e., f_{12n}) is 2^n times the frequency of the note A above middle C (i.e., $f_0 = 440$ Hz).

A8. We have $d_1 = 2(5)$ ft, $d_2 = 2(2.5)$ ft, and in general, $d_n = 2y_n$, with (y_n) the sequence from problem A1. Therefore, $d_n = 2[10(0.5)^n] = 20(0.5)^n$. The series $d_1 + d_2 + d_3 + \cdots$ is a geometric series with first term $a = d_1 = 20(0.5) = 10$ and $r = 0.5$. Therefore, it converges to $\frac{10}{1-0.5} = 20$. Interpretation: Total vertical distance covered by the ball through all its bounces is 20 feet.

A9. (a) $T(1-x)^n$ (b) We need to calculate $\sum_{n=1}^{\infty} T(1-x)^n$. This is a convergent geometric series with $a = T(1-x)$ and $r = 1-x$, so it sums to $\frac{T(1-x)}{1-(1-x)} = \frac{T(1-x)}{x}$.

(c) Setting $\frac{T(1-x)}{x} > T$ yields $\frac{1-x}{x} > 1$ or $1 - x > x$ or $1 > 2x$, which yields $x < \frac{1}{2}$. Interpretation: If the citizens spend more than they save, the initial \$T infusion generates more than \$T in spending.

A10. (a) For $f(x) = \ln(1+x)$, the first Maclaurin polynomial is $M_1(x) = (\ln 1) + x = x$. Therefore, $\ln(1+r) \approx r$. Thus, $T \approx \frac{\ln 2}{r}$.

(b) Since $\ln 2 \approx 0.69 \approx 0.7$, then $T \approx \frac{0.7}{r} = \frac{70}{100r} = \frac{70}{R}$. When $r = 0.03$, $R = 3$, so $T \approx \frac{70}{3} \approx 23$ years.

A11. The first Maclaurin polynomial for $f(x) = \sin x$ is $M_1(x) = \sin 0 + x = x$. Therefore, $\sin x \approx x$. Squaring both sides yields $\sin^2 x \approx x^2$, and squaring again yields $\sin^4 x \approx x^4$. Using this in the $g(x)$ formula yields $g(x) \approx a(1 + bx^2 - cx^4)$, so that $g(1) \approx a(1 + b - c)$. Interpretation: One degree north of the equator, the acceleration of gravity is approximately $a(1 + b - c)$. (Note: $a(1 + b - c) > a$, based on the values of a, b, and c given in the problem. Conclusion: The acceleration of gravity is *greater* $1°$ north of the equator than at the equator.)

A12. (a) Writing $T(v) = t(1 - v^2/c^2)^{-1/2}$, since $v \ll c$, then the binomial expansion (4.43) implies $T(v) \approx t\left[1 - \left(-\frac{1}{2}\right)\frac{v^2}{c^2}\right] = t\left(1 + \frac{v^2}{2c^2}\right)$.

(b) Since $1 + v^2/(2c^2) > 1$, then $T(v) > t$, which means that the time measured by your (moving) clock is less than the time measured by the stationary external observer.

(c) Since the speed of light $c \approx 186{,}282$ miles *per second*, the everyday speeds we travel at (e.g., 60 miles *per hour* in a car) are *soooo much* smaller than c. Therefore, $v^2/c^2 \approx 0$ and so $T(v) \approx t$.

A13. Write $r(\theta) = a(1 - e^2)(1 + e\cos\theta)^{-1}$ and use the binomial expansion (4.43) to get $(1 + e\cos\theta)^{-1} \approx 1 - e\cos\theta$ (since e is close to zero).

D1. (a) Using the quadratic equation on $x^2 - x - 1 = 0$ yields $x = \frac{1\pm\sqrt{5}}{2}$. The positive root is φ, and the negative one is τ. Finally, $\tau = \frac{1-\sqrt{5}}{2} \cdot \frac{1+\sqrt{5}}{1+\sqrt{5}} = \frac{1-5}{2(1+\sqrt{5})} = -\frac{2}{1+\sqrt{5}} = -\frac{1}{\varphi}$.

(b) $x^3 = x \cdot x^2 = x(x+1) = x^2 + x = (x+1) + x = 2x + 1 = F_3 x + F_2$. $x^4 = x \cdot x^3 = x(F_3 x + F_2) = F_3 x^2 + F_2 x = F_3(x+1) + F_2 x = (F_3 + F_2)x + F_3 = F_4 x + F_3$, using the Fibonacci recurrence relation from example 4.3.

(c) $\quad \varphi^n - \tau^n = F_n\varphi + F_{n-1} - (F_n\tau + F_{n-1}) =$
$F_n(\varphi - \tau)$. Dividing by $\varphi - \tau = \frac{1+\sqrt{5}}{2} - \frac{1-\sqrt{5}}{2} = \sqrt{5}$ yields the F_n formula.

(d) We only need to observe that $\tau^n = (-\varphi^{-1})^n = (-1)^n\varphi^{-n}$.

(e) First, note that $\varphi^{-n} = (\varphi^{-1})^n \to 0$ by theorem 4.5 because $r = \varphi^{-1} \approx 0.618 < 1$. Then, since

$$\frac{F_{n+1}}{F_n} = \frac{\varphi^{n+1} - (-1)^{n+1} \ \varphi^{-(n+1)}}{\varphi^n - (-1)^n \ \varphi^{-n}}, \quad \text{as } n \to \infty \text{ the}$$

highlighted terms tend to zero, so $\frac{F_{n+1}}{F_n} \to \frac{\varphi^{n+1}}{\varphi^n} = \varphi$.

D2. (a) $f(x) = ar^x$ grows exponentially if $r > 1$, is constant if $r = 1$, and decays to zero exponentially if $0 < r < 1$. Using this in theorem 4.2 proves the claim.

(b) The function $g(x) = as^x$ has the same growth, constancy, and decay properties as f in (a). Using this in theorem 4.2, and noting that when $r = -1$ the sequence $((-1)^n a)$ diverges, proves the claim.

D3. (b) From (a), $\lim_{n\to\infty} a_n = \lim_{n\to\infty} (S_n - S_{n-1})$. If the left-hand limit is nonzero, then so is the right-hand one.

(c) $\lim_{n\to\infty} (S_n - S_{n-1}) = \lim_{n\to\infty} S_n - \lim_{n\to\infty} S_{n-1} = S - S = 0$

D4. $S_n \to S$ from the definition of series convergence. As $n \to \infty$ so does $n - 1$, and so $S_{n-1} \to S$ too. Solving (4.9) for a_n and taking the infinite limit yields $\lim_{n\to\infty} a_n = \lim_{n\to\infty} (S_n - S_{n-1}) = S - S = 0$ (using a limit law).

D5. The series $\sum_{n=0}^{\infty} e^{nx} = \sum_{n=0}^{\infty} (e^x)^n$ is a geometric series with $a = 1$ and $r = e^x$. It sums to $\frac{1}{1-e^x}$ provided $|r| < 1$, which yields $-1 < e^x < 1$, or $x < 0$.

D6. (c) $\sum_{n=1}^{N} b_n = \sum_{n=1}^{k} b_n + \sum_{n=k+1}^{N} b_n$. So

$$S_N = \sum_{n=1}^{k} b_n + \sum_{n=k+1}^{N} b_n - \sum_{n=1}^{N} b_{n+k}.$$

Using the result from (b), this becomes

$$S_N = \sum_{n=1}^{k} b_n + \sum_{n=k+1}^{N} b_n - \sum_{n=k+1}^{N+k} b_n.$$

And since

$$\sum_{n=k+1}^{N+k} b_n = \sum_{n=k+1}^{N} b_n + \sum_{n=N+1}^{N+k} b_n,$$

solving for the highlighted term and substituting it into the S_N expression above proves the claim.

D7. The last series law is being applied, but it can't be, because that law requires that each ingredient series converge, and neither $\sum \frac{1}{n}$ nor $\sum \frac{1}{n+1}$ converges. We fix the error by noting that $\sum \left(\frac{1}{n} - \frac{1}{n+1}\right)$ is a telescoping series, with $S_N = 1 - \frac{1}{N+1}$. Therefore it *converges* to 1.

D8. (a) $-|a_n| \le a_n \le |a_n|$; adding $|a_n|$ to each side yields the desired inequality.

(b) The second series law

(c) The direct comparison test

(d) The third series law

D9. (a) The resulting series is $1 - \frac{1}{2} - \frac{1}{4} + \frac{1}{3} - \frac{1}{6} - \frac{1}{8} + \cdots$.

(b) Factoring out $\frac{1}{2}$ from the series yields $\frac{1}{2} \sum \frac{(-1)^{n+1}}{n} = \frac{S}{2}$.

D10. For the first series, $L = \lim_{n\to\infty} \frac{n^2}{3n^2 + 1} = \frac{1}{3} < 1$, so the series converges. For the second, $L = \lim_{n\to\infty} \frac{n}{2} = \infty$, so the series diverges.

Chapter 5 Exercises

1. $S_N = 1 - \frac{1}{2N+1}, S = 1$

2. $S_N = 1 + \frac{1}{2} + \frac{1}{3} + \frac{1}{4} - \left(\frac{1}{N+1} + \frac{1}{N+2} + \frac{1}{N+3} + \frac{1}{N+4}\right)$, $S = \frac{25}{12}$

3. $S_N = 7 - \frac{7}{N+2}, S = 7$

4. $S_N = 3 - \frac{3}{(N+1)^2}, S = 3$

5. (a) $\sum_{n=0}^{\infty} (-1)^n \frac{x^{3n+1}}{3n+1}, R=1$

(b) Using the Alternating Series Remainder Theorem, setting $\frac{1}{3n+4} < 0.001$ yields $n \geq 333$.

6. (a) $\sum_{n=0}^{\infty} (-1)^n \frac{x^{4n+1}}{(4n+1)(2n)!}, R=\infty$

(b) Using the Alternating Series Remainder Theorem, setting $\frac{1}{(4n+5)(2n+3)!} < 0.001$ yields $n \geq 1$ (by trial and error).

7. (a) $\sum_{n=0}^{\infty} \frac{(-1)^n}{n!2^n} \frac{x^{2n+1}}{2n+1}, R=\infty$

(b) Using the Alternating Series Remainder Theorem, setting $\frac{1}{(n+1)!2^{n+1}(2n+3)} < 0.0005$ yields $n \geq 3$.

8. (a) Yes, one limit of integration is infinite; (b) converges to $\arctan(\pi/2) = 1$.

9. (a) Yes, one limit of integration is infinite; (b) converges to 1.

10. (a) Yes, one limit of integration is infinite; (b) converges to 1.

11. (a) Yes, one limit of integration is infinite; (b) converges to 2.

12. $\frac{\pi}{6}$ **13.** (a) $\frac{1}{3}$; (b) $\frac{\pi}{6}$

14. (a) $2\pi \int_0^{\infty} xe^{-x} \, dx$; (b) 2π

15. (a) $\int_1^{\infty} \frac{\sqrt{x^4+81}}{x^2} \, dx$; (b) the graph of $f(x)$ goes on forever as $x \to \infty$, so its arc length is infinite.

16. 2

17. (a) Yes, the integrand is discontinuous at $x = 0$; (b) converges to 1/3.

18. (a) Yes, the integrand is discontinuous at $x = 1$; (b) converges to $\pi/2$.

19. (a) Yes, the integrand is discontinuous at $x = 2$; (b) diverges (tends to ∞).

20. (a) Yes, the integrand is discontinuous at $x = 0$; (b) converges to 1.

21. Converges   **22.** Diverges   **23.** Converges

24. Converges   **25.** Diverges   **26.** Diverges

27. (a) $E_3 \leq \int_3^{\infty} \frac{1}{x^4} \, dx = \frac{1}{81} \approx 0.012$; (b) $n \geq 7$

28.   (a) $E_3 \leq \int_3^{\infty} xe^{-x^2} \, dx = \frac{1}{2e^9} \approx 0.000062$; (b) $n \geq 3$

29. (a) $E_4 \leq \int_4^{\infty} \frac{1}{x^2+1} \, dx = \frac{\pi}{2} - \arctan(4) \approx 0.245$; (b) $n \geq 1{,}001$

A1. (a) Write $\frac{1}{\sqrt{1-k^2 \sin^2 x}} = (1 - k^2 \sin^2 x)^{-1/2}$, and use the first two terms in the binomial expansion, (4.43): $(1+z)^m \approx 1 + mz$, with $z = -k^2 \sin^2 x$ and $m = -1/2$. This approximation is valid because $|z| = |k^2|| \sin^2 x| \leq |k^2| \leq 1$, since $k^2 = \sin^2(\theta/2)$.

(b) $T \approx \frac{4}{\omega} \int_0^{\pi/2} \left(1 + \frac{k^2 \sin^2 x}{2}\right) dx = \frac{\pi(k^2+4)}{2\omega}$, which simplifies to the quantity in the prompt.

A2. Use integration by parts twice.

A3. (a) Use the u-substitution $u = x/t$.

(b) Use integration by parts.

(c) $P(20 \leq T < \infty) = \int_{20}^{\infty} \frac{e^{-x/10}}{10} \, dx = \frac{1}{e^2} \approx 0.135$ (13.5% chance)

D1. From the table of Maclaurin series (figure 4.9) we get $\frac{1}{1+x^2} = 1 - x^2 + x^4 - x^6 + \cdots$, valid for $|x| < 1$. From theorem 5.1 we get that $\int_0^1 \frac{1}{1+x^2} \, dx =$

$$\left[x - \frac{x^3}{3} + \frac{x^5}{5} - \cdots\right]_0^1 = \sum_{n=0}^{\infty} (-1)^n \frac{1}{2n+1}.$$ This series converges by the alternating series test.

D2. (b) Hint: Use integration by parts on the integral in part (a), *and* simply the integral in part (b).

D4. For the explanation part, note that $\lim_{b \to \infty} \int_1^b \frac{1}{x} \, dx$ diverges.

D5. (a) Two of the integral test's hypotheses are that f is positive and decreasing. This means that $\int_1^n f(x) \, dx$ is positive for each $n \geq 1$, and that as n gets bigger so does the value of the integral (more of the area under the graph of f is added, so the definite integral's value increases).

(b) Since $\int_1^n f(x) \, dx \leq S_{n-1}$, as $n \to \infty$ the partial sums S_{n-1} increase without bound. The series they represent, $\sum_{n=1}^{\infty} f(n)$, therefore diverges.

D6. (a) $\lim\limits_{b\to\infty} \int_1^b \frac{1}{x}\,dx = \lim\limits_{b\to\infty} [\ln b - \ln 1]$. This limit tends to ∞, and so the improper integral diverges. Since $f(x) = \frac{1}{x}$ is continuous, positive, and decreasing on $[1,\infty)$, it follows by the integral test that $\sum\limits_{n=1}^{\infty} \frac{1}{n}$ diverges too.

(b) Use the fact that $\int_1^b \frac{1}{x^p}\,dx = \left[\frac{x^{1-p}}{1-p}\right]_1^b = \frac{b^{1-p}}{1-p} - \frac{1}{1-p}$.

(c) If $0 < p < 1$, then b^{1-p} is an increasing function. The limit in part (b) therefore tends to infinity, and so the improper integral diverges. If $p > 1$, then b^{1-p} has a negative exponent. The limit in part (b) therefore tends to $-\frac{1}{1-p}$. This is finite, so the improper integral converges. Since $f(x) = \frac{1}{x^p}$ is continuous, positive, and decreasing on $[1,\infty)$, it follows by the integral test that $\sum\limits_{n=1}^{\infty} \frac{1}{n^p}$ diverges for $0 < p < 1$ and converges for $p > 1$.

D7. (a) The expansion is $\frac{x}{e^x - 1} = 1 + \left(-\frac{1}{2}\right)x + \left(\frac{1}{6}\right)\frac{x^2}{2!} + \left(-\frac{1}{30}\right)\frac{x^4}{4!} + \cdots$. Therefore, $B_0 = 1$, $B_1 = -\frac{1}{2}$, $B_2 = \frac{1}{6}$, $B_3 = 0$, $B_4 = -\frac{1}{30}$.

(b) $\sum\limits_{n=1}^{\infty} \frac{1}{n^2} = \zeta(2) = \frac{(2\pi)^2 |B_2|}{2(2)!} = \frac{\pi^2}{6}$,

$\sum\limits_{n=1}^{\infty} \frac{1}{n^4} = \frac{(2\pi)^4 |B_4|}{2(4)!} = \frac{\pi^4}{90}$

(c) $\sum\limits_{n=1}^{\infty} \frac{1}{n^3} = \zeta(3) = \frac{1}{\Gamma(3)} \int_0^{\infty} \frac{x^2}{e^x - 1}\,dt$

Appendix F Exercises

1. (a) Yes: $\frac{\sin x}{2x} \to \frac{0}{0}$; (b) $\frac{1}{2}$

2. (a) No: $\frac{\cos x}{\sin x} \to \frac{1}{0}$; (b) ∞

3. (a) Yes: $\frac{1-e^{2x}}{2x} \to \frac{0}{0}$; (b) -1

4. (a) No: $\frac{\sin x}{x}$ decays to zero as $x \to \infty$

(b) 0

5. (a) Yes: $\frac{\ln(1+e^x)}{x} \to \frac{\infty}{\infty}$; (b) 1

6. (a) Yes: $\frac{3^x}{x^3} \to \frac{\infty}{\infty}$; (b) ∞

7. $\frac{1}{2}$ **8.** 0   **9.** $-\frac{1}{\pi}$ **10.** 2

11. 1

12. 1   **13.** 1   **14.** 1   **15.** $\ln \frac{5}{4}$

16. 0 (L'Hôpital's Rule doesn't apply.)

17. ∞

18. ∞ (L'Hôpital's Rule doesn't apply.)

19. 5   **20.** 0   **21.** e^2 **22.** 1

23. 1

24. ∞ (L'Hôpital's Rule doesn't apply.)

Bibliography

[1] Child, J. M. (1920). *The Early Mathematical Manuscripts of Leibniz.* Chicago: Open Court.

[2] Fernandez, O. E. (2019). *Calculus Simplified.* Princeton, NJ: Princeton University Press.

[3] Marples, C. R., and Williams, P. M. (2022). The Golden Ratio in Nature: A Tour across Length Scales. *Symmetry*, 14, 2059. https://doi.org/10.3390/sym14102059.

[4] Arias, E., Xu, J. Q., Tejada-Vera, B., Murphy, S. L., and Bastian, B. (2022). U.S. State Life Tables, 2020. *National Vital Statistics Reports*, 71(2). DOI: https://dx.doi.org/10.15620/cdc:118271.

Index of Applications

Physical Sciences

Einstein's theory of relativity (time dilation), 151
Electric field of a line of charge, 37
Gravitational acceleration as a function of latitude, 151
Laminar flow, 37
Length of a catenary, 76
Measuring time using a pendulum, 172
Shape of planetary orbits, 152
Speed of molecules in an ideal gas, 173

Life, Behavioral, and Social Sciences; Sports

Cardiac output of the heart, 36
Geometric sequences in basketball, 149
Geometric series in basketball, 150
Life expectancy, 36
Life span inequality, 37
Population density, 37
Population growth: exponential growth, 149

Business and Economics

Balance of savings account that pays interest n times a year, 150

Cost of monthly subscription service, 149
Income inequality, measured via the Gini coefficient, 75
Inflation, 150
Multiplier effect, 150
Rule of 70 (approximates doubling time of savings account balance), 151
Waiting times for customer service, 173

Other

Archimedes's method for finding the area of a circle, 150
Distribution of exam scores, 36
Fibonacci sequence, 1, 91, 152
Gamma function, 174
Geometric sequences in music, 150
Leibniz's formula for π, 173
Riemann zeta function and its relationship to prime numbers, 174
Volume of a mathematical bundt cake, 76
Volume of a mathematical doughnut, 75

Subject Index

Alternating Series Remainder Theorem, 121

antiderivative(s): definition of, 22n6, 190, 195; general form of, 190, 197; relation to indefinite integral, 190, 197 (*see also* integral(s)); usage in the Evaluation Theorem, 10, 190, 196; usage in the Fundamental Theorem of Calculus, 190, 194

applications of differentiation (review), 187–89

Archimedes's method of exhaustion, 150

arc length, 67–70

area: between two curves, 46–52; of a circle, expressed as a limit, 150; formulas for various shapes, 181

Bernoulli numbers, 174

Binet's formula, 152

binomial series: applied to approximating numbers, 143; definition of, 142

calculus: as a dynamics mindset, 1–3, 5–6, 88, 92, 98, 122, 177–78, 193; as the mathematics of infinitesimal change, 1–3, 8, 8n1, 187

Calculus 2 workflow, 4–7, 77, 87, 98–99, 129, 145, 162, 177

cardiac output, 14, 36

catenary, 70, 76

Cauchy, Augustin-Louis, 8n1

compound interest, 150

continuity (review), 185–86

differentiation (review), 186–87

disk method: comparison with washer method, 63; about noncoordinate axes, 82–83; Riemann sums approach to, 79 (*see also* Riemann sum(s)); about the x-axis, 58; about the y-axis, 59

distance formula, 181

Einstein, Albert: theory of relativity (time dilation), 151

electric field of a line of charge, 37

equipartition (of interval), 10. *See also* Riemann sum(s)

error(s): in approximating a function by its Taylor polynomial, 127 (*see also* Taylor polynomial); in approximating the sum of an alternating series with its partial sums, 121 (*see also* infinite series); in approximating the sum of a p-series with its partial sums, 170 (*see also* infinite series); in Midpoint Rule,

21; in Riemann sum approximations, 20; in Trapezoidal Rule, 21

Evaluation Theorem, 190, 196

exponential and logarithmic functions (review), 184

factorial: definition of, 91; relation to gamma function, 174 (*see also* gamma function); nesting property of, 91–92

Fibonacci sequence, 1–2; definition of, 1; relation to the golden ratio, 1–2, 152; as a recurrence relation, 88, 91

Fundamental Theorem of Calculus (review), 190, 194–95

gamma function: definition of, 174; relation to sums of p-series, 175 (*see also* infinite series)

Gini coefficient, 75

golden ratio, 1–2, 91, 152. *See also* Fibonacci sequence

gravitational acceleration as a function of latitude, 151

Growth Order Theorem, 93

improper integrals: convergence/divergence of, 162, 164–65; definition of, 161

indeterminate forms, 209–10

infinite series

—approximation of via the integral test, 170; definition of, 97; definition of absolute and conditional convergence of, 115; definition of convergence/divergence of, 99; laws of, 106–7; notation for, 97; partial sums of, 98–99, 100; procedure for testing for the convergence/divergence of, 118–19; repeating decimals as, 105–6; summation of via Taylor series, 144 (*see also* Taylor series)

—convergence/divergence results: absolute convergence test, 115, 153; alternating series test, 113; direct comparison test, 108; divergence test, 100–1, 153; for geometric series, 105; integral test, 167; limit comparison test, 111; for p-series, 102, 174; ratio test, 116; root test, 154; summary of, 119; for telescoping series, 103

—special types: alternating, 113–14; geometric, 104–106; p-series, 101–2; telescoping, 102–3, 107, 153, 155–56

infinitesimal rectangle(s), 8–9, 56, 60, 63, 66–67, 81–83. *See also* integral(s)